Palgrave Studies in the Future of Humanity and its Successors

Series Editors
Calvin Mercer
East Carolina University
Greenville, USA

Steve Fuller
Department of Sociology
University of Warwick
Coventry, UK

Humanity is at a crossroads in its history, precariously poised between mastery and extinction. The fast-developing array of human enhancement therapies and technologies (e.g., genetic engineering, information technology, regenerative medicine, robotics, and nanotechnology) are increasingly impacting our lives and our future. The most ardent advocates believe that some of these developments could permit humans to take control of their own evolution and alter human nature and the human condition in fundamental ways, perhaps to an extent that we arrive at the "posthuman", the "successor" of humanity. This series brings together research from a variety of fields to consider the economic, ethical, legal, political, psychological, religious, social, and other implications of cutting-edge science and technology.

The series as a whole does not advocate any particular position on these matters. Rather, it provides a forum for experts to wrestle with the far-reaching implications of the enhancement technologies of our day. The time is ripe for forwarding this conversation among academics, public policy experts, and the general public.

For more information on Palgrave Studies in the Future of Humanity and its Successors, please contact Amy Invernizzi, Senior Editor for Philosophy: Amy.Invernizzi@palgrave-usa.com

Georg Theiner • Alcibiades Malapi-Nelson
Editors

Technologies of the Future Self

An Ethics for Transhuman Flourishing

palgrave
macmillan

Editors
Georg Theiner
Villanova University
Villanova, PA, USA

Alcibiades Malapi-Nelson
Humber Polytechnic
Toronto, ON, Canada

ISSN 2945-6592 ISSN 2945-6606 (electronic)
Palgrave Studies in the Future of Humanity and its Successors
ISBN 978-3-031-99901-7 ISBN 978-3-031-99902-4 (eBook)
https://doi.org/10.1007/978-3-031-99902-4

This Palgrave Macmillan imprint is published by the registered company Springer Nature Switzerland AG.
The registered company address is: Gewerbestrasse 11, 6330 Cham, Switzerland

Georg Theiner's dedication: To my wife Xiaozhen Huang,
mother of our newborn twins, with love

Alcibiades Malapi-Nelson's dedication: For my mother,
without whom I'd never have pursued philosophy.

Preface

In October 2024, the MIT Technology Review published an "AI hype index" to afford their readers a quick, at-a-glance summary of latest developments in what is surely a boundary-pushing industry.[1] The index consists of visual icons on a two-dimensional grid, arranged in terms of their respective likelihood of occurrence ("hype vs. reality") and perceived moral salience ("utopia vs. doom"). It would be interesting to extend this index to gauge public reactions to the panoply of transhumanist ideas and proposals.[2] Such a survey of public opinion would include vexed questions such as: Should mind uploading be mandated as a safety precaution for interplanetary travel? Would you prefer to achieve biological, cultural, or digital immortality? What's a more wholesome cause to donate your money—technogaianism, space seeding, or animal uplifting?[3]

All jesting aside, make no mistake: the transhumanist impetus is here to stay. Humans dream their science long before doing it. Our species has a pretty formidable track record for putting things we can imagine, however unlikely they may seem at first, into action. Transhumanist narratives of hope and apprehension have become part of the collective imagination through which we envision, rehearse, and enact or prevent possible futures of the human condition. It comprises an untidy, eclectic

[1] https://www.technologyreview.com/2024/10/23/1105192/ai-hype-index-nov-dec-2024/

[2] https://th-pedia.org/wiki/Comparison_of_futurist_related_ideas_and_positions

[3] For some preliminary testing, see: https://social-epistemology.com/2022/02/09/transhumanism-in-the-schools-redux-steve-fuller/

mix of science, fact and fiction, philosophical speculation and religious yearning, cultural tropes and ideological posturing. (How confident are you, really, in your own assessment of transhumanist ideas?) Lest we confuse ethics with feasibility, an ethical assessment of transhumanist aspirations must always be reflective of this self-propelling force of technological imaginaries.

The editors of this volume are prepared to take transhumanism seriously, from an academic standpoint, which reflects our training and expertise. A few years ago, when we first conceived this book project, we were struck by a mismatch between the exuberant technological idiom on the one hand, and the rather anodyne (mostly utilitarian and/or libertarian) moral idiom in which transhumanists talk about their collective imaginaries on the other. Commonly, transhumanists cast their agenda as a quest for "overcoming human limits", "transcending biological limitations", and "unlocking human potential". Even though they clearly advocate the use of advanced science and technology for the betterment of humankind, we find their goals and aspirations seldom if ever expressed in the language of "human flourishing". In the eyes of transhumanists and their critics alike, central tenets of transhumanism are seen as clashing with ethical conceptions of the "good life" that are substantively defined in terms of *human nature*, which transhumanists seek to transform if not overcome.

But surely, we sensed, this cannot be the final word on the matter. The present volume thus seeks to explore these philosophical tensions and, at the same time, suggest avenues that point towards a rapprochement. In 2022, one of us (Malapi-Nelson) published an article (reproduced in this volume) exploring virtue ethics as an alternative ethical framework for transhumanism. The article was an extension of earlier investigations into the compatibility between Christianity (particularly Catholicism) and the transhumanist ethos based upon a shared concern for *human flourishing*. Its reception revealed a dearth of literature engaging transhumanism from this general ethical vantage point (going well beyond virtue ethics and Christianity). In conversation with the other one of us (Theiner), who had been working on the history of transhumanist thought and the ethics of emerging technologies from the perspective of the "extended mind" thesis, a joint decision was made to fill this scholarly vacuum.

With a nod to the late Foucault,[4] our aim of nudging transhumanism towards a concern for *transhuman flourishing* is to shape its techno-progressive agenda in psychologically meaningful, ethically responsible, and socially desirable ways. At the same time, we take the religious subtext of transhumanist discourse as an opportunity to foreground the nexus between science and religion in developing a "post-secular[5]" ethics for emerging technologies. We thus invited contributions from a relatively wide range of leading scholars whose views might productively coalesce around these themes. The result is what you hold in your hands. We hope it will encourage others to further expand our moral imagination on how to channel the transhumanist impetus towards a "better than good" future for all of us.

We would like to thank the following people for providing valuable input, feedback, and opportunities informing the present work (in alphabetical order): Andy Clark, Michael Collins, Rob Clowes, Ilia Delio, Chris Drain, Travis Dumsday, Brett Frischmann, J.M. Fritzman, Steve Fuller, Shaun Gallagher, Klaus Gärtner, Tertia Gillett, Gregory Hansell, Andrew Jacoby, Philip Larrey, Erickson Lee, Aleksandra Łukaszewicz Alcaraz, Stephen Napier, Orestis Palermos, Christopher Quintana, Francis Remedios, Mark Rowlands, Matt Sanders, Gregory Sandstrom, Susan Schneider, Sally Scholz, Max Shmidheiser, Daniel Shussett, Mark Shiffman, Deborah Tollefsen, James Wetzel, and all our students who've had to endure our musings on transhumanism. We are grateful to Abby Maroun and Thomas McGlone for their editorial assistance. Finally, thanks also to Steven Fassioms, our Production Editor at Palgrave Macmillan, and Jana Yagnavaragan, our Production Supervisor, for their patience and diligent aid in preparing our manuscript for publication.

Villanova, PA, USA — Georg Theiner
Toronto, ON, Canada — Alcibiades Malapi-Nelson

[4] Foucault, Michel. 1988. *Technologies of the Self: A Seminar with Michel Foucault.* Edited by Luther H. Martin, Huck Gutman, and Patrick H. Hutton. First Edition. Amherst, MA: University of Massachusetts Press.

[5] Habermas, Jürgen et al. eds. 2010. *An Awareness of What Is Missing: Faith and Reason in a Post-Secular Age.* Hoboken: Polity Press.

Contents

Notes on Contributors

Debashish Banerji PhD, is the Haridas Chaudhuri Professor of Indian Philosophies and Cultures and the Doshi Professor of Asian Art at the California Institute of Integral Studies (CIIS), San Francisco. He has authored and edited several books on figures of "the Bengal Renaissance," Critical Posthumanism, Integral Yoga Psychology and on several creative and art-related projects. He has curated exhibitions of Indian and Japanese art and has written and produced a documentary film, *Darshan: The Living Art of India* (2018). A monograph *Time-Steps of the Cosmic Horse: The Contemplative Philosophy of the Great Forest (Brihadāraṇyaka) Upaniṣad* and a co-edited volume *Posthumanism and India: A Critical Cartography* are his latest published book projects. A monograph on a theoretical approach to Indian Art, *Visual Imagination of India: History, Theory, Method* is now in press.

Christopher Benek is known internationally for his social and religious analysis and commentary. Rev. Benek is the pastor of the oldest organized church in Miami, First Miami Presbyterian Church. He is also the CEO of The CoCreators Network—an institution that nurtures technological innovation by teaching people how to formationally cultivate their creativity. CoCreators' motto is: Better People, Better Tech, Better World. Additionally, Benek is notably the co-founder and founding Board Chair of the Christian Transhumanist Association. Learn more at ChristopherBenek.com.

Ilia Delio OSF, PhD, holds the Josephine C. Connelly Chair in Theology at Villanova University. Her area of research is Systematic-Constructive theology, with an emphasis on Science and Religion. She is the author of twenty-five books including *The Not-yet God: Carl Jung, Teilhard de Chardin and the Relational Whole, The Hours of the Universe,* winner of the 2022 Gold Nautilus Book Award, and *Making All Things New: Catholicity, Cosmology and Consciousness,* a finalist for the 2019 Michael Ramsey prize. She lectures nationally and internationally on topics in Science and Religion and is founder of the Center for Christogenesis, an online spiritual and educational resource for the integration of science, religion and culture.

Chris Drain works on the "Vygotsky School" of Soviet psychology as well as contemporary philosophy of mind and cognitive anthropology. He holds a PhD in Philosophy from Villanova University and has published on L.S. Vygotsky and Material Engagement Theory, A.N. Leontiev and Enactivism, and issues in social epistemology and technological mediation. He is a lecturer in the Institute for Writing and Rhetoric at Dartmouth College (Hanover, NH).

Francesca Ferrando PhD, teaches Philosophy at New York University (USA), NYU-Program of Liberal Studies. Ferrando is a leading voice in the field of Posthuman Studies and the author of several publications, including *The Art of Being Posthuman* (Polity) and *Philosophical Posthumanism* (Bloomsbury); their work has been translated into a dozen languages. Ferrando was awarded the philosophical prize "Sainati" by the president of Italy; named "One of the 100 Top Creatives Making Change in the World" by ORIGIN Magazine, and defined as "the Philosopher Poet of our Times". More info: www.theposthuman.org

Steve Fuller is Auguste Comte Professor of Social Epistemology in the Department of Sociology at the University of Warwick. Originally trained in history, philosophy and sociology of science, Fuller is the author of nearly 30 books, which most recently have covered visions of a trans- and post-human future (or 'Humanity 2.0'), the current post-truth condition and the normative structure of the university. His latest books are *Post-Truth: Knowledge as a Power Game* (Anthem, 2018), *Nietzschean Meditations: Untimely Thoughts at the Dawn of the Transhuman Era* (Schwabe, 2019), *A Player's Guide to the Post-Truth Condition: The Name of the Game* (Anthem, 2020), *Back to the University's Future: The Second Coming of Humboldt* (Springer 2023), *To Judge and To Justify: Profiles of* *the Academic Vocation* (Springer 2024) and *Media and the Power of Knowledge* (Bloomsbury 2025).

Brian Patrick Green is the director of technology ethics at the Markkula Center for Applied Ethics at Santa Clara University. His work is focused on the ethics of technology and various aspects of the impact of technology and engineering on human life and society including the relationship between technology and religion, particularly the Catholic Church. Green has written more than 30 academic articles and chapters, is co-author of the *Ethics in Technology Practice* corporate technology ethics resources, and is author, co-author, and co-editor of five volumes including *Space Ethics, Ethics in the Age of Disruptive Technologies: An Operational Roadmap, Religious Transhumanism and Its Critics, Encountering AI: Ethical and Anthropological Investigations,* and a special issue of the *Journal of Moral Theology* on AI. He also helps to convene several academic discussion groups on AI for the Vatican's Dicastery for Culture and Education, as well as working with various international organizations and corporations on issues related to technology and ethics.

Alcibiades Malapi-Nelson (PhD, York University) currently teaches at the Faculty of Liberal Arts and Sciences at Humber Polytechnic, and at the Faculty of Humanities and Social Sciences at Sheridan College—in the Greater Toronto Area. He pursued studies in philosophy and theology in Latin America, and philosophy in French Canada and Toronto. Alcibiades published his first book (*The Nature of the Machine and the Collapse of Cybernetics: A Transhumanist Lesson for Emerging Technologies*) for Palgrave Macmillan in 2017. Since then, he has been working on the implications of transhumanism and emerging technologies for the future of humanity, particularly in what regards human identity, ethics, and religion. You can find some of his publications at https://www.researchgate.net/profile/Alcibiades-Malapi-Nelson

Joanna Pascoe is a posthuman feminist arts-based researcher/writer/maker/philosopher and lecturer in the School of Education at Auckland University of Technology (AUT). Her research interests include posthuman pedagogy, public pedagogy, the affirmative ethics of joy, hopepunk possibilities, speculative fiction and poetic inquiry. Her doctoral thesis *The Posthuman Learner: Mothers, Monsters and Machines* explores how engagement with speculative fiction texts in light of posthuman theory may open lines of flight for an affirmative ethics of joy in education. Her background is in creative writing, theatre and applied linguistics, with a focus on curiosity, creativity, and a deep love for the world.

Ted Peters (PhD, University of Chicago) teaches systematic theology and ethics at the Graduate Theological Union in Berkeley, California, USA. Along with Robert John Russell, he co-edited the journal *Theology and Science,* at the Center for Theology and the Natural Sciences. He is the author of *God–The World's Future* (Fortress, 3rd ed., 2015) and author of *Playing God? Genetic Determinism and Human Freedom* (Routledge, 2nd ed., 2003). He recently edited a new volume on artificial intelligence, *The Promise and Peril of AI and IA: Technology Meets Religion, Theology, and Ethics* (ATF 2025). Along with two colleagues, Arvin Gouw and Brian Patrick Green, he co-edited, *Religious Transhumanism and its Critics* (Lexington 2021). He is the author of a fiction thriller with a Transhumanist plot, *Cyrus Twelve,* with Aprocryphile Press [http://tedstimelytake.com/series/leona-foxx-thriller/]. Visit his website: TedsTimelyTake.com and his blog site on Public Theology, https://www.patheos.com/blogs/publictheology/.

Ljiljana Radenovic (PhD, York University, Toronto) is a full professor in the Department of Philosophy, University of Belgrade. She has published in the philosophy of developmental psychology, cognitive science, and the history of emotions. Her current research interests lie at the intersection between history, psychology, and religion.

Micah Redding is founder of the Christian Transhumanist Association, where he facilitates public dialogue on artificial intelligence, transhumanism, and the future of humanity. He has contributed to several books on technological futurism, and his work has been covered in places like BBC World Radio, Channel News Asia, and Christianity Today. Micah Redding holds a Masters in Philosophy, Science and Religion from the University of Edinburgh.

Stefano Rozzoni PhD, is a postdoctoral research fellow and lecturer at the University of Bergamo, Italy. He holds a PhD in *Transcultural Studies in Humanities* from the University of Bergamo and a PhD in *Literary and Cultural Studies* (cotutelle) from Justus-Liebig-University Giessen. His research focuses on forms of human-nonhuman ethical relationality in Anglophone literature and culture across various periods, media, and themes (including the pastoral, rural and urban spaces and places, and pollution). Rozzoni adopts a transdisciplinary approach, integrating frameworks from Philosophy (Posthumanism), Economics (Civil Economy), Education Studies, and Literature (Ecocriticism, Postcoloniality) within the context of the Environmental Humanities. He is also a co-founder of the Italian Posthuman Network.

Stefan Lorenz Sorgner is a philosophy professor at John Cabot University in Rome, Director and Co-Founder of the Beyond Humanism Network, fellow at the Institute for Ethics and Emerging Technologies (IEET), research fellow at the Ewha Institute for the Humanities at Ewha Womans University in Seoul, academic advisor of Humanity+, and visiting fellow at the Ethics Centre of the Friedrich-Schiller-University in Jena. He is editor of more than ten essay collections, and author of several monographs on transhumanism, including *On Transhumanism* (Penn State University Press 2020), *We Have Always Been Cyborgs* (Bristol University Press 2022), *Philosophy of Posthuman Art* (Schwabe 2022). In addition, he is Editor-in-Chief and Founding Editor of the "Journal of Posthuman Studies". As a public intellectual, his views have been featured in media outlets like *Die Zeit*, *Philosophy Now*, and *Il Sole 24 Ore*.

Georg Theiner (PhD, Indiana University) is Associate Professor of Philosophy at Villanova University, where he also teaches for the Cognitive Science Program and at the Law School. He works mostly in philosophy of mind and cognitive science, with a focus on technologically and socially distributed cognition, and has an abiding research and teaching interest in examining the impact of AI on the future of humanity. He has published numerous articles and book chapters in these areas, and is the author of a monograph, *Res Cogitans Extensa: A Philosophical Defense of the Extended Mind Thesis* (Peter Lang, 2011). He served as editor-in-chief of the journal *Social Epistemology* (Taylor & Francis) from 2019–2023, where he also guest-edited a forthcoming special issue on "The Mind-Technology Problem in the Age of GenAI." Visit http://villanova.academia.edu/GeorgTheiner to find out more about his work.

Tracy J. Trothen is Professor of Ethics at Queen's University, jointly appointed to the School of Religion and the School of Rehabilitation Therapy. Trothen is the author or editor of numerous articles, chapters, and ten books including *Religion and the Technological Future: An Introduction to Biohacking, A.I., and Transhumanism,* coauthored with

Calvin Mercer. She co-chairs the American Academy of Religion's (AAR) Artificial Intelligence Exploratory Session and is a fellow of the International Society for Science and Religion (ISSR).

David W. Wood D.Sc., was one of the pioneers of the smartphone industry. He is now a futurist writer, consultant, and speaker. He has been organising and supporting transhumanist events and projects based around London since 2005. As Chair of London Futurists, David has organised over 300 public meetings since March 2008 on futurist and technoprogressive topics. David's most recent book (published July 2023) is *The Death of Death*. His previous books include *The Singularity Principles*, *Vital Foresight*, *Smartphones and Beyond*, *The Abolition of Aging*, and *Sustainable Superabundance*. David has a triple first-class mathematics degree from Cambridge and undertook doctoral research in the Philosophy of Science. He has an honorary Doctorate in Science from Westminster University. He is Executive Director of Longevity Escape Velocity Foundation, a fellow of the Royal Society of Arts (FRSA), and sits on the Board of the IEET (Institute for Ethics and Emerging Technologies).

List of Figures

Who's Afraid of Transhuman Flourishing: Introduction

Georg Theiner

Transhumanism is a growing intellectual and cultural movement promoting the use of advanced science and technology to enhance human beings—their bodies, minds, and the worlds they inhabit—far beyond biologically familiar modes of existence. The movement is often cast as a technologically intensified continuation of the Enlightenment, much in the spirit of the original "humanist" conception of humanity as an unfinished project that stands in need of completion by self-directed means—be that in the form of education and training, cultural reform and political organization, or medicine and technology. Transhumanists lean into this progressive-minded impulse, banking on major technoscientific breakthroughs to ameliorate seemingly ineluctable features of human life which traditionally have been viewed as lying beyond the scope of human agency. In the prescient words of Hannah Arendt, transhumanism can be seen as a radicalized "rebellion against the given":

G. Theiner (✉)
Department of Philosophy, Villanova University, Villanova, PA, USA
e-mail: georg.theiner@villanova.edu

G. Theiner, A. Malapi-Nelson (eds.), *Technologies of the Future Self*, Palgrave Studies in the Future of Humanity and its Successors,
https://doi.org/10.1007/978-3-031-99902-4_1

> The human artifice of the world separates human existence from all mere animal environments, but life itself is outside this artificial world, and through life man remains related to all other living organisms. For some time now, a great many scientific endeavors have been directed toward making life also 'artificial,' toward cutting the last tie through which even man belongs among the children of nature. [...] This future man, whom the scientists tell us they will produce in no more than a hundred years, seems to be possessed by *a rebellion against human existence as it has been given* (our emphasis), a free gift from nowhere (secularly speaking), which he wishes to exchange, as it were, for something he has made himself. (Arendt 1958, 2–3)

Transhumanists are indeed a fundamentally optimistic bunch. Their self-professed ambition to level up the human condition by trading "the given" for "the self-made" is far-flung and comes in many varieties. They target such things as the radical enhancement of our physical, intellectual, and emotional capacities; the cessation of disease, suffering, and death, with the prospect of radical life extension; the cybernetic merging between humans and machines;[1] the attainment of digital immortality through uploading our minds into supercomputers; and the creation of artificial superintelligence and a concomitant transition to post-scarcity economics. Such lofty goals reveal a techno-utopian outlook on human life which draws upon ideals and aspirations which can be viewed as secularized religious motifs. As James Hughes (a leading proponent of social-democratic transhumanism) noted: "Much of what the Transhumanists long for is already available to Christians: eternal life and freedom from pain, suffering, and the burden of a frail body" (Hughes 2006, 297). To be sure, Hughes was quoting a distinctly Christian, bioconservative *critique* of transhumanism, but in the context of our volume, we propose to take the quip seriously as an inclusive expression of *approval*.

The transhumanist impetus for "building better persons" is clearly animated by an overarching concern with improving (trans)human well-being—or, to use a more ethically charged term, *(trans)human flourishing*. However, explicit treatments of transhumanist ethics—notably, by main proponents of the movement—seldom if ever express their aspirations in the idiom of "human flourishing". Instead, despite pronouncements that transhumanism is in principle open to alternative ethical and political

[1] For an account of the historical connection between classical cybernetics and transhumanism, see Malapi-Nelson (2017, Chap. 8).

frameworks, the default ethical outlook of transhumanism has been utilitarian, and its default political outlook (especially in a North American context) libertarian. Thus, we often find a two-pronged strategy by which transhumanists advocate their techno-progressive agenda. The first prong pitches the aims of transhumanism in terms of "overcoming human limits", "transcending biological limitations", and "unlocking human potential", and appeals to freedom of choice over one's body to make the case for letting it happen. The second prong, with an eye on the outcomes of transhumanist interventions, defends their moral desirability in terms of utilitarian principles, on the grounds that in the long run, they are destined to produce "the greatest good for the greatest number of people".

Once the transhumanist agenda is framed in this manner, it is all too easily typecast, by its critics, as the dangerous outpouring of a puerile "boundary-busting" mentality, whose veneer of legitimacy derives from an ill-guided "long-termist" ethical calculus passionately loathed by its detractors. Transhumanists have mostly dug into their heels, refusing to entertain ethical alternatives to their libertarian-utilitarian outlook. To prompt alternative ways of imagining the transhumanist project, we believe the notion of *transhuman flourishing* deserves to be given greater philosophical weight. This is easier said than done, because to many ears, the notion of transhuman flourishing sounds close to a *contradictio in adjecto*. It is widely believed that central tenets of the transhumanist creed clash with ethical conceptions of the "good life" that are substantively defined in terms of "human nature". Without intending to be exhaustive, several tensions are frequently noted in this regard.

1 Clashes Over the "Good Life"

Within normative ethics, "human flourishing" is the term of choice for expressing Greek *eudaimonia*—also translated as happiness, well-being, or the purposive realization of human nature, understood as the actualization of potentialities that make one human. For ethical frameworks of a broadly "Neo-Aristotelian" or eudaimonist bent,[2] the ultimate good of *living well* consists in the development and virtuous exercise of capabilities that have

[2] We use "eudaimonist" as an umbrella term to include not only ethical frameworks (such as virtue ethics and ethical perfectionism) but also related views of human well-being (such as the capabilities approach or positive psychology) that are broadly consonant with the former.

intrinsic value for us because they are part of human nature. Of course, humans still need to make the right choices for these potentialities to become actualized. However, the fact that one is human, and thus objectively possesses these species-specific potentialities, is not a matter of human choice. This conflicts with the declared objective of transhumanists to harness the power of emerging technologies to transcend the natural, biological limitations of the human organism—such as mortality, senescence, and vulnerability. For transhumanists, an appeal to "human nature" does not harbor positive connotations of the good, the wholesome, and the desirable, but is associated with undesirable features such as arbitrary limitations, frailty, and needless suffering.

Closely related to their different moral valuations of "human nature" are several other theoretical tensions. First, for eudaimonist ethics, preserving the integrity of the biological body is essential to human flourishing, and thus valuable as an end in itself, not merely as a means. Thus, radical technological alterations to one's body are seen as violations of human nature and thus morally problematic, since human dignity is viewed as inextricably entwined with our biological nature. Transhumanists, on the other hand, contend that there is nothing sacrosanct about the biological make-up of the human body. Their appeal to "morphological freedom" asserts a person's right to reconfigure their bodies in ways they find desirable, regardless of whether those alterations deviate from biologically evolved or culturally accepted norms.

Second, the virtuous use of practical wisdom to discern the proper weighting or balance between a plurality of ends and goods that are constitutive to human flourishing is an integral component of eudaimonist ethics. In the absence of universal rules dictating *a priori* how a moral agent ought to conduct herself, it requires careful consideration of particular and contingent circumstances. In the eudaimonist tradition, the virtue of *moderation* is thus seen as providing a moral compass to guide the proper use of practical wisdom. In stark contrast, the slogan of striving for "better than well-being"—adopted by the (former) World Transhumanist Association—clearly evinces a refusal to let customary biological constraints dictate what might be humanly attainable levels of flourishing. From a eudaimonist perspective, the prospect of radical enhancement of human capabilities is thus likely to appear as an untempered departure from a harmonious balance, bound to result in excess or deficiency, and thus morally undesirable.

Third, eudaimonists often rebuke the culture of neutrality, central to the liberal-democratic tradition in Western political thought, as impeding public conversations about how technology contributes (or not) to the good life. In their eyes, an appeal to human flourishing provides a normative foundation of the political and legal order, lending shape and direction to public policy making. While this attitude is historically associated with the bioconservative stance characteristic of the natural law tradition, liberal minded exponents of eudaimonist ethics have embraced a similar viewpoint. For example, Nussbaum (1992) has argued that an ethical commitment to human flourishing implies the political aim to secure the development of certain universal capabilities hypothesized to underlie human flourishing. But what exactly are those capabilities, and how does one validate such a list? How should we go about resolving conflicts between competing conceptions of human flourishing? Are there ethical norms (such as the liberal ideals of tolerance, diversity, and freedom) that should be counterbalanced with the promotion of human flourishing? Although the connection between ethical and political perfectionism is tenuous, it is certainly strong in the eyes of many transhumanists—especially (but not only) those of a libertarian persuasion. For them, the liberal ideal of neutrality is a bulwark for protecting individual rights and liberties from state interference by paternalistic regimes (however well-intended). Perhaps ironically, then, while liberally minded transhumanists submit that research and development of radical enhancement technologies should be supported, pointing to their value in improving quality of (trans)human life, they nevertheless tend to defer considerations about the moral permissibility of adopting them to the private decisions of individuals.

It should be clear from these *prima facie* tensions that an appeal to "human nature", arguably the nub of ethical controversies over transhumanism, can't be reduced to empirical questions about "biologically inborn" capacities and their evolutionary origin. The meaning of "human nature" is deeply bound up with normatively salient metaphysical conceptions about what (trans)human futures are possible, desirable, and good. For the purpose of our volume, we propose to treat the term "human nature" akin to what political scientists, with a nod to W.B. Gallie (1956), call an "essentially contested concept"—i.e., a concept whose meaning is widely recognized to be subject to deep theoretical rifts, giving rise to disputes which "cannot be settled by appeal to empirical evidence, linguistic usage, or the canons of logic alone" (Gallie 1956, 169). To be sure, by adopting this posture, we don't mean to imply that there are no truths to be unveiled about the meaning of "human nature", nor that the fruits of scientific labor have no bearing on this question. And we certainly don't mean to deny that conflicting viewpoints on the matter can be subjected to rational debate. Rather, the starting position we adopt—espoused by Karl Popper (among many others)—stems from a recognition that all of science rests on a "metaphysical research program", a cluster of presuppositions about the structure of the world, and our place within it, which are neither testable nor "falsifiable" empirically, but nonetheless play an essential role in the progress of science (Dupuy 2007, 234). Readers of our volume should thus be prepared to wade into the metaphysical thicket.

2 Clashes Over "Human Nature"

In the Judeo-Christian tradition, the metaphysics of "human nature" has been elaborated, against the backdrop of ancient philosophy, through the notion that humans are created in the image and likeness of God (*imago Dei*). The historically dominant "structural" understanding of this notion locates the divine likeness of human beings in the faculty of reason, a set of qualities associated with the intellect and the will.[3] On this view, the faculty of reason, housed by an immaterial soul, is a category-defining feature of humanity which sets us apart from (non-human) animals and the rest of material creation. It secures human beings a privileged spot on that "great chain of being" (Lovejoy 1936), precariously perched somewhere between divinity and animality. Like angels, humans are endowed with an immaterial soul; however, unlike the purely immaterial nature of angelic intelligences, the human soul is tied to the human body and thus bears an essential relation to the material world. Within this picture, the precise grounds of human uniqueness were historically worked out under the influence of the two main ancient Greek models of how the human soul relates to the body.

In abstract terms, the metaphysical difference between the two hinges on whether the duality of matter and form is conceived as wholly divided (Plato's dualism) or as fully integrated in the individual (Aristotle's hylomorphism). Applied to the peculiar amalgamation of mental and physical qualities in human beings, they yield two competing conceptions of human nature. Whereas the Platonic model sees humans as *embodied souls*,

[3] For a succinct comparison between "structural", "functional", and "relational" interpretations of *imago Dei*, and their history, see Cortez (2010, Chapter 2).

the Aristotelean model views us as *ensouled bodies*. The difference in emphasis is subtle but consequential. In the Platonic model, the locus of human uniqueness, and thus moral value, is identified entirely with our existence as an immaterial soul. On this view, our possession of animal bodies is generally regarded with suspicion if not animosity and contempt. Bodies are but expedient vessels for the human soul and epistemically debilitating encumbrances which humans should be eager to discard, or at least trade for environments that are more hospitable to the soul. The Aristotelean model of humans as rational animals, likewise, places a heavy premium on the faculty of reason as the immaterial apex of the human soul, as opposed to lesser forms of ensoulment found in other animals and plants. However, in contrast to Plato, the human soul is not a pure spirit but necessarily tethered in its operation to the body, with which it forms an individual substance. Thus, it is the proper *integration*—not *separation*—of mental and physical qualities which humans should seek to achieve.

Enshrined in these two rival conceptions of "human nature" are conflicting assessments of what humans can ideally hope to become. To illustrate this point, consider a crucial feature of the Porphyrian Tree—an ancient taxonomic device, much indebted to Aristotle's theory of predication, for representing the "scale of nature" (Parrochia 2024; cf. Fig. 1). Aristotle's theory rests on an opposition between the *essence* of a thing—the ingredients which define it, make it the kind of thing it is, and which can't be changed; versus its *accidents*, qualities that can be added or subtracted without modifying the essence of a thing. In this taxonomic system, the "real definition" of a category—its essence—is given in terms of its *proximate genus* and *specific difference*. For instance, the category of "living (things)" is defined as "animate body" and so forth; until we descend to the category of humanity, the *infima species* beyond which no real definition (in the Aristotelean sense) can be given. Importantly, the Porphyrian Tree aims to classify things (*being*) into categories that are mutually exclusive and jointly exhaustive at a given level of generality. For example, anything is either a material or an immaterial substance, but not both; any material substance is either animate or not; and so forth.

It follows from this Porphyrian partitioning of being (substance) that no specific difference can appear in more than one genus. For example, it is impossible that there could be a category-defining division, among minerals, into those which can think (say, a computer) and those which don't. Only humans can think, and humanity is subordinate to being alive.

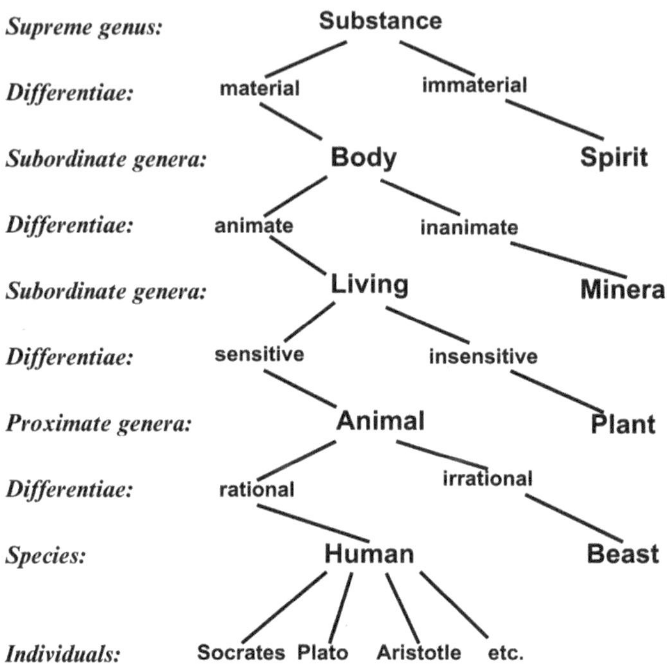

Fig. 1 The Porphyrian Tree. Reproduced after: https://iep.utm.edu/classification-in-science/

Likewise, there can be no such thing as a flesh-eating plant. It goes against the nature of an entity endowed with only a vegetative soul (plant) that it might succeed in entrapping, and devouring, a more capable being (animal) endowed with a sensitive soul (Mancuso and Viola 2015). It is equally impossible that non-human animals ("beasts") might exhibit mental traits associated with rationality—such as self-awareness, freedom of will, or creativity of expression—simply because they lack an immaterial soul. There is no room for upwards or downwards mobility in the Porphyrian Tree. Its logic constrains the realm of metaphysical possibilities, by defining the outer limits of what a thing may or may not become in virtue of its essential nature. No amount of creative ingenuity—human or divine, for that matter—can change that.

The contrast between the Platonic vs. Aristotelean model of human nature can thus be cast as a metaphysical disagreement over the set of transformations that are deemed "ontologically admissible" for humans—i.e., (idealized) operations that in principle could be performed on members of the class humankind without violating any of its category-defining features. This way of looking at things is reminiscent of Edmund Husserl's method of "eidetic variation", devised (in *Ideas* I and III) as a purely "ideational" method for the inquiring phenomenologist to discern the essential structures of consciousness (Drummond 2007). Husserl's method starts by contemplating an exemplary instance of the phenomenon in question, and then freely varies its features in the imagination to elicit an intuitive grasp of what stays invariant throughout the entire series of variations—ranging not only over actually occurring but all *conceivable* (i.e., imaginable) cases. Through applying his methodically refined form of eidetic variation, Husserl believed that ideation is apt to yield infallible knowledge about the nature of consciousness that is impervious to change, since all evidence (both in its favor or against) would presumably be given in the form of conscious experience. However, Husserl's supposedly clean separation between eidetic and empirical knowledge has long been challenged on a variety of grounds (Belt 2022). In the context of transhumanism, the epistemic quandaries besetting Husserl's method of "eidetic variation" have turned into ethical quandaries about the future of being human, as questions about human nature are increasingly beholden to the technological futures we can imagine. Husserl's method has left the ideational realm of pure consciousness and become a technoscientific project of human self-determination.

We are on the verge of achieving unprecedented levels of technological mastery over the very conditions which factor into our conception of "human nature"—such as the range of what our minds and bodies can do, the range of communities we can create, and the range of environments we can live in. Our technologically imagined futures are laden with existential risks and opportunities which our ancestors would not have dared to dream of. They project a far greater diversity of ideas about what it means to be human, what makes human lives worth living, and what human beings might aspire to become. It would be a huge mistake to abdicate responsibility for deciding which of these futures we'd like to create for ourselves by treating technology as a quasi-autonomous agent of change, rendering moral questions about the "good life" (lived in whatever form(s) a future humanity might assume) as subsidiary to considerations of what's technologically feasible. The technological futures we can imagine, however dimly, feed back into our present, creating urgent moral quandaries that can only be addressed if we expand the traditional horizons of our shared moral imagination. Transhumanism stands in dire need of a fundamental *ethics of (human) transformation* which has yet to be formulated. The present volume should be seen as a contribution to this project.

2.1 *Quaestio mihi factus sum*

From this perspective, the rift between the Platonic and the Aristotelean model of human reason should really be seen as a boundary dispute over the limits of human transformation. A brief examination of an illustrative debate in the history of transhumanism will help drive this point home. In his "History of Transhumanist Thought", Nick Bostrom (2005) refers to the conspicuous opening passages of Pico della Mirandola's *Oration on the Dignity of Man* (2016). Bostrom presents the treatise as a signature manifesto of Renaissance humanism, foreshadowing the advent of Enlightenment rationality which he regards as the birthing place of contemporary transhumanist thought. The passage at hand, worth quoting beyond the short fragment relayed to us by Bostrom, depicts the nature of human beings as not yet having a determinate form, and thus responsible to fashion one for themselves:

> Adam, we give you no fixed place to live, no form that is peculiar to you, nor any function that is yours alone. According to your desires and judgment, you will have and possess whatever place to live, whatever form, and whatever functions you yourself choose. All other things have a limited and fixed nature prescribed and bounded by our laws. You, with no limit or no bound, may choose for yourself the limits and bounds of your nature. We have placed you at the world's center so that you may survey everything else in the world. We have made you neither of heavenly nor of earthly stuff, neither mortal nor immortal, so that with free choice and dignity, you may fashion yourself into whatever form you choose. To you is granted the power of degrading yourself into the lower forms of life, the beasts, and to you is granted the power, contained in your intellect and judgment, to be reborn into the higher forms, the divine. (Pico della Mirandola 2016).

Pico portrays the exceptional status of humanity, as creatures of a uniquely "indeterminate nature", as God's response to a dilemma he faced in the act of creation. On the last day, as God was about to create a rational being capable of apprehending the fullness and beauty of his divine handiwork, he recognized that he had already exhausted every possible nature in the process. Seeing as there wasn't any fixed essence left after which human nature might be fashioned, God decided to give humans an indeterminate nature, making it Adam's prerogative to create one for themselves. The popular interpretation of this passage—air-brushed through post-Kantian history of philosophy and adopted by Bostrom—considers it an important prelude to "modern" conceptions of morality that are grounded in freedom and autonomy as opposed to obedience to God's commands. Indeed, conservative Christians have long chastised Pico's radically optimistic view of the self-transcending powers of the human mind as an epitome of Renaissance Pelagianism, with its seeming disregard for the redemptive role of divine grace (Passmore 1970, Chapter 5).[4]

Reading beyond the opening passages of the *Oration*, though, we find that Pico's "humanistic" concern for education—the *studia humanitatis*—is animated by a desire to reunite with God through the pursuit of spiritual excellence (Copenhaver 2024). When Pico describes the ultimate purpose of education (*paideia*)—Pico's technique of choice for instilling in humans the power to assume "whatever form we choose"—it turns out that his aspiration is for humans to become like angels. This is because in order to be metaphysically fit for a reunion with the divine—in other words, for human nature to be divinized—Pico thought that whatever ties the human soul to its earthly existence must be gradually discarded. His ideal of *paideia* thus involves a stepwise process of spiritual purification, beginning with discursive types of knowledge such as ethics, logic, natural philosophy and theology; then ascending through esoteric forms of knowledge such as magic and (Christianized) Kabbalah, and ending in ecstasy and pure contemplation. The final goal of this mystical ascent is not only to shred our mortal bodies, but also to dissolve our individual personalities and identities which separate us from each other, and also from God. Only after formerly embodied humans have been fully transformed into pure spirits, fully purged of matter, and thus angel-like, can our human journey back into the mind of God be completed.

The radical transformation of human nature to which Pico's ideal of *paideia* aspires reveals his sympathies for the Platonic model of human

[4] For a critique of the presumptive identification of transhumanism with Gnosticism, see Malapi-Nelson (2016)

nature.[5] On the Aristotelean model, a divine call for humans to emulate angelic perfection would literally be unintelligible, and thus not one we could be rationally expected to heed. This incompatibility can be brought out with reference to Aquinas' conception of the difference between human and angelic intelligence (McInerny and O'Callaghan 2009). Aquinas' discussion of the matter occurs in the context of his repudiation of the so-called "Plurality of Forms" doctrine in the *Summa*, which implied (inter alia) a separation between a form of corporeity for the human body and a rational form of the human soul in addition to it. Aquinas rejects the "Plurality" doctrine, which he says induces an ontological fissure between the mental and physical features of humanity, along the lines of the Platonic model. For Aquinas, rationality *is* the form that intelligence takes in the life of the human animal "acting knowingly and willingly". Reason distinguishes humans *as* animals, not *from* animals.[6] It would thus be metaphysically improper to attribute rationality to immaterial substances.

In the twentieth century, it was Julian Huxley (1887–1975), evolutionary biologist of "modern synthesis" fame and first director general of UNESCO, who parlayed Pico's exalted vision of humanity's ontological indeterminacy into the technologically inflected, naturalistic idiom of the present age. In a series of short essays, Huxley (1957) grounded his idea of "transhumanism"—a "metaphysical research program" in the Popperian sense mentioned earlier—in a cosmic vision of the unique custodial responsibility humans are destined to shoulder. The essays were written on the heels of major scientific breakthroughs and technological innovations, which Huxley believed have put us on a course to rechart human nature as we know it. He expressed his vision by saying that in the evolution of the human person "the universe has become conscious of itself" (1957, 13). Huxley's delightful turn of phrase suggests a "doubly reflexive" conception of humankind. On the one hand, it retains a sense of human uniqueness anchored in the powers of reason, imagination, and self-awareness, but their genesis is now given a broadly naturalistic grounding. Splitting the difference between materialism and dualism, Huxley viewed the special features of the human mind as emergent cognitive outcomes of

[5] In his discussion of Bostrom, Sorgner (2016) also points out that Pico's Neoplatonism makes him a rather unrepresentative figure of Renaissance Humanism, which was dominated by Aristotelianism.

[6] We borrow this felicitous phrase from McInerny and O'Callaghan (2009).

a deep evolutionary chain of which humans form part but which we nevertheless have managed to transcend and eventually redirect.[7] At the same time, the dawn of human reason also represents an inflection point for evolution on a cosmic scale, as it equips the universe with a sensorium to peek into its own history and catch a glimpse of its future.[8] Having arrived at this critical juncture, transhumanism was Huxley's way of posing the Augustinian *quaestio mihi factus sum* ("a question I have become for myself").

Huxley embraced science and technology as powerful media for reimagining what it means to be human but regarded them as subservient to a deeper concern for (trans)human flourishing. In a passage one might consider Huxley's "this-worldly" journey of a self-improving *paideia*, he emphasized the recognition "above all, that there are two complementary parts of our cosmic duty – one to ourselves, to be fulfilled in the realization and enjoyment of our capacities, the other to others, to be fulfilled in service to the community and in promoting the welfare of the generations to come and the advancement of our species as a whole" (1957, 17). In the service of creating more favorable social environments, Huxley stressed the central value of education and spiritual development in bringing about the envisioned goals of his transhumanism (ibid.) True to his training as a scientist, he viewed the scientific method as our most reliable intellectual tool for understanding our place in the natural world. But his conception of science was by no means narrowly restricted to the "natural sciences". Recapitulating the "reflexive turn" of evolution, Huxley asked for the development of a comprehensive "psychosocial science" (1959, 403) that would turn the traditional subject of scientific inquiry—the human— into its privileged object. Such a science would be both descriptive (a science of human nature) and normative (a science of human possibilities)—in other words, a progressive educational method for humanity to represent and intervene into its own nature in ameliorative ways.[9]

[7] For a book-length discussion of Huxley's "evolutionism" within the history of transhumanist thought, see Dunér (2025).

[8] See Delio and Redding (both this volume)

[9] With apologies to Ian Hacking (1983).

3 Coping with "Legacy Code"

Our readers will have noticed (e.g., in our treatment of Pico de Mirandola and Julian Huxley) that we are quite comfortably moving between the secular and the sacred—that is, between describing transhumanism as a scientific and as a religious phenomenon. Indeed, we believe a characteristic feature of transhumanist discourse is to afford thematic transpositions between secular and religious motifs. Should we consider transhumanism as a religious phenomenon, inspiring a secular vision? Or, as a secular phenomenon, inspired by a religious vision[10]? In our view, these questions rest on the flawed presupposition to demarcate science and religion into two separate realms of human existence—as "nonoverlapping magisteria" (NOMA), in the language of Stephen Jay Gould (1997). Especially (but not only) when it comes to the history of technology, the relationship between the two is far more intimate and contentious—that is to say, more interesting than NOMA would have it. To borrow a notion from evolutionary biology, it would be far more accurate, historically speaking, to consider the entanglement between science, technology, and religion in the Western tradition as a *homology*, united by a shared intellectual ancestry yet subsequently taking on different forms and functions.[11] This puts transhumanism on the map as a site of their potential reconvergence.

Here, we'd like to suggest an alternative way of framing the exchange between the secular and the sacred, specifically in the context of transhumanism. It would be to treat religious motifs as a kind of "legacy code", inscribed in the technologies we create, that we would do well to learn how to decipher, preserve, or—if appropriate—modify for future use. In the field of software engineering, from which the metaphor is borrowed, "legacy code" was long used as a pejorative term to describe the cumbersome task of maintaining code that has become hopelessly obsolete due to changes in or of the platform for which it was originally designed. This negative view mirrors the skeptical posture of secular transhumanists who make it a point to distance themselves from religious sensibilities which they view as mired in irrational dogma and superstition.

[10] Our formulation is indebted to Redding (this volume).

[11] See, e.g., discussions by Funkenstein (1986), Jaki (1988), Noble (1997), Brooke (2005), Harrison (2009, 2010, 2015), Hurlbut and Tirosh-Samuelson (2016) and Gouw et al. (2022).

In recent years, however, there has been a veritable "transvaluation" (in the Nietzschean sense[12]) of "legacy code" (McGee 2005). To wit, a more positive meaning has been attached to the deliberate practice of inserting code into new software for the purpose of maintaining older functionalities which otherwise would no longer be supported by current technology.[13] This view puts the coded creation of "legacy systems" on a par with other legacies we cherish in our lives and take pride in passing along from one generation to the next. We might call this a "precautionary" attitude, invested in the preservation of what has worked in the past for future contingencies.[14] A more "proactionary" attitude is to treat the technological decisions we made in the past, which have cumulatively shaped the decisions we face today, as a learning context to avoid past mistakes and guide future improvements. Going beyond the practical benefits of preventing malfunction, the more proactionary approach to legacy code management would stress the innovative potential of understanding recurrent patterns and problems in our dealings with technology.[15] We invite our readers to test the suggested metaphorical framework as they are browsing through the chapters to follow.

4 Chapter Overview

As propaedeutic to our volume, it will be helpful to restate three prominent leitmotifs of ethical debates of transhumanism that are played out in different keys in the present work, ranging over scientific, philosophical, theological, and even poetic registers. The first leitmotif, *identity*, concerns the question of who we are—what it means to be human. The second leitmotif, *embodiment*, concerns the significance and value of physicality to human existence. The third leitmotif, *perfection*, concerns

[12] We're using the Nietzschean term "transvaluation" loosely, in the spirit of Fuller (2019). On assessing the relationship between Nietzsche's ideas and transhumanism, see the collection of essays in Tuncel (2017).

[13] An example would be the insertion of code into an operating system to support a serial interface even though most modern computers no longer have a built-in serial port.

[14] We borrow the terms "precautionary" and "proactionary" from contemporary debates about the ethics of risk-taking, especially in the context of transhumanism; see Fuller and Lipinska (2014, Chapter 1).

[15] To be sure, we're not saying that a positive valuation of legacy code necessarily leads to an endorsement of transhumanism (in fact, it may lead some to conclude the opposite!). However, we do believe it's a stance which encourages a critical yet constructive exploration of the nexus between science and religion.

the question of what we can ideally hope to become. One way or another, each of the essays takes a definite stance on these fundamental questions. This gives them diagnostic value in comparing the wide spectrum of visions of (trans)human flourishing that are expounded in this volume, and the metaphysically diverse views of "human nature" on which they rest. A reader with skin in the transhumanist game may want to use our three leitmotifs as a scorecard to gauge the plausibility of the different treatments they receive. We hope our thumbnail sketch of what's to come will help.

At this point, it is worth acknowledging that our contributors have their (institutional) home in different fields. Their respective approaches to the central theme of our volume reflect these differences in standpoint and training, including their choice of register to engage the three leitmotifs that are guiding our inquiry. It may take some adjustments from the reader to discern the core melodic ideas and how they transpose between different essays, but we believe the rewards are well worth the extra effort. With some hesitation, we have grouped our chapters into two parts: "Transhumanism and Human Flourishing" (Part 1) and "Transhumanism and Christian Philosophy" (Part 2). This division is not meant to reinforce the unhelpful categorical divide between philosophical and theological anthropology. To wit, our authors from Part 1 are grappling with notions that reveal a complex religious legacy, and our authors from Part 2 use a theological hermeneutic to grapple with developments in advanced science and technology. Moreover, since we strongly believe that the religious "legacy code" of transhumanism can be deciphered from the perspectives of any of the world religions, we felt the need to flag the noteworthy restriction of our present sample (in Part 2) to Christian philosophy. Within this ambit, our volume still gathers a variety of religious faiths, including Roman Catholic, Presbyterian, Lutheran, Eastern Orthodox, and Unitarian denominations.

5 Part 1: "Transhumanism and Human Flourishing"

Part 1 juxtaposes a variety of philosophical frameworks from which the transhumanist ambition of reconfiguring and elevating the human condition has been imagined. In the opening essay, **David Wood** plots a course towards "a transhumanist state of sustainable superabundance for all" based on his ethical framework of "active transhumanism". Organized around 13 principles, Wood describes it as a delicately balanced "*mixture* of brakes and accelerators" to steer the development and deployment of NBIC technologies closer towards the intended destination. He breaks down the multifaceted notion of "sustainable superabundance for all" into the "five supers": *superlongevity*, *superintelligence*, *superhappiness*, *superdemocracy*, and *supernarrative*. They refer to a concurrent series of profound transformations of human life which Wood alternatively dubs "the five rejuvenations"—with the proviso that we're talking not about a restoration to some previous state but hitherto unattainable levels of health and wellbeing. Religiously minded observers will see in the imagery of "rejuvenation" or "healing" an oblique reference to the physical and mental ailments besetting "human nature" as a result of the Fall. From this vantage point, what might seem like a radical enhancement relative to our postlapsarian predicament would in fact be considered "therapeutic" by prelapsarian standards (Harrison 2009).

Wood's chapter clears the ground for the discussions to follow. It is aimed at exposing "naïve" conceptions of transhumanism which cling to simple slogans and ideologically motivated distortions which side-step real-world complexities of the interplay between humans and technology. They cast what he calls a "transhumanist shadow"—a collection of attitudes and practices that have gained currency because they are promulgated by prominent "influencers" who are not, however, representative of the larger community. Their circulation in popular media can make the movement an easy target for thoughtful observers—a vicious circle, for Wood, which in effect prevents a deeper engagement of the public with its core ideas. Part of the "transhumanist shadow" are risk complacency, the lure of determinism (both technological and genetic), the loathing of biology, scientism, techno-solutionism, anti-regulation fallacies, anti-money fundamentalism, "trickle down" transhumanism, and others. Wood expands on his 13 ethical principles, with references to the Transhumanist Declaration, to dispel the magical sway which the "shadow" seems to hold over the public perception of transhumanism. He embraces a diversity of conceptions of what (trans)human wellbeing involves, how to measure it, and how to improve it; provided they are constrained by two non-negotiable principles: "the fundamental importance of *individual* wellbeing, and the observation that our individual wellbeing in turn depends on the responsible exercise of *mutual welfare*". In the remainder of his chapter, Wood points to five initiatives that are integral to his "active transhumanism": the uplifting of education, collective intelligence, emotional intelligence, politics, and creative intelligence.

Alcibiades Malapi-Nelson directly challenges the prevailing narrative that transhumanism is necessarily tied to utilitarianism, and argues that alternative ethical frameworks—most notably, principles of virtue ethics—ought to be taken seriously as contenders. To make his case, he first hones in on two ethical quandaries widely acknowledged to challenge a purely utilitarian justification for transhumanism. First, framed as an issue of justice, transhumanism threatens to exacerbate not only socio-economic inequalities between individuals as well as countries, but also between generations, unfairly inhibiting the young from integrating themselves into a society dominated by "temporal super predators" (Fuller 2019) blessed with indefinitely long health spans. Second, framed as an issue of choice, morphological freedom leaves transhumanists with a seemingly endless proliferation of lifestyle possibilities to ponder, yet wondering "what moral beacon are we to follow for designing this human 2.0"?

Underlying these challenges, Malapi-Nelson argues, is that transhumanism is caught in an impasse between traditional understandings of morality based upon the formation of "character" through the practice of "virtue" (from Aristotle and Aquinas to Nussbaum and MacIntyre), pitted against an "ethics of modernity" (Bentham's utility and Kant's duty) which prizes the freedom of human agency above all else. This way of drawing the fault lines seemingly leaves transhumanists little choice but to embrace the latter. "Thus, whereas for virtue ethics, human nature determines the sphere of possibilities for the moral act, for utility and duty, this nature is the springboard, or raw material, from which human agency reaches for new heights of moral and personal accomplishments." In Steve Fuller's assessment (Fuller 2011), the virtue-ethical standpoint can be elaborated in several ways though all effectively lead to the same, transhumanism-averse outcome. Following the Thomistic principle of *agere sequitur esse* ("action follows being"), a cornerstone of the natural law tradition, humans are destined to grow into their divinely decreed niches as given by the *scala naturae*, which is presumed "fixed and immutable". To be sure, after Darwin, the normative horizons of human development are often "re-naturalized" by appeals to our so-called "shared evolutionary heritage" as the foundation of human flourishing. Either way, when their hands are forced, transhumanists facing the ethical quandary: "*Does function follow form or form follow function?*" (Fuller 2013,

51) will happily embrace the latter, rejecting any implicit appeal to the fixity of human form.

As a way out of this impasse, Malapi-Nelson suggests there is room for an "Aristotelean transhumanism". One way to see this relatively unexplored option is to distinguish between morally valuable *ends* and morally permissible *means* for achieving them. For traditional virtue ethics (VE), the ultimate moral end of human flourishing is rather concrete, and our paths for getting there are fairly heavily constrained by "human nature". For the ethics of modernity (ME), moral ends are conceived more abstractly in terms of duties or utilities, while our means in this pursuit can be relatively freely chosen. A hybrid account of "transhuman flourishing" would combine the greater specificity of moral *ends* mandated by VE with the greater latitude of *means* permitted by ME. To develop this suggestion, Malapi-Nelson argues that radical enhancements to *either form or function* are bound to necessitate corresponding upgrades to the other. Going off the Thomistic *agere sequitur esse* principle, a genetically or prosthetically modified human being will need to develop an altered set of competences and responsibilities to fulfill its moral end—along the lines of "to those who much is given, much is required" (Luke 12:48). At the same time, if "transhuman" is taken to refer to an intermediary *transition* between current and future humans (rather than a *replacement*), there must be some structural continuity of *being* in the human form which survives the envisioned alterations. Malapi-Nelson identifies this locus of structural continuity with the possession of a mind, the rational aspect of our species-being which is directly tied to human action, and hence its *telos* of (trans)human flourishing.

Taking up Malapi-Nelson's challenge, **Brian Patrick Green's** chapter examines the prospects for transhuman flourishing by investigating the relationship of transhumanism and Aristotelean thought. Aristotle's work is of particular interest to this volume not only because it is foundational to eudaimonist ethics, but because it reveals a "tension between the normativity of nature and the normativity of transcending nature", which Green's chapter places in fruitful dialogue with transhumanism. Transhumanists have claimed a kinship with Aristotle's "perfectionism" which seeks the attainment of intellectual and moral excellence—the actualization of inherently valuable potencies of the human soul—as the highest forms of human perfection, and thus inherently valuable. At the same time, this aspiration is held in check by an equally fundamental principle of Aristotle's ethics: that nature is teleological and normative, that "nothing which is contrary to nature is good". Within the context of Aristotle's hylomorphism, these two principles seem to pull into opposite directions. The most normatively salient, uniquely human potencies that must be actualized are the transcendent virtues of science, understanding, and wisdom, but those are precisely the ones towards which the human animal, in its "natural" state, seems relatively ill-disposed.

Following the philosopher of technology Carl Mitcham, Green draws a distinction between *cultivation* and *construction* to relieve this tension: "Cultivation draws a living entity towards its natural perfection; while construction constrains nature into a form it would not otherwise produce." As Green notes, the distinction derives from the different roles humans play vis-à-vis the system being completed. In the case of construction, humans are altering systems they have engineered themselves (*techne*), whereas in cultivation, they are merely tweaking "someone else's work"—in Aristotle's case, nature (*physis*). Building on this distinction, Green develops an Aristotelean typology of transhumanism in terms of what type of perfection it is reaching for, what means are to be used to attain this end, and to what extent transhumanist alterations would count as "natural" in an Aristotelian sense. His typology first divides transhumanism into two strands: embodied transhumanism, which maintains a human body, and disembodied transhumanism, which does not. Embodied transhumanism is further subdivided into transhumanism which cultivates and fulfills human nature and transhumanism which replaces or reconstructs human nature. Green morally evaluates each strand of transhumanist thought through an Aristotelian lens, concluding—specifically with respect to the second subdivision—that "[t]here remains a genuine tension between our natural rational *telos* of divine knowledge and our identity as human beings, made of organic matter."

Christopher Drain focuses on the work of Soviet psychologist L.S. Vygotsky (1896–1934) to articulate a "Vygotskian transhumanism" which straddles (or perhaps subverts) the Aristotelean division between cultivation and construction, siding with Marx (as opposed to Aristotle) on the question of human nature. For Marx, the nature of human "species-being" (*Gattungswesen*) is neither the expression of a timeless essence, nor the actualization of biologically preformatted potentialities. Instead, the process of becoming human is "always already" socially and culturally mediated in ways that are inherently connected to the human transformation of nature, and thus contingent on historically specific modes of production. Vygotsky built up Marx' anthropology into his scientific research program of "cultural-historical" psychology, jockeying for influence in the nascent socialist academy. Central to his program is the thesis that the so-called "higher mental functions" (unique to humans) are transformed versions of "lower mental functions" (shared with non-human animals). On both an evolutionary and a developmental scale, the developmental transitions along this path are mediated through the internalization of sociocultural practices (such as labor, tool use, and language) which, redirected internally, enable humans to achieve increasingly higher forms of self-regulation (parallel to higher levels of external mastery over nature). Vygotsky's opus has been a bellwether for the contemporary medley of ideas travelling under the banners of distributed cognition, the extended mind thesis, cultural-cognitive niche construction, neural constructivism, and their ilk—epitomized by Andy Clark's evocative image of humans as "natural born cyborgs" (Clark 2003). This makes it attractive to transhumanists, since Vygotsky's thesis effectively "normalizes" technological self-enhancement as the continuation of a species-defining activity we have been doing all along.[16]

The focal point of Drain's chapter are Vygotsky's writings from the 1920s in which he extrapolates his cultural-historical conception of human becoming into the future. Borrowing amply from Trotsky's *Literature and Revolution*, and centralizing the Marxian point of self-mastery, Vygotsky's "peak psychology" (a counterpoint to Freud's retrograde "depth psychology") indulges an exalted vision of what the human body (and therefore mind) may as yet be capable of. He anticipates a socialist future with a materially higher type of human, a "superman", that "will resemble the old kind of man, 'the old Adam', in name only" (Vygotsky 1994, 183). Drain reconstructs Vygotsky's position as a brand of "eudaimonist transhumanism" where human excellence is understood as radically contingent on historical, cultural, and technological factors. In contrast to those (not infrequent among the Marxist left of the early twentieth century) who took Nietzsche's musings about the "superman" as an inspiration for state-sponsored eugenics, Drain emphasizes that "Vygotsky, and those who currently find inspiration in Vygotsky, thus take not biogenetic but ontogenetic engineering as the key to human transcendence".

The second part of Drain's chapter investigates the ethical and political implications of Vygotskian transhumanism, with its characteristic insistence that any valuable transformation of the biological organization of

[16] See Theiner (2021a, 2021b), Sorgner (2021) for an elaboration of this theme.

man must result in "*the social liberation of man*" (Vygotsky 1994, 183). Vygotsky's program recalls Marx' developmental perfectionism, which looks towards a future organization of society in which "fully developed individuals" freely and creatively exercise their human capacities to the fullest, in collective realization of their finally liberated "species-being". This puts Vygotsky in conversation with Aristotle, virtue ethics, and Nussbaum's capabilities approach. Drain concludes that despite overlapping commitments, their rapprochement remains limited, harking back to the key distinction between *physis* and *techne* that was also noted by Green: "Marx's 'productivist' anthropology maintains that human capacities are self-cultivated. Aristotle, on the other hand, recognizes no such thing; though he acknowledges the history of political association, this crucially does not entail the historicity of human potentials."

Steve Fuller continues his long-term campaign against the "Linnean equation[17]", understood as the identification of human nature with biological membership in the species *Homo Sapiens* (the "crime against the concept of humanity", as per the title of his chapter). Going back to his first systematic treatment of transhumanism (Fuller 2011, Chapter 2),[18] Fuller has developed his conception in dialogue with Foucault's "disappearance of man" thesis and continues to do so in his present essay. As Fuller puts it, he and Foucault agree that the "Linnean equation" is a historically contingent construct with an expiration date, but they differ "in both substance and spirit" about what comes before and after. For Foucault, the early modern taxonomic construction of the "human" hinged on adopting an emerging conception of "intentionality"—a distinctly human capacity for meaning-making—as the hallmark of humanity's privileged status as "rational ape". That foundation was finally dismantled, according to Foucault, in the nineteenth century, by the "masters of suspicion" Marx, Nietzsche, and Freud. Fuller challenges Foucault's larger narrative arc: "The biggest difference between us is that I regard humanity primarily as a concept—not an object. Whereas for me *Homo Sapiens* marks the moment when the concept of humanity coalesced around a concrete object, for Foucault that is the only time in which humanity has ever existed."

[17] See Fuller (2021, 2022)

[18] For a review of Fuller's first (2011) and most recent (2020) book on transhumanism, see Malapi-Nelson (2013) and (2021) respectively.

Contra Foucault, Fuller suggests to "think of *Homo Sapiens* as a prototype or 'proof of concept' for *humanity*, but not the ultimate standard-bearer of the value—or set of values—represented by 'humanity'". In the idiom of Quine, Fuller asserts that "to be human is to satisfy the open predicate, 'X is human'"—with transhumanists claiming that there can be instantiations of X which exemplify humanity better than current specimens of *Homo Sapiens.* Under this common umbrella, different variants of transhumanism offer competing visions of which features in the constitution of today's *Homo Sapiens* should be retained or amplified in the construction of tomorrow's "Humanity 2.0", and which should be diminished or eliminated. The resulting indeterminacy presents us with a new "riddle of induction" (Goodman 1955)—with "human" as a "projectible predicate" admitting of an open-ended variety of transhumanist futures which register somewhere between "the '*god*' and '*ape*' poles", aka the "great *carbon-silicon* divide" (cf. Fuller 2011).

As Fuller notes, this new riddle creates an ethical "meta-problem" due to "a lack of agreement over the range of beings entitled to be treated 'humanely'". Fuller links this question, which we already start to see playing out in debates over animal vs. machine rights, to the ancient art of *theodicy*—the tricky business of trying to figure out "what to believe and how to act in a state of profound ignorance and uncertainty. The original object of this ignorance and uncertainty was the 'Mind of God'." In the context of transhumanism, where humans are acting as both creature and creator, the problem of theodicy resurfaces in the logic of recognition, "as defined by two complementary questions: How do you recognize a human? How do you become recognized as a human?" Fuller concludes his chapter by considering a "reflexive application" of the Turing Test to adjudicate these matters and lays out four key questions for future research on the topic.

Stefan Lorenz Sorgner has been a leading proponent of an alternative version of transhumanism—dubbed "Euro-Transhumanism" in his chapter[19]—developed in conversation with kindred spirits "whose thinking primarily bears constructive interferences with the Continental European cultural tradition from Heraclitus via Spinoza and Nietzsche into our times". Sorgner's programmatic statement of twelve differences between (what he calls) "Classical Transhumanism" (CT) and

[19] For a recent essay collection discussing Sorgner's work, see Schussler & Balistreri (Eds., 2024), *Metahumanism, Euro-Transhumanism, and Sorgner's Philosophy.*

"Euro-Transhumanism" (ET) is animated by a concern reminiscent of Wood's "transhumanist shadow"—that a particular strand of transhumanist theorizing, rooted firmly in the Anglo-American cultural tradition and championed by the likes of Ray Kurzweil, Elon Musk, and Peter Thiel, is wrongly taken in the public eye and by academic critics alike as representative of transhumanism in general. Sorgner is worried that many of the professed goals and aspirations of CT such as mind-uploading, digital immortality, or the abolition of human suffering are in fact neither physically realistic nor ethically desirable, giving transhumanism a bad reputation. In his view, the central, actionable concern of transhumanism ought to be formulated as follows: "It is in our interest to push our current boundaries in a significant way using the latest techniques, as this will promote the likelihood of individuals achieving a better quality of life." This formulation highlights the importance of health as a precondition for leading a fulfilled life, however diverse the plurality of goals we choose to associate with it. Thus, when transhumanists advocate the use of advanced technology to increase people's health span, their position should find widespread support, since most (though not all) people associate health with a successful life one way or another. The fight for affordable schemes of universal health care should thus be a major transhumanist cause (Sorgner 2021).

Running down the list of differences between ET and CT, the reader will find that some differences are a matter of emphasis, others are more substantive and/or "meta-philosophical" in nature, and yet others turn on a difference in political orientation. For example, whereas CT flirts with a techno-utopian outlook (e.g., with its language of "paradise engineering"), ET eschews utopian promises which can all too easily turn into political apologetics for sacrificing the "deficient" present for a perfect future. CT affirms its continuity with Enlightenment rationalism, whereas ET is highly critical of the logocentric Enlightenment tradition (from Descartes to Kant) yet supportive of the body-centric Enlightenment tradition (from de Sade to Nietzsche). This means ET has an ambivalent relation to the Enlightenment: it rejects the categorical mind-body dualism which upholds the primacy of Reason (with a capital "R") as a source of ultimate truth, but without negating the special significance of reason for the embodied human mind. From this starting point, several of the other differences can be deduced. Sorgner contends that without metaphysical dualism to back it up, a non-formal notion of "the good" is philosophically indefensible. Thus, ET embraces a radical diversity of different concepts of the good, in opposition to (e.g.) Bostrom's "hyper-Renaissance ideal of perfection" and Savulescu's paternalistic "commonsensical" account of the good life. Unlike CT, ET resists the reduction of ethics to a glorified "cost/benefit" type of calculus, aka "utilitarianism", and embraces a hermeneutic understanding of ethics (philosophically influenced by Sorgner's former academic mentor, Gianni Vattimo).

In the final essay of Part 1, **Debashish Banerji, Francesca Ferrando, Stefano Rozzoni,** and **Joanna Pascoe** take us on a "mythical journey towards posthumanist ethix" to deal with the complex and ever-evolving challenges created by unparalleled forms of technological and environmental transformation in the Anthropocene. Their chapter is written from the perspective of Cultural Posthumanism (CP). In many ways, CP is an inversion of transhumanism insofar as it aims to *decenter* (rather than intensify) the qualities of humanity which form the metaphysical core of classical humanism (Hayles 1999; Braidotti 2013; Ferrando 2019). Thus, CP is post-anthropocentric in rejecting the exceptional status of the human as morally superior to other species, which it considers a blanket license for imposing its will to mastery and exploitation of the earth and all its creatures. CP is post-dualist in rejecting the complex legacy of a rationalist ontology which divides up the world into hierarchized binaries such as spirit over matter, life over death, subject over object, man over female, or colonizer over colonized. Instead, it emphasizes the deep interconnectedness between humans and non-human entities, such as animals, plants, technology, and ecosystems, as part of a complex, co-evolving system.

Importantly, though, CP shares with transhumanism the idea that technology is a prime vehicle—conceptually but also practically—to reconfigure our posthuman existence for good, by creating new "transversal alliances" which relate the posthuman subject "at the same time to the Earth—land, water, plants, animals, bacteria—and to technological agents—plastic, wires, cells, codes, algorithms" (Braidotti 2013, 117–118). In short, technoscientific transhumanism and cultural posthumanism both embrace an ethos of remaking of the human as a creative, agential project (Tirosh-Samuelson 2014; Haraway 2016). Along post-dualist lines, the authors develop in their chapter a posthuman "ethix" (fusing the terms "ethics" and "praxis") which is "to be embraced collectively, through a pluralistic, perspectivist, and affirmative glance, functioning as a *phronesis* for the 21st century". Echoing the etymological origin of *êthos* (ἦθος)—from which ethics derives—as denoting "character" and "personal disposition", the chapter is a kaleidoscope of different genres and writing styles (including poetry, photography, diary entries, and essays) which reflect the lived experiences of those who contributed. It is held together by a narrative centered on a human being discovering their posthuman subjectivity, self-consciously composed in the form of a self-similarity between style and content which recalls the ancient Greek trope of the *nostos* ("homecoming").

6 Part 2: "Transhumanism and Christian Philosophy"

Part 2 engages various traditions in Christian philosophy to decipher the transpositions between secular and religious motifs which are emblematic of transhumanist discourse, to provide vision and guidance for developing a comprehensive ethics of transhuman flourishing.[20] Leading the way, **Micah Redding** frames the question of a Christian transhumanist ethics in terms of three central concepts of Christian theology that map directly onto the three leitmotifs of our volume: the notion of imago Dei (*identity*), the incarnation of Christ (*embodiment*), and the purposes and proprieties of Christian eschatology (*perfection*). Redding retraces the religious moorings of transhumanist thought, from the scientific theology of Francis Bacon, to the eschatology of Nikolai Fyodorov, to the evolutionary vision of Teilhard de Chardin which inspired the aforementioned "father of transhumanism," Julian Huxley. He shows that even Huxley, who viewed transhumanism as a "secularized religion", did so with the intent of reintegrating ideas and concerns of a broadly religious character into the naturalistic world view of modern science. Redding's recommendation is "[f]or Christians [...] to recognize the offshoots of our own theology, and address them as such. In doing so, we can learn to recognize the ways our own theology can lead to excesses, and the ways our traditions have learned to mitigate those risks. [...] For transhumanists, [...] they should lean into these themes and their origins, and embrace them as the rediscovered theological frameworks they are."

Redding reaffirms the centrality of *imago Dei* as foundational to human dignity, as a moral bedrock to limit and check purely utilitarian considerations. He agrees with the evolutionary vision (expressed by Huxley and

[20] Within this context, Malapi-Nelson (2022) argues for an increased alignment between the views of the Catholic Church and transhumanist values in virtue of their shared goal of promoting human flourishing.

Teilhard) of humanity as a cosmic "inflection point", an emergent evolutionary force uniquely equipped with some God-like capacities and thus sharing in special responsibilities vis-à-vis the created world. Going well beyond the ambit of redirecting *biological* evolution, Redding appeals to David Deutsch's notion that human minds, *qua* being endowed with "creative universality", can in principle realize any possibilities not precluded by the laws of physics, including the ability to fuse our cognitive horizons with a future superintelligence. Discussing the eschatological leanings of transhumanism, Redding responds to concerns by Moos (2016) that a technological "immanentization of the eschaton" runs the risk of losing sight of the strategic insights of "religious reason" to cope with the "secular gap" between the immanent and the transcendent. Those include developing a sophisticated vocabulary for acceptable "God-talk" or managing human expectations in the face of eschatological delay. Redding agrees with Moos that transhumanists ignore these insights at their own peril, but points to a variety of thought experiments (such as "Roko's Basilisk", the "Simulation Argument", or the role of the "Singularity" in coming to understand "meaning of life") through which transhumanists have developed their own ways of reckoning with the "secular gap" between immanent realities and transcendent demands.

Finally, Redding aptly introduces the significance of the Incarnation for transhumanism: "In the majority tradition, Christ is 'God become man', and also, though sometimes spoken more quietly, 'man become God'. God has taken on human flesh, and rather than discarding it again, has raised it up to partake of his absolute transcendence." Within the Christian tradition, the theological discourse about the person of Jesus having a "duplex" nature (combining human and divine attributes) has been a vehicle for exploring the ontological precariousness of the human, suspended between being "animal" and aspiring to be an "ideal" (Fuller 2011, Chap.2). The ramifications of Christology for transhumanism—and the ethics of human transformation in general—are profound. On the one hand, religiously minded transhumanists have endorsed a relatively literal-minded interpretation of the Christian teaching—common among the church fathers—that "the Word of God became man, that you may learn from man how man may become God". As Redding notes: "This is *theosis*, and in a sense is the *imago dei* played back in a higher key." Critics, in turn, rebuke the transhumanist misappropriation of this notion to conceal a hubristic desire for achieving god-like perfection, underhandedly paving the way towards idolatry. Redding suggests that this theological dispute will partly turn on the choices we make in years to come: "*Who is the agent of transformation—us or the Holy Spirit?* A Christian can employ the language of synergism, where our wills cooperate with the Spirit. *How do we attain transformation—technology or faith?* A Christian can employ the language of salvation by faith, expressing itself through works, manifesting itself in tangible blessing."

Rev. Christopher Benek picks up the Christological theme in his chapter, which frames Christian transhumanism as a technological hermeneutic for humans to understand how they might more fully participate in Christ's redemptive purposes. Benek is acutely aware that a flawed (or at least severely limited) understanding of what it means to be human is bound to yield pathological realizations of "transhumanness" in the future. Ten years ago, he took to task Zoltan Istvan's "transhumanist hero" who seeks to overcome himself, Zarathustra-style, by thrusting aside conventional morality in the name of unlimited, technologically supercharged self-realization (Benek 2014). Benek raised the concern that Istvan's unfettered idolatry of the individual (I+) would lead to a forfeiture of our humanity (H-) instead of a bettering of it. "For as soon as we make gods of ourselves in the way Istvan advocates, our self-absorption dehumanizes us by separating us from our larger relational community." For Benek, the coming technological journey ought to be one "to enhance humans collectively and, as such, our understandings of what it fundamentally means to be human". For Benek, the explicit focus on community betterment means that religious traditions with centuries of prior experience in this regard surely have much to offer to ethical discussions of transhumanism.

In the present chapter, the main target of Benek's critique is very much an antipode of Zoltan's "transhumanist hero": namely, the bioconservative view of Christian evangelical author Andy Crouch, whose diatribe against transhumanism (Crouch 2019) Benek describes as a "religious techno-hesitancy and anti-enhancement" position that has taken root in evangelical circles. It was written on the heels of Crouch's (2017) *The Tech-Wise Family*, a book aiming to give practical advice on how to use technology intentionally in the home to build stronger family relationships and live more fully by putting technology in its "proper place". Benek agrees that Christians, like everyone else, should be concerned with how we have come to rely on technology in our daily lives. Yet, he contends that Crouch's practical skepticism has turned into a deeper, theologically motivated indictment of technology in general, demonizing its adverse impact on human personhood. Summing up his argument, Crouch claims "that true human flourishing is found in relationships, community, and embodied presence, not in technological convenience or superpowers" (Crouch 2022), and that "instead of exponential advancements – Christians should focus on the example of Christ, whose incarnation teaches the value of ordinary, embodied human life" (Crouch 2019).

Against the backdrop of "the life Andy Crouch is looking for", Benek develops an alternative Christian hermeneutic which puts technology at the very heart of creation, and thus not at odds with God's redemptive transformation of humanity but rather an integral part of it. Benek draws on work in digital physics and related approaches which suggest that the universe, at bottom, is a giant computer, bringing forth a reality which at its most fundamental level is informational. This means everything in existence, including all biological life, can on some level be viewed as digital information technology—and thus as divine technology, from the standpoint of the Christian Creator-God. Within this universe, humans created in *imago Dei* are God's own AI ("Alternative Intelligence"), destined to act as technological stewards "to support the advancement and application of technology with the intentional aim of participating in Christ's redemptive work for humanity and all creation". As Benek notes, "[t]his should encompass all forms of digital and technological expression, whether it's crayon drawings on walls, playing in the mud, marveling at bumblebees, learning the violin, engaging online or in virtual reality, gene editing, performing miracles, or any other God-inspired activity we are called to undertake. In every creative act, we should strive for the collective well-being of *all* humanity."

Benek goes on to wed his technological hermeneutic to the Christian conception of the triune God. He interprets Christ's resurrection and ascension as "technological events via digital information" which "point to the reality that humans will change and upgrade". The transformed nature of the Risen Christ challenges Crouch's fixation on the value of "ordinary, embodied life": "He is effectively Jesus 2.0, unrecognizable to disciples who know him, potentially like a shapeshifter Jesus, until he performs ritual action—i.e., an identifiable (digital, technological) informational pattern." In sum, Benek agrees with Crouch (and holds it against Istvan) that "part of being human means living in a relationship with others and God" and to the extent that our use of technology alienates us from this calling, it dehumanizes us. But against Crouch, Benek reminds Christians to recognize "that our current relationship to God is via the

Holy Spirit—an informational [...] person of the trinitarian God embodied in individual human persons with specific gifts enacted corporately in the community who collectively represent Christ." Through advanced technological stewardship, we thus participate in Christ's redemptive purposes.

Ted Peters pours cold water on the notion that there could be a convergence between the transhumanist plan for cyber-enhancement—be that through intelligence amplification (IA) or artificial intelligence (AI)—and the scriptural promise of spiritual transformation of the "Old Adam into the New Adam". Peters makes it clear from the outset that the two projects take off from different launching pads—secular transhumanists prize *intelligence* and followers of Jesus value *love* as their respective *summum bonum*. That said, the real target of his chapter are ongoing attempts by Christian transhumanists to find some substantive overlap between these two projects. While the goal of his chapter is to demonstrate their incompatibility, it goes to Peters' credit that he sincerely tries to position the two projects in ways that at least make their comparison worthwhile. For those less familiar with the Christian tradition, we recapitulate a few of Peters' points of departure.

Setting out the problem, Peters cites methodist John Wesley's description of Christian perfection:[21] "Pure love reigning alone in the heart and life", and the words of St. John's that "Love has been perfected among us ... We love because [God] first loved us" (1 John 4:17–19). He makes these points to emphasize that: "To transform the self from self-centeredness to God-centeredness has been the goal of spiritual discipline for two millennia." The Christian aspiration for a transformed or "sanctified" self, then, fundamentally involves a recognition of *relationality*: "the self re-centers itself in what transcends the self. This makes possible the pursuit of virtue and loving the neighbor for the sake of the neighbor." Peters continues with Christian soteriology: "The *Adam* of Genesis, to whom we are presently heir, was subject to tragedy in a way that the new *Adam*, who participates in the resurrection of Jesus Christ, will not be." Building on Irenaeus, Peters suggests "we think of Christ's work as *pre*capitulation. Christ establishes ahead of time what it is that will define who we as humans shall be". As the first appearance of the New Adam, "Christ as the divine image is our prototype. We live now as the *imago Dei* insofar as we live in him, insofar as we participate in the reality of the future self

[21] On the history of Christian perfectionism, see Passmore (1970, Chapter 7).

that we will be." Aided in this quest, it is "God's Holy Spirit that binds our self-centered self today with our sanctified self of tomorrow." For Christian believers, these are central articles of faith. Thus, the stage is set for Peters' inquiry. Can Christian transhumanism deliver? Might not Christian perfection and technological perfection go hand in hand?

Peters points to two key shortcomings in the transhumanist position which severely limit any amount of overlap between what he thinks are two very different views of human transformation. The first concerns a background supposition regarding the notion of "intelligence" which transhumanists seek to amplify. Peters argues that an essential trait of human intelligence, inherited from its role in biological evolution, is "self-generated intentionality manifest as agency". As of this date, says Peters, no AI can set its own goals. We humans set those goals. Relatedly, Peters argues that human intelligence implies self-awareness—having a first-person perspective on the world around us and using it reflectively in our actions. He samples five extant models of the self which transhumanists might try to cyber-enhance. He finds them severely lacking in what to a Christian should matter most: namely, the ability of a self to *center itself.* But without the ability of re-centering oneself in God, the road to Christian perfection—and thus to genuine *eudaimonia*—is closed off. Citing Buber: "We become a fulfilled self only when we treat God as a Thou and we treat our neighbor as a Thou." The second weakness is closely related to the first: "the human self that would persist into superintelligent techno-utopia would remain the untransformed or unsanctified self. In short, the current human propensity for selfishness accompanied by competition and violence would be perpetuated in the transhumanist scenario." Notably, Peters specifically objects to the transhumanist appropriation of Eastern orthodox *theosis* ("divinization" or "deification"). It prompts a mistaken reliance on technology as a vehicle of human self-transcendence towards the divine, a role that is reserved only for the sanctifying grace of God's Holy Spirit.

The chapter by **Ljiljana Radenovic** picks up right where Peters' left off. She explores the concept of *theosis* in Eastern Orthodoxy, with particular emphasis on the philosophy of nature in the work of the Russian Orthodox polymath Pavel Florensky (1882–1937). Radenovic articulates her version of Christian transhumanism as a counter to Wesley Smith's (2022) recent critique: that transhumanism rests on a reductive materialist metaphysics, with a blithely self-centered and egoistic agenda that puts it at odds with the Christian values of compassion, charity, and spiritual development for the sake of *theosis*. Most importantly, Smith argues that transhumanism is "a wrong-headed project because it aims at transforming our God-given human nature in the wrong way". Radenovic retorts that the belief in a "fixed human nature" is a distinctly Aristotelian aspect of Western Christianity (see also Fuller 2019). Drawing on Eastern Orthodox traditions, she argues, *contra* Smith, that a strict ethical bifurcation between "good ways" of reforming our nature from *within* (e.g., through moral efforts or self-discipline) as opposed to "bad ways" of reforming our nature from the *outside* (e.g., through technological enhancement) cannot be maintained.

Radenovic starts from the Eastern Orthodox emphasis on the credo that God created humans with the capacity to reunite with God. Espousing a so-called "realistic" reading of *theosis* (cf. Jacobs 2009), this implies that human nature was essentially incomplete at its creation—rather than "autonomous and self-sufficient" (in the Aristotelean sense)—and is yet to be fully completed in the reunion. As Radenovic emphasizes, the "incompleteness" of human nature does not simply refer to an inherently human imperative to strive for moral perfection or a human psychological need to be close to God. Instead, it reveals a substantive ontological gap which human beings, to become *fully human*, must fill by reducing their distance from reunion with God. To repudiate the supposed difference between "spiritual" vs. "physical" means for doing so, Radenovic appeals to Florensky's metaphysics of matter. Opposing the atomistic picture of a "mechanistic worldview", Florensky held that "matter is never entirely devoid of the Divine, so it should never be treated as something despicable, something that should be discarded in our ascent to God." Florensky's metaphysics is a version of Platonist panpsychism, which asserts the ontological priority of the forms; but importantly for him, forms are not only spiritual but *individualized*. Citing Florensky (2015, 31): "If an organism is a metaphysical form and not just periphery of our body, then the form permeates every particle, and since the form is individual, it follows that every particle is individual and personal." Radenovic puts this point saying that "the form will pick its particles on its own just as the organism is able to select sugar and discard saccharine". It follows that human nature is both physical and spiritual at the same time since *all* bodies—including human bodies—are individualized instances of spiritualized matter. Florensky himself appealed to this principle to make sense of the resurrection—and metaphysically underpinning the technological enthusiasm for radical life extension and restoration of life to the dead which were *en vogue* among the Russian cosmists (Young 2012). The same principle sustains Radenovic's argument to bridge the gap between "therapeutic" versus "enhancing" uses of technology for salvific purposes. By improving or extending our bodies, we "bring more order (form) to matter", and thereby, according to Florensky, make matter more spiritualized. Florensky's well-known advocacy for the artificial reconstitution of the human organism to better equip future generations to live beyond earth should be viewed in this light.

In the final section of her essay, Radenovic stresses that an ethically appropriate use of enhancement technologies requires that we first develop an adequate understanding of our place in the universe. For transhumanists aspiring to be Christian, the criterion of ethical acceptability of a proposed alteration of human nature rests on the right motive: whether it's done with empathy, compassion, and charitable regard for the suffering of others (as a direct expression of our love for God). Here, she underscores the ethical significance of a "realistic" understanding of *theosis*: "human beings are not individual atoms but their identity as persons is co-constituted by their relationship to other human beings. Who we are is created in relation to others: in other words, in this world, we always come 'in pairs'."

Tracy Trothen aims to reorient the moral discourse surrounding the use of human enhancement technologies by engaging a feminist Christian theological perspective. Her perspective is anchored in a "fourth wave" feminist approach which emphasizes the theological values of relationship, interdependence, and social justice. Many of the ethical concerns about transhumanism that she raises in her chapter overlap with the posthumanist critique of the logocentric, rationalist subject as a normative category in the Western tradition. This is by no means a coincidence. The philosophical "turn to relationality" has prompted renewed theological interest in relational conceptions of *imago Dei* which define the concept of the human person as constituted in relationship, rather than as an autonomous, independently existing individual (LeRon Shults 2003; Herzfeld 2002). The "relational" model of *imago Dei* is often contrasted with the more traditional "structural" and "functional" models which featured prominently in previous chapters. It centralizes the interconnectedness between humans and God, other humans, and indeed all of creation—a conception which also reflects the relational nature of a triune God. Feminist theologians have utilized the relational model to show how certain historical articulations of the *imago Dei* have served as a tool for oppressing socially marginalized "others", but also as a basis for developing more inclusive theologies of the ecological and technological environments which mediate our social relationships (Ruether 1983; Fisk 2014). Trothen insightfully extends this relationalist approach in her discussion of human enhancement technologies, bringing it into fruitful conversation with the leitmotifs of our volume.

For example, concerning our self-understanding as human beings, Trothen suggests that our inflated trust in AI-powered decision-making, e.g., in the context of healthcare, betrays a "hyper-valuing of efficiency, utility, and supposed rational intelligence" which downplays the significance of embodied, relational ways of knowing. She connects this bias to the legacy of dualist thinking to split the world "into absurd dualities in which the first element is valued at the expense of the second", such as spirit/body, reason/emotion, or male/female. From a theological standpoint, this includes the traditional privilege of male-associated types of knowing, such as the divine "Word" over human emotion, or the transcendent/immanent dualism erected to separate God from the life of the World. If theology operates with exclusively human white male images of God, and we then conceive of AI in anthropocentric terms, "we keep encountering ourselves instead of the Other in our human-human and human-machine relationships, but also in relationships with God". For example, the familiar ethico-religious discourse centered on the human risks of "playing God" implicitly shifts "the focus on sin as individual pride" and "draws attention away from communal sin, and also from the possible value of pride." Moreover, it also obscures how technology might systematically impact *all* of life, including ecology and relationships.

Trothen recalls the centrality of the embodied incarnation of Jesus in the Christian tradition, which signifies that "the enfleshed body is confessed to be good". At the same time, she warns that Christianity has a complicated track record of dealing with embodiment: "the binary model of Mary as redeemed and pure, is contrasted with the sexual, disobedient and sinful Eve." Such tropes are all too easily extended into the world of chatbots and AI girlfriends. More generally, Trothen emphasizes that "embodiment offers salvific possibilities of connection and relationship. Embodiment also illuminates bias and prejudice". This leads her to consider to what extent we might be able to mitigate social injustice through an erasure, or at least radical transformation, of our physicality, such as via uploading into minds into a computer or by becoming cyborgs. For example, she refers to research suggesting that imaginary gender transfer in sexual harassment scenarios through virtual reality representations can reduce judgmental bias. However, the desired changes in attitudes and values are unlikely to result from technological transformations alone, unless they are explicitly ***intended***, as indeed they were in the study. More generally, she worries that "a shift away from our vulnerable fleshy embodiment" would promote increased social disconnection and a loss of caring.

In the final chapter, **Ilia Delio,** O.S.F., attempts a captivating synthesis between many of the dualities we have run up against in our volume, such as those between spirit vs. matter, emergence vs. panpsychism, science vs. religion, evolution vs. technology, and divine transcendence vs. immanence. Her essay begins by sharpening the contrast between the discordant conceptions of "human nature" associated with transhumanism vs. (cultural) posthumanism. For Delio, transhumanists long for "a post-biological future where super informational beings will flourish and biological limits such as disease, aging and death will be overcome". This could in principle be achieved in a variety of ways, including the creation of superintelligent AI into which we integrate ourselves, the uploading of our minds into a digital platform, or the gradual cyborgization of our bodies.[22] A commonly held assumption among transhumanists regarding the envisioned transition, "from the archaic biological species of *homo sapiens* into a new technologized species of *techno sapiens,*" is that it "brackets nature, as if biology is a mere passing phase in the overall evolution of life". The posthuman alternative, as portrayed by Hayles (1999), "does not really mean the end of humanity. It signals instead the end of a certain conception of the human"—a turn away from its Eurocentric image and toward the embedded, hyperconnected person. "Hayles' cultural achievement lies in a critical perspective on technology in which the human species limits itself to that of a 'co-evolving' partner in the relationship and against the technical will to disembodiment and immateriality."

Delio deftly connects the posthumanist perspective to the work of Jesuit scientist-priest and philosopher-theologian, Pierre Teilhard de Chardin (1881–1955). Teilhard anticipated that the next phase of human evolution, brought about by computer technology, will give rise to a new type of person: the ***ultrahuman***. Delio describes the ultrahuman as "a new type of person of extended cognition or complexified intelligence but also

[22] These three scenarios are mentioned in Kurzweil (2005), 197–202.

one marked by more spiritual energy. The ultrahuman is one who has consciousness of relational wholeness and seeks to advance in love." How do we become ultrahuman? Teilhard shared Huxley's belief that the human person is "evolution become conscious of itself". As Delio notes, Teilhard eagerly appropriated Huxley's phrase to characterize humanity as "a unique biological, collective, and global phenomenon, whose past, present, and future is intimately bound up with the formation, life and ultimate transformation of the earth". More specifically, "Teilhard's evolutionary paradigm is a development of consciousness through various levels of material complexity. Life evolves toward greater wholeness and higher consciousness by deepening the three levels of matter, mind and spirit." While Teilhard only caught a glimpse of today's information and communication technologies, he foresaw the emergence of a globally interconnected network which he called the noosphere (Greek *nous* = mind)—"a psycho-social process, a planetary neo-envelope *essentially linked with the biosphere* in which it has its root yet is distinguished from it". With the formation of the noosphere, thought is elevated from the level of the individual to the level of the collective, towards "greater spiritual energy through a maximization of consciousness and a complexification of relationships".

Pulling forward the universal cosmic trend towards progressive convergence is a "principle of centration" which Teilhard calls *Omega*—in theological terms, the emergent evolution of God who draws all things into himself (Delio 2011). Delio is careful to explain: "Omega is not merely the point that would eventually be reached by the process of evolution itself; rather, Omega is 'already in existence and operative at the very core' of evolution. That is, Omega is not a final cause of the evolutionary process as it is a formal cause; the energetic centric presence that undergirds the unitive nature of evolution toward more complexified wholeness. It is supremely present and supremely personal, a universal center that influences individually centered human personalities without annihilating those centers." For Teilhard, spirit and matter, but also God and the created world, are not separable. Just as spirit evolves *in* and *through* matter, Teilhard also believed that "God is not only *in* all things but *as* all things, which means God emerges with complexified matter". In contemporary parlance, we might say Teilhard aimed to developed a Christian version of panentheism that synthesizes emergentism and panpsychism.

Delio summarizes the ethical upshot of Teilhard's "ultrahumanism" as a corrective to transhumanism: "The significance of Teilhard's paradigm is that computer technology and artificial intelligence are inserted into a much broader process of cosmic evolution or big history. To view AI as a cosmic phenomenon, rather than merely a human one, eliminates AI as a threat to human existence. Rather, as Teilhard realized, technology becomes the next phase of human evolution toward Omega." Teilhard's vision for the future rests on a "*homo progressivus*" with faith in the spiritual energy of the human phenomenon, who sees the power of technology as a call "to reconsider the infinite potential of the human person to become a fully relational person, one who lives by the life of the whole".

References

Arendt, Hannah. 1958. *The Human Condition*. Chicago: University Of Chicago Press.

Schussler, Aura Elena, and Maurizio Balistreri, eds. 2024. *Metahumanism, Euro-Transhumanism and Sorgner's Philosophy: Technology, Ethics, Art*. Budapest: Trivent Publishing.

Belt, Jaakko. 2022. Eidetic Variation: A Self-Correcting and Integrative Account. *Axiomathes*32(2):405–434.https://doi.org/10.1007/s10516-021-09611-1.

Benek, Christopher. 2014. Humanity+ vs. Individualism+. *Hplus Magazine*. https://web.archive.org/web/20140716170912/http://hplusmagazine.com/2014/07/14/humanity-vs-individualism/.

Bostrom, Nick. 2005. A History of Transhumanist Thought. *Journal of Evolution and Technology* 14 (1): 1–25.

Braidotti, Rosi. 2013. *The Posthuman*. Cambridge, UK: Polity.

Brooke, John Hedley. 2005. Visions of Perfectibility. *Journal of Evolution and Technology* 14 (2): 1–12.

Clark, Andy. 2003. *Natural-Born Cyborgs*. New York: Oxford University Press.

Copenhaver, Brian. 2024. Giovanni Pico Della Mirandola. In *The Stanford Encyclopedia of Philosophy*, ed. Edward N. Zalta and Uri Nodelman. https://plato.stanford.edu/archives/fall2024/entries/pico-della-mirandola/.

Cortez, Marc. 2010. *Theological Anthropology: A Guide for the Perplexed*. New York; London: T&T Clark.

Crouch, Andy. 2017. *The Tech-Wise Family: Everyday Steps for Putting Technology in Its Proper Place*. Grand Rapids: Baker Books.

Crouch, Andy. 2019. Transhumanism and the Cult of 'Better, Faster, Stronger.' *Christianity Today*, March 15. https://www.christianitytoday.com/2019/03/transhumanism-image-god-modern-technology-human-future/.

Crouch, Andy. 2022. *The Life We're Looking For: Reclaiming Relationship in a Technological World*. New York: Convergent Books.

Delio, Ilia. 2011. *The Emergent Christ: Exploring the Meaning of Catholic in an Evolutionary Universe*. Orbis Books.

Drummond, John J. 2007. *Historical Dictionary of Husserl's Philosophy*. Lanham, MD: Scarecrow Press.

Dunér, Ingrid. 2025. *Julian Huxley, Evolutionism and the History of Transhumanism*. London: Palgrave Macmillan.

Dupuy, Jean-Pierre. 2007. Some Pitfalls in the Philosophical Foundations of Nanoethics. *The Journal of Medicine and Philosophy* 32 (3): 237–261.

Ferrando, Francesca. 2019. *Philosophical Posthumanism*. London: Bloomsbury Academic.

Fisk, Anna. 2014. *Sex, Sin, and Our Selves: Encounters in Feminist Theology and Contemporary Women's Literature*. Eugene, OR: Pickwick Publications.

Florensky, Pavel. 2015. In *At the Crossroads of Science & Mysticism: On the Cultural-Historical Place and Premises of the Christian World-Understanding*, ed. Boris Jakim. Semantron Press.

Fuller, Steve. 2011. *Humanity 2.0: What It Means to Be Human Past, Present and Future*. Basingstoke: Palgrave Macmillan.

Fuller, Steve. 2013. *Preparing for Life in Humanity 2.0*. Basingstoke: Palgrave Macmillan.

Fuller, Steve. 2019. *Nietzschean Meditations: Untimely Thoughts at the Dawn of the Transhuman Era*. Basel: Schwabe.

Fuller, Steve. 2021. Cyborg Persons: Humanity Played in a Different Key. *Postdigital Science and Education* 3 (3): 668–677. https://doi.org/10.1007/s42438-020-00214-4.

Fuller, Steve. 2022. The Mind-Technology Problem. *Postdigital Science and Education* 4 (2): 247–252. https://doi.org/10.1007/s42438-021-00226-8.

Fuller, Steve, and Veronika Lipinska. 2014. *The Proactionary Imperative*. Basingstoke: Palgrave Macmillan.

Funkenstein, Amos. 1986. *Theology and the Scientific Imagination from the Middle Ages to the Seventeenth Century*. Princeton, NJ: Princeton University Press.

Gallie, Walter B. 1956. Essentially Contested Concepts. *Proceedings of the Aristotelian Society* 56 (1): 167–198.

Goodman, Nelson. 1955. *Fact, Fiction, and Forecast*. Cambridge, Mass.: Harvard UniversityPress.

Gould, Stephen J. 1997. Nonoverlapping Magisteria. *Natural History* 106:16–22.

Gouw, Arvin M., Brian Patrick Green, and Ted Peters, eds. 2022. *Religious Transhumanism and Its Critics*. Lanham, MD: Lexington Books.

Hacking, Ian. 1983. *Representing and Intervening: Introductory Topics in the Philosophy of Natural Science*. Cambridge, UK: Cambridge University Press.

Haraway, Donna J. 2016. *Staying with the Trouble: Making Kin in the Chthulucene*. Durham, NC: Duke University Press.

Harrison, Peter. 2009. *The Fall of Man and the Foundations of Science*. Cambridge: Cambridge University Press.

Harrison, Peter, ed. 2010. *The Cambridge Companion to Science and Religion*. Cambridge: Cambridge University Press.

Harrison, Peter. 2015. *The Territories of Science and Religion*. Chicago: University of Chicago Press.

Hayles, N. Katherine. 1999. *How We Became Posthuman: Virtual Bodies in Cybernetics, Literature, and Informatics*. Chicago: University of Chicago Press.

Herzfeld, Noreen. 2002. *In Our Image: Artificial Intelligence and the Human Spirit*. Minneapolis, MN: Fortress Press.

Hughes, James. 2006. Human Enhancement and the Emergent Technopolitics of the 21st Century. In *Managing Nano-Bio-Info-Cogno Innovations: Converging Technologies in Society*, ed. William Sims Bainbridge and Mihail C. Roco, 285–307. Dordrecht: Springer.

Hurlbut, J. Benjamin, and Hava Tirosh-Samuelson, eds. 2016. *Perfecting Human Futures: Transhuman Visions and Technological Imaginations*. Wiesbaden: Springer VS.

Huxley, Julian. 1957. *New Bottles for New Wine*. London: Harper & Brothers.

Huxley, Julian. 1959. The Future of Man. *Bulletin of the Atomic Scientists* 15 (10): 402–409.

Jacobs, Jonathan. 2009. An Eastern Orthodox Conception of Theosis and Human Nature. *Faith and Philosophy* 26 (5): 615–627.

Jaki, Stanley L. 1988. *The Savior of Science*. Grand Rapids, MI: Eerdmans Publishing.

Kurzweil, Ray. 2005. *The Singularity Is Near*. New York: Viking.

Lovejoy, Arthur O. 1936. *The Great Chain of Being: A Study of the History of an Idea*. Cambridge, MA: Harvard University Press.

Malapi-Nelson, Alcibiades. 2013. Review of Steve Fuller (2011), *Humanity 2.0: What It Means to Be Human Past, Present and Future. International Sociology 28* (2): 240–243.

Malapi-Nelson, Alcibiades. 2016. Transhumanism, Christianity and Modern Science: Some Clarifying Points Regarding Shiffman's Criticism of Fuller. *Social Epistemology Review and Reply Collective* 5 (2): 1–5.

Malapi-Nelson, Alcibiades. 2017. *The Nature of a Machine and the Collapse of Cybernetics: A Transhumanist Lesson for Emerging Technologies*. Cham, Switzerland: Palgrave Macmillan.

Malapi-Nelson, Alcibiades. 2022. Unlikely Allies? Transhuman Dignity and Catholic Doctrine on Human Flourishing. In *Postdigital Theologies: Technology, Belief, and Practice*, ed. John Reader and Maggi Savin-Baden, 75–91. London, UK: Springer.

Malapi-Nelson, Alcibiades. 2021. Review of Steve Fuller (2019), *Nietzschean Meditations: Untimely Thoughts at the Dawn of the Transhuman Era. Postdigital Science and Education* 3:1008–1016.

Mancuso, Stefano, and Alessandra Viola. 2015. *Brilliant Green: The Surprising History and Science of Plant Intelligence*. Translated by J. Benham. Washington: Island Press.

McGee, Jim. 2005. Legacy Systems: Why History Matters. *Enterprise Systems*. https://esj.com/Articles/2005/10/11/Legacy-Systems-Why-History-Matters.aspx.

McInerny, Ralph, and John O'Callaghan. 2009. Thomas Aquinas. In *The Stanford Encyclopedia of Philosophy*, ed. Edward N. Zalta and Uri Nodelman. https://plato.stanford.edu/ARCHIVES/WIN2009/entries/aquinas/.

Pico della Mirandola, Giovanni. 2016. *Oration on the Dignity of Man*, ed. Francesco Borghesi, Michael Papio, and Massimo Riva. Translation edition. New York: Cambridge University Press.

Moos, Thorsten. 2016. Reduced Heritage. In *Perfecting Human Futures: Transhuman Visions and Technological Imaginations*, ed. J. Benjamin Hurlbut and Hava Tirosh-Samuelson, 159–178. Wiesbaden: Springer Fachmedien. https://doi.org/10.1007/978-3-658-11044-4_8.

Noble, David F. 1997. *The Religion of Technology*. New York: Penguin Books.

Nussbaum, Martha C. 1992. Human Functioning and Social Justice: In Defense of Aristotelian Essentialism. *Political Theory: An International Journal of Political Philosophy* 20 (2): 202–246.

Parrochia, Daniel. 2024. Classification. In *Internet Encyclopedia of Philosophy*. https://iep.utm.edu/classification-in-science/. Accessed 8 Dec 2024.

Passmore, John Arthur. 1970. *The Perfectibility of Man*. London: Duckworth.

Ruether, Rosemary Radford. 1983. *Sexism and God-Talk: Toward a Feminist Theology*. Boston: Beacon Press.

Shults, F. 2003. *LeRon. Reforming Theological Anthropology: After the Philosophical Turn to Relationality*. Eerdmans Publishing.

Smith, Wesley J. 2022. The Impossibility of Christian Transhumanism. *First Things*. https://www.firstthings.com/web-exclusives/2022/03/the-impossibility-of-christian-transhumanism.

Sorgner, Stefan Lorenz. 2016. Three Transhumanist Types of (Post)Human Perfection. In *Perfecting Human Futures: Transhuman Visions and Technological Imaginations*, ed. J. Benjamin Hurlbut and Hava Tirosh-Samuelson, 141–157. Wiesbaden: Springer Fachmedien. https://doi.org/10.1007/978-3-658-11044-4_7.

Sorgner, Stefan Lorenz. 2021. *We Have Always Been Cyborgs: Digital Data, Gene Technologies, and an Ethics of Transhumanism*. Bristol: Bristol University Press.

Theiner, Georg. 2021a. The Extended Mind: A Chapter in the History of Transhumanism. In *The Mind-Technology Problem: Investigating Minds, Selves and 21st Century Artefacts*, ed. Robert Clowes, Klaus Gärtner, and Inês Hipólito, 275–322. Berlin, Heidelberg, New York: Springer.

Theiner, Georg. 2021b. Uniquely Human – Without Exception. *Culture e Fede (Cultures and Faith)* XXIX (3 (Supplement)): 21–26.

Tirosh-Samuelson, Hava. 2014. Religion. In *Post- and Transhumanism*, ed. Robert Ranisch and Stefan Lorenz Sorgner, 49–71. Frankfurt/Main; New York: Peter Lang.

Tuncel, Yunus, ed. 2017. *Nietzsche and Transhumanism: Precursor or Enemy?* Newcastle upon Tyne: Cambridge Scholars Publishing.

Vygotsky, Lev Semyonovich. 1994. The Socialist Alteration of Man. In *The Vygotsky Reader*, ed. Rene Van der Veer and Jaan Valsiner, 175–184. Oxford: Wiley-Blackwell.

Young, George M. 2012. *The Russian Cosmists: The Esoteric Futurism of Nikolai Fedorov and His Followers*. Oxford; New York: Oxford University Press.

Transhumanism and Human Flourishing

Human Flourishing Requires Transhumanist Flourishing

David W. Wood

1 The End Is Nigh?

In every generation, there have been people who thought they might take part in some kind of culmination of history. Perhaps they observed frightening trends of floods, famines, wars, or pestilence, and deduced that more (*much* more) was to follow. Perhaps they believed that contemporary events had been foretold in apocalyptic prophecies in holy scripture. Or perhaps they were captivated by the mesmeric orations of charismatic preachers with their proclaimed miracles. In each case, they thought the end was nigh.

So far, every time, these expectations proved to be overblown. There may have been tumults and revolutions, but history has continued. The end was *not* nigh.

D. W. Wood (✉)
London Futurists, London, UK

Delta Wisdom, Surrey, UK
e-mail: davidw@deltawisdom.com

G. Theiner, A. Malapi-Nelson (eds.), *Technologies of the Future Self*, Palgrave Studies in the Future of Humanity and its Successors,
https://doi.org/10.1007/978-3-031-99902-4_2

But this time, things really might be different. Within a few decades—perhaps within just a few *years*—human life may be fundamentally altered. In that scenario, ages-old limitations will be vanquished. Humans will be liberated as never before. Even death, our most ancient foe, will be overthrown. The earth will be transformed into a literal paradise.

The agent of this radical change? New waves of technology.

Let me spend a few paragraphs building the case for that remarkable possibility, and the rest of this chapter exploring reactions to that idea—reactions that include scepticism, bioconservative opposition, naive accelerationist embrace, and active transhumanism.

2 The Possibility of Rapid Radical Progress

Thanks to technological innovations arising from the first industrial revolution onward, lifespan expectancy has already doubled within only a few centuries. The average person can now travel much more easily to many more locations than in past times. Information used to be scarce—available in limited numbers of hand-produced manuscripts—but can now be accessed and queried via the touch of a button or a spoken message to an attentive software assistant. Women now have much greater choice over the size and timing of their families. Digital storage and cloud connectivity enable us to explore vast libraries of music, art, and entertainment.

Now consider that the technological innovations of the coming decades could dwarf even these giant leaps of the last few centuries:

- Nanotech (N)—the systematic re-engineering of molecules—is poised to enable further advances in energy, manufacturing, and recycling.
- Biotech (B)—the systematic re-engineering of the molecules, cells, and tissues of living organisms—is poised to enable further advances in agriculture, food preparation, and healthcare.
- Cognotech (C)—the systematic re-engineering of aspects of our brains and minds—is poised to enable further advances in education, entertainment, and creativity.
- Infotech (I)—the systematic re-engineering of collections of data, in which machine learning reaches conclusions beyond those available to unaided human intellect—is poised to accelerate the N, B, and C transformations listed above.

(A brief aside: the term "Cognotech" in this list is a deliberate variation on the term "cognitive science" used when William Bainbridge and others first introduced (in 2002) the concept of NBIC.)

Importantly, the I transformation is also poised to accelerate itself, with new AI tools being used to help create even better AI tools. Moreover, there are many positive feedback loops between these different transformations. For example, the more insight that cognotech sheds on what's happening inside our own brains, the more ideas can be applied within the I transformation to improve AI. Again, the new computing chips created by improved nanotech result in better AI, which in turn helps with the design of even better computing chips.

If handled wisely, the outcome of the NBIC convergence would be what I have called (Wood 2019) a "sustainable superabundance for all". In broad terms, five areas of human experience await transformation. The outcomes can be called "the five supers":

1. *Superlongevity*: Transcending current human tendencies towards physical decay and decrepitude; enabling dramatically improved biological health, including (Wood 2016) the abolition of aging
2. *Superintelligence*: Transcending current human tendencies towards mental blind spots and collective stupidity (cognitive biases); enabling dramatically improved all-round thinking capabilities
3. *Superhappiness*: Transcending current human tendencies towards egotism, depression, alienation, self-loathing, and needless internal suffering; enabling dramatically improved states of consciousness
4. *Superdemocracy*: Transcending current human tendencies towards xenophobia, divisiveness, deception, and the abuse of power; enabling dramatically improved social inclusion and resilience, whilst upholding diversity and liberty
5. *Supernarrative*: Transcending current human tendencies towards small-mindedness or naive fantasies; enabling dramatically improved engagement and collaboration on a credible, compelling transformational journey

These transformations could also be called "the five rejuvenations": the rejuvenation of physical health, the rejuvenation of all-round intelligence, the rejuvenation of emotional health, the rejuvenation of social relationships, and the rejuvenation of vision. However, to be clear, the transformations I'm describing won't simply *restore* a previous state of health and wellbeing. *They'll result in levels of health and wellbeing never before possible.*

3 A Question of Timescales

Regarding timescales for the kind of radical transformation I've just described, four broad ideas need attention.

The first is the sceptical response that such a transformation *could never happen*. That response assumes a rigidity in the human condition, whereas in reality many steps towards a reconstitution of human experience have already taken place. The *potential* of NBIC technologies to improve health, all-round intelligence, emotional health, and social relationships cannot be denied. What can be debated is the *extent* to which these improvements are feasible. My observation is that there is nothing in nature itself which dictates that living creatures need to age and die (indeed, many species show "negligible senescence"), that intelligence is capped at the current level of the human brain, that heights of positive emotion need always be accompanied by depths of depression, or that societies must always feature deception and tribalism.

A second response is that such a transformation *will* never happen, because we humans will bungle things along the way to that outcome. Mismanaged, NBIC will lead to huge increases in crime, environmental damage, terrorist outrage, deadly pathogens, and intended or unintended wars. Rather than the respect for science, and for openness, which are crucial for the development and wide deployment of new aids to human wellbeing, society will lose awareness of the difference between science and quackery, and will elevate into power autocrats that forbid alternative ideas.

That's a response to which I have more sympathy. Indeed, alongside this statement of the first of what I call the thirteen principles of active transhumanism,

- *Radical opportunity*: The near future can be *much* better than the present situation. The human condition can be radically improved, compared to what we've inherited from evolution and history.

I state this as the second of these principles:

- *Existential danger*: The near future can be much *worse* than the present situation. Misuse of powerful technology can have catastrophic consequences.

Denial of that possibility is one of the main flaws in the "transhumanist shadow" that I will describe in more detail later. *Acceptance* of that possibility requires that very serious consideration be given to options for avoiding any such existential disaster.

The third response to the question of timescales is that this kind of momentous transition may happen at some point in the future, but not any time soon. It's akin to a large meteorite causing havoc in the future by crashing into the earth with astronomical momentum: we'll watch out for any evidence of such an impending impact, but there's none visible in any of the detailed models that track all known large meteorites in the solar system.

But that's where I disagree. That's why I strongly support the *fourth* response to the timescale question, namely, that whilst there are lots of grounds for uncertainty, we cannot rule out the possibility of an NBIC explosion by, say, 2030. Things are unlikely to unfold quite so quickly, but they might. Accordingly, we need to put a great deal more attention onto that possibility.

Here are the reasons why NBIC convergence might dramatically accelerate:

- The factor that makes the biggest difference with technological progress is the scale of the intelligence applied in research and development, that is, the ability to spot patterns, formulate theories, make predictions, recommend interventions to change the course of events, and to learn from the results of experiments.
- AI already surpasses many aspects of the intelligence that human researchers and engineers bring to the task of improvements in various NBIC fields.
- There are powerful incentives—both financial and geopolitical—for teams to discover and deploy enhancements in AI.
- There's a long string of ideas that are being explored for how AI could be improved, including alternatives to the transformer architecture, neuromorphic computing, combinations of deep learning with symbolic and logical versions of AI, AIs that understand the difference between correlation and causation, AIs that exploit quantum hardware or new quantum algorithms, data from sources like YouTube videos and radio broadcasts being cleaned and fed into deep learning systems, and much more.

- Various features of human intelligence that previously seemed almost magical have been duplicated unexpectedly easily by new AI systems in the last few years.

My conclusion is that it's possible that NBIC convergence could surge forward faster than commonly expected, at any time in the next few decades. Perhaps even before the end of the 2030s.

Ahead of that happening, what steps should humanity take, to increase the probability of the "radical opportunity" scenario of sustainable superabundance for all, and to decrease the probability of "existential risk" scenarios?

That's where I point to the remainder of the thirteen principles of active transhumanism.

4 No Easy Options

Given the likely chaos that lies ahead, "more of the same" won't be sufficient. We won't reach a better future, avoiding the numerous landmines along the way, by driving on cruise control.

I summarise this as follows:

- *No easy options*: If humanity gives too little attention to these radical future options, on account of distraction, incomprehension, or intimidation, there's a high likelihood of a radically bad outcome.

Instead, we humans will need to bring our "A game" to this task:

- *Human agency*: The difference between these two radical future options depends critically on human agency: wise human thinking and concerted human action.

Rather than distraction, we'll need to apply *focus*. Rather than tribal divisiveness, we'll need *constructive collaboration*. Rather than myths and illusions, we'll need to apply the best principles of *science and data analysis*. And rather than fear or lethargy, we'll need *an uplifting vision*.

5 Resisting Change?

I've given my objections to the sceptical response to the scenario of radical opportunity. It's now time to consider a response that can be called bio-conservative opposition. This response says that:

- Human nature is something to be embraced and preserved, rather than altered.
- Humanity already has sufficient material goods.
- Because technology can be a terrible two-edged sword, we must decisively slow down its development and deployment.

However, I don't believe that response is credible. Today's medical treatments are *not* sufficient: far too many people are struck down in the primes of their lives (or even before), when they still have so much more to contribute and experience. Today's energy systems are, likewise, *not* sufficient, as humanity stares ineffectively at the prospect of runaway climate change when the accumulation of greenhouse gases reaches tipping points. Today's economic distribution systems are, yet again, *not* sufficient, as increasing numbers of people find they cannot afford adequate housing, healthcare, or education. In all these cases, NBIC has the potential to transform our systems from substandard to abundance. Who will really stand against that possibility?

What *is* credible, however, instead of a blanket "no" to more technology, is a combination approach, in which much greater wisdom is applied to determining *which* technologies are prioritised, and *how* their deployment should be overseen. Rather than just pushing on the brake pedal, it's vital to use a *mixture* of brakes and accelerator. Whilst keeping our collective hands firmly on the steering wheel.

6 Against Naive Transhumanism

I'll say it again: it's a *mixture* of tools that's needed. Society as a whole won't tolerate too much use of brakes, given the marvellous potential of NBIC technologies. But too much use of the accelerator pedal will propel matters out of control.

There are many within the broad transhumanist community who see things differently. They express various points of view that I regard as opposed to the deeper wisdom within the same movement. Rather than dealing with the full complexity of the human situation—the complexity of how technology is developed, of how humans respond to that technology, of the distorting effects of one-dimensional incentives, and of the interplay of technology with the environment, with economics, and with politics and geopolitics—they prefer to uphold simple slogans.

With their simplicity, these slogans may appeal to aspects of human psychology, but I point instead to a variant of the saying attributed to Einstein: we should make our messages as simple as possible, *but no simpler*. Over-simplification brings major drawbacks.

That's why I speak of a "transhumanist shadow"—a collection of attitudes and practices that obscure the most important insights of transhumanism. When thoughtful observers notice these beliefs and actions, it makes them recoil, and this prevents them from engaging more fully with the community.

Transhumanism is by no means the only movement with a shadow—a dark underbelly which can cause it problems. Indeed, the shadow is often cast by people who are outside the community, but who are (wrongly) believed by observers to be reliable representatives of that community. I'm thinking of various prominent figures within Silicon Valley, or people who proudly call themselves "accelerationists" without having much to say about transhumanism.

The first part of the shadow is *risk complacency*—a focus on the potential large upsides of NBIC technologies, whilst giving only lip service to their potential large downsides. In this case, the transhumanist enthusiasts in question may accept intellectually that new technologies do bring some risks, but they tend to express an easy confidence that these risks will be relatively straightforward to manage. Accordingly, they wish to minimise how much attention is given to any potential dangers.

In contrast, let's consider the nearest thing in the transhumanist community to a canonical document: the Transhumanist Declaration. It contains eight clauses. Four of these clauses (numbered 3, 4, 5, and 6) make nuanced statements regarding the spectacular possibilities of technological change:

3. *We recognize that humanity faces serious risks, especially from the misuse of new technologies. There are possible realistic scenarios that lead to the loss of most, or even all, of what we hold valuable. Some of these scenarios are drastic, others are subtle. Although all progress is change, not all change is progress.*

4. *Research effort needs to be invested into understanding these prospects. We need to carefully deliberate how best to reduce risks and expedite beneficial applications. We also need forums where people can constructively discuss what should be done, and a social order where responsible decisions can be implemented.*
5. *Reduction of existential risks, and development of means for the preservation of life and health, the alleviation of grave suffering, and the improvement of human foresight and wisdom should be pursued as urgent priorities, and heavily funded.*
6. *Policy making ought to be guided by responsible and inclusive moral vision, taking seriously both opportunities and risks, respecting autonomy and individual rights, and showing solidarity with and concern for the interests and dignity of all people around the globe. We must also consider our moral responsibilities towards generations that will exist in the future.* (Baily et al. 2009)

Risk complacency stands in opposition to the "existential danger" principle of active transhumanism. The other seven aspects of the transhumanist shadow, that arise from time to time, likewise oppose other elements of these principles. Let's review them now, one by one.

7 Against Technological Determinism

Consider *technological determinism*: the (mistaken) view that technological development will control what happens in the future, more or less regardless of the actions of individual humans. In this view, bigger forces of history are at play, dictating outcomes, ensuring the ultimate transcendence of humanity. Transhumanists therefore are reduced to cheering from the sidelines, awaiting what has been heralded as "the inevitable victory of transhumanism". This cheerleading sees itself as potentially speeding matters up slightly, but not as changing any overall direction of history.

As an example, the Facebook group "Scientific Transhumanism" displayed for several years on its front page a banner that proclaimed "Victory of transhumanism is inevitable".

This mindset has some similarities with Marxist determinism, which foresaw as inevitable the rise of the working class and the overthrow of capitalism. There are similarities, too, with religious visions of forthcoming messianic interventions and apocalyptic vindication for the faithful.

However, a wider reading of history opposes any such technological determinism. There can be no guarantee, in advance, of how NBIC technologies will develop and be deployed. The outcomes will depend critically on numerous human actions. The uncertainty is at four levels:

1. The *pace* at which key breakthroughs are made, governing when new products and services become available
2. The *use* to which these new products and services are put—that is, whether the impact is beneficial or deleterious
3. The *reactions* against the use of these new technologies
4. Other unforeseeable *interactions* between various trends and countertrends.

It may be psychologically appealing to deny the relevance of human action, and to indulge in wishful "rah-rah" thinking regarding the smooth arrival of sustainable superabundance. It's understandable that people who arrive in the transhumanist community with battered psyches, on account of their personal histories, may be attracted to that kind of vision. But any such "hands off" mentality makes it more likely that the trajectory of technological development will lead us to a dreadful landmine encounter. The only way to navigate through these landmines is to remain vigilant and active. Cheerleading is *not* what is needed.

That's why I advocate this (as already stated) as one of the principles of active transhumanism:

- *Human agency*: The difference between these two radical future options depends critically on human agency: wise human thinking and concerted human action.

In other words, technological determinism should give way to human determinism. It should be humanity, as a whole, that determines the uses to which technology is applied—taking into account and weighing up the needs of non-human sentients and any recommendations made by synthetic agents, but resisting any suggestions that the future is already pre-determined.

8 Science and Data

It's one thing to say that humanity should determine how technology is developed and deployed. It's another thing to offer more specific guidance—to move beyond platitudes to concrete policies.

Three of the principles of active transhumanism offer help in answering that "how" question. The first of them is this:

- *Science and data*: To improve our decisions, it's important to seek, publish, and review objective data, respecting the best scientific principles, rather than accepting the say-so of would-be pundits or celebrities.

This sometimes provokes the claim that transhumanism is guilty of scientism. That criticism has two prongs:

1. Some vital aspects of human life lie outside the understanding of present-day science, and stand to be trampled or destroyed by transhumanists who, allegedly, take present-day science as their sole guide as to what should be done
2. Transhumanists think they are applying the best insights of science, but their thinking is allegedly dominated by ideas from "the hard sciences", such as physics (as exemplified by the adoption of the term "singularity"), and lacks awareness of other parts of science.

I acknowledge that certain people have a narrow conception of science. They are blind to the wide range of approaches used within and alongside science. As the philosopher of science Karl Popper said, they are guilty of "the aping of what is widely *mistaken* for the method of science" (Popper 1979, 185).

It's true that problems can arise when people are mistaken about what the methods of science involve, or when they apply these methods incorrectly. But that's not a problem of transhumanism. It's a problem of *lack of openness*—of *too much dogmatism*.

One way this shows up in the transhumanist shadow is as *solution hyping*. Rather than following the principles of science, this takes an overly rosy attitude towards various entrepreneurs or would-be visionaries. It tolerates, or even promotes, unwarranted marketing claims.

The issue arises when people seek to make money from particular products, without any reliable evidence of the efficacy and safety of what they are selling. Examples include various pills or potions supposed to help extend lifespan, increase intelligence, sharpen concentration, or strengthen emotional bonds.

To be clear, it's my belief that various pills and other treatments *can* deliver the kinds of results claimed, with a good evidence base. These products are not the problem. Nor is there a problem when a treatment is made available even though the experimental evidence for it remains unclear, so long as the people who choose to take part in a trial of it are fully aware of the product's uncertain status. The problem, instead, is with rogue products, and when some transhumanists fail to demand appropriate levels of rigour and evidence before endorsing a product.

In some cases, the over-selling is deliberate: the people involved understand that they are over-stating their claims, but they proceed regardless. In other cases, the over-selling arises from naivety and lack of sufficient knowledge by the vendors and their supporters. In all cases, lots of damage can arise:

- To people who purchase the goods or services and suffer actual health problems as a result
- To people who squander lots of money on ineffectual goods or services
- To the reputation of the transhumanist community as a whole, for apparently tolerating, or even praising, the products involved or the people who were pushing them

9 Beware Solution Dogmatism

As a variant of the problem of solution hyping, consider the problem of *solution dogmatism*. It's another aspect of the transhumanist shadow, and, once again, it prevents a deeper discussion of the pros and cons of various approaches to technological innovations.

Examples of solution dogmatism include genetic determinism—the view that the most important changes in our biology will necessarily result from altering our genes. Another example is anti-capitalist fundamentalism—the view that progress towards transhumanist goals will be held up until the capitalist system is overthrown. Yet another example is anti-money fundamentalism, which insists that the first thing that needs to be done is to abolish money (or, perhaps, to abolish the ability of banks to issue new "fiat" money).

Whilst there's definitely room to evaluate some interventions as more significant than others, it's important to keep an open mind. Dogmatism and fundamentalism shutter perceptions, and make it more likely—again—that technological development will go off the rails in unexpected ways.

Whilst genes clearly do influence biological wellbeing, they're by no means the only factor in play. To improve human wellbeing, interventions at many other levels of biology should be considered too—alongside interventions at social, political, or mindset levels. It's the same with the pros and cons of capitalist economics, and with the operation of banks and currencies. In all these cases, having closed minds could cause us to miss the best opportunity to achieve the future we desire.

That's why I state this principle of active transhumanism:

- *Openness*: In deliberations between conflicting insights, no book, thinker, or tradition should be given any absolute priority. Society needs to remain open to the possibility that our current favoured ideas and methods will be superseded. Of course, respect can be shown to books, thinkers, or traditions with good track records as sources of insight. But that respect should be tempered with caution. Runs of success can come to an end—especially in new circumstances or new contexts.

This is sometimes expressed that, in transhumanism, there is no infallible pope, and no infallible scripture.

10 Proactive Vigilance

Another implication of the need to avoid dogmatism and, instead, to be ready for unexpected developments is the principle of proactive vigilance:

- *Proactive vigilance*: We need to expect the unexpected—in other words, to keep monitoring for new risks and new opportunities. Where possible, actions should be preferred that are reversible, lest surprises arise. In anticipation of the possibility of changes in plan, we should design for change, rather than with certainty in mind.

This means collectively keeping an eye on whether plans turn out as well as expected—and also being ready to switch from one plan to another, whenever new insight or new information becomes available.

Importantly, this opposes a narrative which is often advanced—*anti-regulation fundamentalism*:

1. Entrepreneurs and engineers are assessed as being the great heroes of societal progress, creating and applying new technologies.
2. Politicians and regulators are assessed, in contrast, as being the villains of this story, hindering progress for their own short-sighted reasons, such as preserving a status quo from which they benefit.
3. Therefore, all regulation of the economy should be resisted; entrepreneurs and engineers should be allowed a free rein to make progress as fast as they would like.

Along these lines, some members of the transhumanist community blame regulatory organisations such as the FDA (Food and Drug Administration) for what they describe as "hundreds of thousands" of "needless and preventable" deaths, due to the slow approval of new treatments.

Like a number of the other aspects of the transhumanist shadow, this anti-regulation narrative depends on a very selective reading of history. A fuller assessment shows:

- Some public health initiatives, driven by visionary politicians, have actually saved millions of lives (Johnson 2021).
- Products introduced by entrepreneurs and engineers often have significant safety issues, especially when quality has been sacrificed, or when users have been deliberately misled—hence there's a positive role for regulators, to curtail misleading marketing.
- Powerful companies, if left to their own trajectory, often eliminate competitors, and then move from a mode when they frequently introduce innovations into one that restricts disruptions—hence, again, there's a positive role for regulators, this time to ensure market competitiveness.
- The wisest way to solve various problems isn't always by applying new technology; instead, political changes can sometimes prove much more effective.

In other words, there are two problems with anti-regulation fundamentalism:

1. It is in thrall to *techno-solutionism*, always expecting technical fixes, whereas non-technical changes are sometimes a better solution.
2. Even when a technological change would be an appropriate solution, sometimes the best way to ensure such a solution is developed is via positive political direction of the economy, via *industrial strategy*.

Active transhumanism aims to elevate the operating methods of entrepreneurs and engineers, *and* the operating methods of politicians and regulators. To be clear, I am not advocating *more politics* or *more regulation*. Instead, I am advocating *better politics* and *better regulation*. That's part of what I meant by the term "superdemocracy" earlier in this chapter. I'll return to that theme in my concluding remarks.

11 All-round Human Wellbeing

Having considered the "how" question, let's turn again to the "whither" question. What exactly is involved in human wellbeing? In which ways will sustainable superabundance be better than the status quo?

Yet another consequence of the need to avoid dogmatism is retaining an open mind about potential wider dimensions of human wellbeing. I summarise it like this:

- *Human wellbeing*: In all decisions, actions should be preferred that lead to the increase of human wellbeing. As for what wellbeing involves, how to measure it, and how to improve it, these are enquiries that remain open and ongoing. Wellbeing involves vitality, liberty, creativity, health, intelligence, awareness, happiness, collaboration, and bliss—but extends beyond these separate traits. Over time, our understanding of the conditions and possible expression of human wellbeing will surely evolve and improve. That's as it should be.

However, within the broad spectrum of choices, there are two angles that need to be upheld: the fundamental importance of *individual* wellbeing, and the observation that our individual wellbeing in turn depends on the responsible exercise of *mutual welfare*:

- *Individual sanctity*: Individual wellbeing should never be sacrificed or subordinated in favour of collectivist goals. Individuals should never become cannon-fodder in service of some tribal, national, ethnic, religious, or ideological quest. In short, individuals are ends not means.
- *Mutual welfare*: It's in our mutual interest to protect and elevate all members of society. Rather than keeping quiet about impending dangers about to befall someone, or major opportunities they are about to miss, we should find ways to speak up, just as we would ourselves like to be alerted to such dangers or opportunities in an equivalent circumstance. We're stronger together, rather than in isolation or a state of war.

Deprioritising the latter principle leads to another area of the transhumanist shadow, namely, a lack of concern regarding unequal access to the fruits of technology. This can be described as *egocentric transhumanism*: the transhumanist enthusiast celebrates his or her own opportunities for better health and better intelligence, etc., but shows little practical concern regarding whether others also share these benefits. When explicitly asked, the enthusiast might indicate that others should be able to benefit too, but this topic seems to be far from their mind most of the time.

This lack of concern may be accompanied by what can be called a "trickle down" theory of transhumanism: the belief that once technological benefits are available for a vanguard set of early adopters, they will in due course become available to everyone who is sufficiently interested in them.

Some support for that kind of thinking comes from observing that the consequence of the anticipated forthcoming superabundance of goods and services is that there will be sufficient for everyone to enjoy a high quality of life. However, different considerations apply *in the lead-up* to the achievement of that superabundance. During the extreme stress and turmoil of the next decades, if groups of people perceive that they are falling behind from the wave of benefits that other people are receiving—people who seem, perhaps, to be more grasping and self-seeking—the resulting anguish and hostility will magnify the challenges in steering humanity away from major landmines.

A different deviation from the above principles is when some transhumanists give the impression that they disdain certain aspects of human wellbeing, namely those connected to the biological body, in preference for a forthcoming "uploaded" state of being.

Hence the *loathing of biology* area of the transhumanist shadow. These transhumanist enthusiasts find aspects of their biological existence so distasteful or limiting that they concentrate on ideas for transferring their minds as soon as possible out of biology into alternative hardware substrates, such as a cloud of silicon computers. These transhumanists show less interest in options to repair and augment their existing physical bodies, and they speak contemptuously about their bodies as being "meat sacks".

Although that view represents, in my experience, only a small minority of the overall transhumanist community, it has been selected by some researchers as somehow providing the core definition of transhumanism. In this view, transhumanism is all about "uploading minds to the cloud". Thus, in the 2019 BBC/HBO series *Years and Years*, one teenage character, Bethany, comes out to her parents as being "trans", meaning "transhumanist". What shocks Bethany's parents most is her expressed desire to "live forever as information".

It may be true that, at one level, everything that exists is information. It may also be true that analysing reality in terms of information leads to significant new insights. However, that's no reason to diminish the importance of analysing what exists at higher levels. Accordingly, active transhumanism recognises the paramount phenomenon of emergence. We are all atoms, but we are *more than atoms*. Similarly, we are all information, but we are *more than information*. We are all individuals, but we are *more than individuals*, as we interact with mutual responsibilities.

As for our biology, that's not something to be rejected. It's something to be enhanced.

12 Key Generalisations and Extensions

Beyond the ten core principles I've already described, active transhumanism points to three key generalisations and extensions:

- *Sustainability* (generalising from the present into the future): Our plans need to enable, not only wellbeing today, but also wellbeing tomorrow—and the days and years that follow. Hence the need to avoid actions that reduce the possibilities for future wellbeing.

- *Consciousness* (generalising the previous principles beyond present-day humans): In all decisions, actions should be preferred that lead to the increase of the wellbeing of consciousness. To the extent that animal or artificial minds possess core attributes of consciousness, these minds deserve at least some of the same care and support as human minds. This care includes possibilities for growth and development, and the reduction in needless suffering.
- *Diversity*: The above principles leave many questions unanswered. They define a broad envelope that can accommodate a multiplicity of different viewpoints. That diversity is, itself, something to embrace, as a celebration of variety, and for the rich pragmatic benefits of pluralism. Hence the final core principle: within the overall transhumanist framework, diverse opinions and lifestyles should be cherished and nurtured. It's from interactions between diverse subcultures that some of the most important insights and experiences are likely to arise.

A denial or disinterest of the last of these principles—diversity—results in the ***monoculture transhumanism*** aspect that is sometimes seen within the transhumanist shadow. In this case, individual enthusiasts show concerns for other people who are sufficiently similar to themselves, but they are disinterested, or even antagonistic, to other potential supporters of transhumanist principles who have different backgrounds, different primary or secondary aspirations, or different ways of expressing themselves.

Sadly, so long as any one transhumanist monoculture dominates:

- The movement is weaker, because it fails to take advantage of the experience, insights, and talents from more diverse sources.
- The growth of the movement is stymied, because of alienating and deterring potential new supporters with different backgrounds and desires.

A denial or disinterest of the sustainability principle is often justified by referring to the forthcoming superabundance, in which NBIC technologies will provide more than enough for everyone. In this line of thinking, there's no need for concern over potential rapid climate change, since NBIC will enable ample green technologies. Any damage experienced by the environment before then can be reversed in due course. And there's no need for concern about lack of access to adequate healthcare or social support, since NBIC will make high-quality healthcare and social support available at low cost to everyone on the planet.

But I fundamentally disagree. The natural environment could tip over into a state of disaster *before* NBIC has delivered sufficient green technological innovations. Just as there can be positive feedback cycles with huge *positive consequences* (such as AI enabling more AI), so also there can be positive feedback cycles with huge *negative consequences*—such as increased global temperatures changing aspects of the biosphere so that temperatures ratchet up further and faster. When certain transhumanists give the impression of waiting for "pie in the sky, bye and bye"—that is, forthcoming marvellous geoengineering interventions, carbon capture mechanisms, and so on—they risk being too late, namely, allowing an existing problem to accumulate to the point of catastrophe.

And just as there are serious risks of the environment transitioning into a much uglier state of being, so also there are serious risks of the political sphere becoming much uglier, driven by the wide perception by individuals, groups, and entire nation states, that they are being unfairly *left behind* in the distribution of the bounty of modern life. In a state of despair and alienation, voters can fall under the spell of Svengali-like would-be political leaders, with disastrous consequences.

Therefore, rather than a "pie in the sky", what's needed is to show clear signs that a paradise is being constructed, here and now, on the earth, such that everyone will benefit—with access to unprecedented levels of the five supers: superlongevity, superintelligence, superhappiness, superdemocracy, and supernarrative.

13 Transhumanism in Action

To recap: NBIC is unleashing into the world the equivalent of extraordinarily powerful genies. These genies could enable human flourishing on a scale as never before, sustainably, with open trajectories for further growth and discovery, with no one left behind against their will. But these genies will require a different calibre of management and governance. Business-as-usual will not be sufficient. Nor will politics-as-usual. Nor will philosophy-as-usual.

From where will higher calibres of management and governance arise? From the best of humanity's insights and practices, aided and augmented by selective use of the technologies delivered by the early phase of NBIC. In other words, from humans who are becoming transhumans, individually and collectively—with the transformation accelerating as these insights and practices are more fully understood and spread ever more widely

Specifically, I point to five vital transhumanist initiatives: the uplifting, respectively, of education, collective intelligence, emotional intelligence, politics, and creative intelligence.

14 Uplifting Education

The requirements for uplifting education—for people of all ages—can be split into six groups.

First, whereas previous educational models tended to assume that most change is linear (incremental), transhumanist education will raise awareness of exponential change and accelerating feedback cycles. It will focus attention on the potential for emerging disruptive technology to transform human experience in multiple ways that are hard to predict. In particular, it will cover skills of anticipatory foresight and scenario design and evaluation, including the identification and management of both disruption and contingency.

Second, whereas previous educational models assumed a broadly static cultural background, transhumanist education will prepare students for ultra-diversity. People's interactions with each other nowadays aren't just more *numerous* than in the past, but are much more *varied*—because of greater global travel, and because groups of people are increasingly adopting new lifestyles, new bodily modifications, new social structures, and new philosophies. Moreover, transhumanist education will prepare us to coexist, not just with diverse other people, but also with advanced AIs and robots.

Third, whereas previous educational models tended to prioritise the memorisation of facts, transhumanist education will focus on critical thinking and collaborative intelligence. The goal, in effect, will be to inoculate all of us, in advance, against the damaging mental pathogens that are increasingly bombarding us. Indeed, we all are coming into contact with ever-greater amounts of information, in rapidly evolving formats—with much of that information being *designed* to mislead or confuse, and lots more being *unintentionally* misleading or confusing. We must learn to handle all that better.

Fourth, many of the assumptions behind traditional frameworks for ethics and morality are being challenged by new ideas that are circulating, and by unprecedented new possibilities that technology puts in our hands. It will remain important for people to be able to listen to their intuitions, but these intuitions will likely prove misguided more frequently than before. That's a reason for improved fluidity and clarity of thinking. It's also a reason for transhumanist education to elevate not only IQ (measures of raw intellectual power) but also EQ, including emotional resilience and mental agility.

Fifth, whereas most of history has witnessed a competitive struggle for scarce resources, NBIC technologies will increasingly enable a post-scarcity society, in which the majority of the products and services needed for a good standard of living will be available in abundance. Accordingly, transhumanist education will highlight the management of abundance alongside the management of scarcity—noting that both skills will be needed in parallel.

Sixth, as a special case of transitioning from scarcity to abundance, society will need to be ready for a transition in what it is assumed most people will be doing with their time—a transition *from* the assumption that most people should seek and gain paid employment, *to* the assumption that most people will be "post-employment".

How can all these changes be accelerated? Existing schools and universities may adapt in some respects, but the bigger disruptions are likely to take place on the fringes of the mainstream educational establishment. A big role can be played by AI tutors, responsive to the particular backgrounds and thought patterns of each individual learner. Another big role will be played by improved collective intelligence.

15 Uplifting Collective Intelligence

One template for improved collective intelligence is Wikipedia. It's far from perfect, but it shows what can be accomplished once good structures and processes are put in place that enable the merging together of the best insights of different contributors.

YouTube also provides many remarkable resources, but these are often mixed in with materials that are much less trustworthy. Accordingly, sites that curate and review these resources can be particularly valuable. My own "Vital Syllabus" project falls into this category, as does the London Futurists Podcast. Researchers such as Geoff Mulgan (2017) and Thomas Malone (2019) have pointed out other ways in which communities of humans, aided by artificial intelligence, can boost collective intelligence.

Curating resources, and providing short summary information about each resource, is time-consuming and expensive. AI summarisers can handle some of the workload, but the fact remains that most collective intelligence projects are underfunded. Those projects which *are* well funded often promote ideological viewpoints that are partisan or suspect. This is a wake-up call for all of us: finances, whether from private or public sources, need to be preferentially applied to well-run collaborative intelligence projects, which follow the principles of the academy (pursuit of knowledge) more than the principles of activism (pursuit of power).

16 Uplifting Emotional Intelligence

Sadly, we humans often resist learning about things that would challenge our current ideas.

It's as the American writer Upton Sinclair said: "It is difficult to get a man to understand something, when his salary depends on his not understanding it". (Alternative, equally valid versions of that sentence would involve the words "ideology", "worldview", "identity", or "tribal status", in place of "salary".)

We like to think of ourselves as a *rational* species, but we can equally be described as a *rationalising* species. We search eagerly for facts that appear to confirm what we want to believe, and we readily turn away from facts that highlight shortcomings in our beliefs.

This confirms the burning need, not just for more rational intelligence, but for more emotional intelligence—ways of inspiring each other to set aside our dogmatism and to be willing to entertain alternative ideas.

The idea is to seek to change, not only people's *understanding of the world*, but also their *mental dispositions*. Rather than accepting the mix of anger, partisanship, pride, self-righteousness, egotism, vengefulness, deceitfulness, and so on, that we have inherited from our long evolutionary background, we should be using a variety of methods to transform our mental dispositions.

The transformative technology movement documents methods that are already known to assist people with psychological transformation, embracing compassion, humility, kindness, appreciation, and so on. These methods include various drugs, supplements, meditative practices, and support from electronic and computer technologies.

Related, the insights of the movement for improved *cultural intelligence* deserve to be more widely appreciated and uplifted.

Some of these methods have been discussed for millennia, whereas others have only recently become possible. The scientific understanding of these methods (including potential drawbacks as well as potential benefits) is still at an early stage, but it deserves much more focus.

17 Uplifting Politics

As indicated above, many of the projects that will help humanity to make the best out of NBIC involve decisions from the world of politics:

- Wiser allocation of public funds, free from partisan pressures
- Various restrictions on technologies in unsafe circumstances
- Opposing the actions of huge corporations that are unduly motivated by short-term share price movements, market presence, or racing to overtake competing products
- Upgrading core aspects of our collective infrastructure, fixing latent vulnerabilities
- Revisions in the social safety net that assists each of us in harder times
- Incentivising the development of new skills and attitudes

It is in the world of politics that NBIC technologies may have the largest impact. It's not just that politicians (and their advisors) will benefit from uplifted education, uplifted collective intelligence, and uplifted emotional intelligence. It's also that well-understood AIs can augment the political decision-making process (Wood 2018):

1. *Real-time fact checking*, which could highlight almost immediately whether there is evidence to contradict (or verify) a claim being made by a politician or their supporters
2. *Real-time logic checking*, that could highlight any invalid jumps in reasoning—or conversely, confirm the soundness of an argument
3. *Real-time source checking*, to verify whether a quotation, image, or video extract may have been altered before being presented, changing its meaning or any implications arising
4. *Reputation systems*, that keep track of the reliability of people's past contributions to particular topics
5. *The creative generation of new proposals*, that can combine key elements from previous suggestions in ways that address earlier concerns

6. *Simulations*, in huge computational models, of potential consequences of envisioned legislative changes, to help the identification and evaluation, ahead of time, of possible real-world risks and benefits of these changes
7. *Improved electoral systems*, involving elements of proportional representation as well as ranked preference votes, to avoid the present situation in which far too many people feel that their votes will be essentially meaningless, or that they ought to cast a tactical vote for a candidate different from their actual first preference
8. *Liquid voting*, in which citizens can temporarily assign their voting rights in specified topic areas to people they trust to vote on their behalf regarding these topics
9. *Facilitated citizens' assemblies*, that act in similar ways to jury trials, with representative groups of citizens being selected at random, and supported in a task to evaluate possible solutions to given social issues, ahead of the conclusions of these assemblies being forwarded to society as a whole for wider consideration
10. *A "House of AI" revising chamber of government*, that applies AI reasoning to provide feedback on proposed legislation, as well as suggesting possible revisions to that legislation

In this way, AI can assist humanity to take best advantage from various domain experts, without falling foul of experts who fail to recognise the limits of the applicability of their knowledge. Indeed, the fact that someone is an expert in one domain does not entail any special priority applies to their viewpoints in other domains. An expert sailing navigator gains no authority in a different field, such as medical treatments, just by virtue of their sailing proficiency. And if an expert doctor offers suggestions about how to steer an unfamiliar boat in the midst of a storm, we would be foolish to disregard all contrary opinions just because the doctor has an "MD" credential after their name.

What's more, decisions frequently involve the intersection of several different domains. A decision that appears sound from one perspective may be recognised as inadequate when other perspectives are introduced. If we listen only to experts from the first perspective, we risk reaching a bad decision. Accordingly, there are major limitations to the concept of delegating hard decisions to domain experts. The recommendations of such experts need to be incorporated into the broader process that I have called "superdemocracy"—the involvement of representatives from the entire community in a deliberative process to reach decisions.

18 Uplifting Creative Intelligence

The positive developments listed above are unlikely to happen unless enough people believe that they are credible, desirable, and actionable. Otherwise, what will prevail is distraction, fear, confusion, and lethargy. We're up against powerful forces of reaction, egotism, and inertia. These forces cannot be overcome by wishful thinking. Nor by drawing hopeful exponential curves. Nor by reiterating the sermons and essays from previous generations. No: what is needed is a vision that can uplift, inspire, and inform new types of thinking and action around the world.

That's where we need to bring together the best of *critical* intelligence with the best of *creative* intelligence. Instead of Black Mirror scenarios in which technology adversely impacts humanity, we need engaging, diverse scenarios of technology uplifting humanity.

I have offered a number of these in the past in my own writing, such as a speculative "A year-by-year timeline to 2045" (Wood 2022b) and associated materials, including "A day in the life of Patricia, 2045" (Wood 2022c) and "A day in the life of Asimov, 2045" (Wood 2022d). I've seen many other initiatives that can be gathered under the banner of "White Mirror" scenarios, but they frequently fall short in my view:

- They fail to appreciate the full scale of the potential forthcoming changes in the human condition.
- They have a limited view of what humans can soon become.
- They pay insufficient attention to political, economic, and social forces.

That's one reason I strongly welcome the initiative represented in the chapters of this book, *Technologies of the Future Self*.

19 Further Reading

For numerous references to ideas that expand or challenge the narrative in this essay, see Wood (2021) and Wood (2022a), as well as my articles on dw2blog.com.

References

Baily, Doug, et al. 2009. The Transhumanist Declaration. *Humanity+*. https://www.humanityplus.org/the-transhumanist-declaration.

Johnson, Steven. 2021. *Extra Life: A Short History of Living Longer*. New York: Riverhead Books.

Malone, Thomas W. 2019. *Superminds: The Surprising Power of People and Computers Thinking Together*. New York: Little, Brown Spark.

Mulgan, Geoff. 2017. *Big Mind: How Collective Intelligence Can Change Our World*. Princeton, NJ: Princeton University Press.

Popper, Karl R. 1979. *Objective Knowledge: An Evolutionary Approach (Revised Edition)*. Oxford: Oxford University Press.

Wood, David W. 2016. *The Abolition of Aging: The Forthcoming Radical Extension of Healthy Human Longevity*. Delta Wisdom.

Wood, David W. 2018. *Transcending Politics: A Technoprogressive Roadmap to a Comprehensively Better Future*. Delta Wisdom.

Wood, David W. 2019. *Sustainable Superabundance: A Universal Transhumanist Invitation*. Delta Wisdom.

Wood, David W. 2021. *Vital Foresight: The Case for Active Transhumanism*. Delta Wisdom.

Wood, David W. 2022a. *The Singularity Principles: Anticipating and Managing Cataclysmically Disruptive Technologies*. Delta Wisdom.

Wood, David W. 2022b. A Year-by-Year Timeline to 2045. https://dw2blog.com/2022/05/15/a-year-by-year-timeline-to-2045/.

Wood, David W. 2022c. A Day in the Life of Patricia, 2045. https://dw2blog.com/2022/05/15/a-day-in-the-life-of-patricia-2045/

Wood, David W. 2022d. A Day in the Life of Asimov, 2045. https://dw2blog.com/2022/05/15/a-day-in-the-life-of-asimov-2045/

Is Transhumanism Necessarily Utilitarian? Recasting Alternative Ethical Systems Towards a Future Human Flourishing

Alcibiades Malapi-Nelson

1 Introduction

The last decade has witnessed an explosion of ethical narratives regarding the pros and cons of such 'philosophy' of human enhancement. A common feature of most positions ethically defending transhumanism is that these 'have assumed a consequentialist foundation, in the form of various kinds of utilitarian'[1] (More 2013, 13). As Stefan Sorgner points out, '[i]t is true that many transhumanists take a utilitarian standpoint, and the

[1] Perhaps the most radical utilitarian stance in this context is Pearce 1995. Hughes 1996 developed a political version of an envisaged transhumanist democratic society based upon utilitarianism. Levin 2020, ch. 4 criticizes transhumanism in general precisely due to its alleged over-reliance on utility ethics.

A. Malapi-Nelson (✉)
Humber Polytechnic, Toronto, ON, Canada
e-mail: alcibiades.malapi-nelson@humber.ca

G. Theiner, A. Malapi-Nelson (eds.), *Technologies of the Future Self*, Palgrave Studies in the Future of Humanity and its Successors, https://doi.org/10.1007/978-3-031-99902-4_3

bioethics debate in the English speaking [sic] world in general is also dominated by this ethical theory' (2010, 9). The reason for this is fairly straightforward. Utilitarianism in general advocates for the greatest happiness for the greatest amount of people, emphasizing the consequences of the moral action rather than on the intentions fueling it. It would seem logical that this ethical framework is the most naturally related to a philosophy that advocates human enhancements—since the point of these enhancements is the happiness and flourishing of the many. However, as we will see, some problems do stem from the utilitarian angle that transhumanism standardly prefers.

2 Problems with a Utilitarian Transhumanism

Although transhumanism has several staunch and vocal advocates,[2] Steve Fuller arguably is the most prominent *academic* proponent of it.[3] Having experienced a critical embracement of transhumanism about a decade ago, Fuller acknowledges, as most transhumanists, that utilitarianism is the most logical fit for the movement in terms of an ethical framework (Fuller 2011, 150–160, 222–231; 2012: Ch. 4). And yet, he recognizes that there are disarming ethical challenges for transhumanism linked to both the inequality *and* unfairness that it might reinforce. Although Fuller is aware of these problems, some of which without an apparent or foreseeable solution, he does not see that the utilitarian framework itself might be insufficient. Instead, he adds clever theoretical patches.

This dual challenge is instantiated at the socio-economical and intergenerational levels. In the former case, transhumanism may exacerbate political, social, and financial global inequalities, by allowing a wealthy few

[2] While Max More, Natasha Vita-More, Zoltan Istvan, James Hughes, and Ray Kurzweil are perhaps the most vocal advocates of transhumanism these days, neither of them holds academic posts.

[3] Nick Bostrom, founding director at Oxford University's Future of Humanity Institute, and Anders Sanders, senior research fellow at the same institute, are both academicians, but Bostrom's position as of late is increasingly critical of transhumanism to a point that his endorsement is now put into question (Fuller 2019, 72–80). Stefan Sorgner is a late promoter coming from the Nietzschean scholarship. Although he is criticized by Fuller for conflating both transhumanism and posthumanism (Fuller and Sorgner 2019), he might be the other one, next to Fuller, advocating for transhumanism from within their academic positions.

to enjoy the benefits of the enhancing perks while leaving the poor and destitute aside. In the case of intergenerational injustice, transhumanism may irreversibly affect generations of people not yet born, fostering contexts where the youth would not be able to get integrated into society—or do so in a diminished capacity. Let us briefly explore these two socio-economic objections, raised by Finn Collin and David Pedersen, respectively, in their criticism of Fuller's transhumanist utilitarianism.

Collin (2013) notes the insufficiency of the standard utilitarian perspective embedded in transhumanist thought when applied to the various ways in which 'directed evolution' can go. More specifically, he is uncomfortable with the endless lifestyle possibilities becoming available for modified humans, vis-à-vis the absence of a pre-given 'anthropological base' that may guide them through their deep somatic and cognitive alterations. Given a radically augmented freedom to redefine ourselves down to the genetic level, what type of future person are we to expect? What moral beacon are we to follow for designing this human 2.0? Should this individual be kind and loving? Or instead, a capacity for hard and efficient work should be emphasized? Should she rather be a pleasure-seeking entity? Or a chilled and relaxed one? Fuller's alleged lack of 'genuine anthropology' (Collin 2013, 79) would generate a dangerous prospect for humanity, now prey to whatever form it may take following individual whims. Without a clear image of what this human is supposed to be, indeed leaving our future entirely to 'choice', we risk veering towards a dystopian future—so the criticism goes.

A preemptive reply to this objection could be that we should not even aim at having a 'model' of humanity, given that, as Jean Jacques Rousseau (1762, 9) famously observed, '[w]e do not know what our nature permits us to be'—put in other words, 'there is a domain of human freedom not dictated by our biological nature' (Kompridis 2009). In this respect, Fuller reminds us that '[t]raditional natural law appeals to violations of human nature... increasingly lack intellectual currency, given the socially constructed character of humanity' (Fuller 2011, 109). This fluid character of humanity entails leaving the very question regarding our nature always open, and there is an important value in that, since 'preserving its openness allows us more freely to frame alternatives to what heretofore seemed like its only possible answer' (Kompridis 2009).

The genius of modern ethics partly consists in capturing this sentiment. After the development of post-eighteenth century moral theory (i.e., utilitarianism and deontology), the emphasis has been placed instead on the possibilities opened up for human agency—*contra* Collin's objection. Furthermore, it would seem that Collin's discomfort could be soothed by some type of 'character'-based aretaic narrative, so that human nature remains anchored, unalterable, and protected against getting 'lost' while aiming at improving itself.

Leaving the modification of human nature to 'choice' entails even further problems, as David Pedersen (2013) points out. Redistributing the outcomes of the genetic and social 'lotteries' by means of a fair restructuring of society is what John Rawls' (1971) theory of justice was largely about. The biological and socio-economic conditions with which we are born constitute the underserved goods that profoundly shape our entire social lives. So that the distribution of these goods is not left at the mercy of 'natural luck', Rawls devised a state whose duty would be to organize this distribution based upon fairness, relative necessity, and equality of opportunity.

What would happen, however, if these goods would be altered before they get distributed (as transhumanism advocates)? How would a fair society be structured if the lottery is rigged—namely, if the goods that we are to start with, and that society would redistribute fairly, are beforehand modified? Pedersen foresees unequal social conditions nurtured by transhumanism after wealthy individuals get biologically enhanced. Access to genetic and prosthetic improvements by those who can afford them would expectably exacerbate the already sharp divide between the rich and the poor.

To be sure, Fuller's answer to this dystopian prediction is already contained in the 'proactionary imperative' (Fuller and Lipinska 2014): a 'welfare state 2.0' would redistribute these 'goods' among the population *at large*, given that human 2.0s would be benefiting from research on (and taxation of) human 1.0s. Thus, state regulatory agencies would actively intervene and balance out possible enhancement-driven inequalities. However, Pedersen thinks that this redistribution may work at best *ad intra* nation-states. Given its limited geopolitical nature, a welfare state 2.0 could not guarantee that the 'superhumans' share the enhancements with the 'unenhanced' at the *global* level. And in any case, why would one 'species' (human 2.0)—continues Pedersen—feel any kind of solidarity with another one (human 1.0)?

Without dismissing this concern, one wonders whether pointing towards an ongoing moral deficiency can constitute grounds for holding back on accessing unprecedented opportunities for human flourishing.

Indeed, it seems we are before a classical instance of the 'dual-use dilemma' (Rappert and Selgelid 2013). On the one hand, discriminatory treatment of those perceived as 'the other' or 'different' (e.g., racism, ableism, etc.) is an ethical ailment that we aim at overcoming, without precluding us to try in the meantime improving the whole human community in other aspects. And even if another species were to emerge out of this human 2.0, the discrimination or subservience of the 'unenhanced' would likely be widely condemned as an instance of 'speciesism'—i.e., mistreatment of a living entity based upon its species-belonging (Singer 2001). Pointing out potential moral problems in general, which will exist regardless, can hardly validate slowing down technological progress, given that the benefits resulting from this progress will likely outweigh the pains endured by the group in question.

On the other hand, paying attention to how said progress unfolds would not stop humanity from enjoying its benefits. In this context, the nature of the criticism against technologies of enhancement must take into consideration the end result of the equation between risky costs and qualitative overall social improvement. As Fuller retorts, '"principled objections" to Humanity 2.0 should follow the example of resisters of technological innovations, namely, to argue that unregulated innovation is likely to increase already existing inequalities in society' (2011, 109). In other words, it is not clear how a precautionary world, where the human 1.0 continues with its usual ailments at every possible level (including the moral one), would be in any way a better environment for this human.

Nevertheless, Fuller admits that 'uncertainties and risks remain, but in the first instance they are about how costs and benefits are distributed' (2011, 109). Confronted with this reality, his notion of 'superutilitarianism' (Fuller 2012: Ch. 4) aims at soothing the above concerns. Superutilitarianism calls for taking a step back to contemplate the temporal big picture so that we realize that the risks—an inherent aspect of frontier science and innovation—will, *in the long run*, produce a greater good. Since 'evils' (i.e., suffering) will occur, these should be understood more as instances of investments in such greater good, rather than as costs. This framework of 'moral entrepreneurship' is best understood as having the role of 'recycling evils into good' for the betterment of the whole human community (Fuller 2012, 63). Why for the *whole* and not a few, one may ask? Because these benefits are derived from the population at large via taxation and experimentation—as indicated above. The welfare state 2.0

referred to above, putting policy tools in place (e.g., hedgenetics[4]) would fairly redistribute these benefits throughout the community. Thus, the utilitarian dictum of the most happiness for the many (fulfilling both Bentham's calculus and Mills' proviso of sacrificing personal gain for the common good) gets applied in an extended time spectrum.

In contrast to all these, the issue of intergenerational unfairness is, in Fuller's mind, quite possibly the most intractable one in what concerns social problems potentially brought up by transhumanism (Fuller 2019, 187–195). For instance, if humans get to have access to substantially longer and healthier lives, what would be the point of bringing children to the world, since one could *eventually* do most of the life pursuits themselves? Further, if the implications of having sustained better health include staying cognitively sharp, and hence, avoiding the need to retire at the 'usual' age (i.e., 65), how would that affect the youth in what regards them entering the workforce?[5] Perhaps most importantly—indeed mulling over from a superutilitarian perspective—given that generational change brings an influx of new ideas to all strata of society, how would a world of stagnated minds, slow to change and naturally inimical to fresh ideas, affect humanity at large? A world without the youthful risk that sparks innovation and out-of-the-box thinking would hardly advance towards a better future for the whole.

So far, these 'temporal super predators' (Fuller 2019, 188) get a pass. Fuller admits that there is no foreseeable solution to this potential grave problem and that it is largely ignored, acknowledging the denial in which the movement is currently locked: 'Contemporary transhumanism, despite its much-vaunted "futurist" vision, remains largely oblivious to the intergenerational consequences of its quest for indefinite longevity' (Fuller 2019, 107). Now, facing this seemingly unsurmountable conundrum, with no solution in sight, are we to despair and start jumping off the transhumanist ship?

Not so fast—at least not yet. Sorgner remarks that 'many transhumanists take a utilitarian standpoint, and the bioethics debate in the English speaking world in general is also dominated by this ethical theory.

[4] A legal framework devised by Fuller and Lipinska, where the citizen, aware of its genetic makeup, exercises property rights, backed by the state, to whatever may come out of using its genetic information—by profit or non profit entities (Fuller and Lipinska 2014, 111–128).

[5] A silver-lining might be that since someone healthier would retire at, say, 85 instead of 65, younger generations would not have to for 40, but 20 years in pensions for the elderly via taxation.

However, this does not mean that other ethical theories are irrelevant in this respect or have to be inconsistent with transhumanism' (Sorgner 2010, 9). Indeed, as Max More (2013, 13) reminds us, transhumanism is ultimately value system neutral, given that in the broad transhumanist landscape, 'the meta-ethical basis for making moral decisions and according rights can be consequentialist, deontological, or virtue-based'. To be sure, some non-negotiable transhumanist mandates, such as the defense and promotion of 'morphological freedom' (More 1993)—the liberty to radically alter our bodies as we see fit—would have to be maintained. In this context, it is at least conceivable that divergent value systems could successfully carry such mandate and others while shaping moral dilemmas raised by transhumanist implementations. Further, it is not unthinkable that they could dissect these vexing problems in such a way that these would not even emerge in the alternative framework. With this possibility in mind, let us explore the divergent moral route.

3 A Kantian Transhumanism?

In as much as diverse human moral problems can be addressed from alternative ethical systems, it stands to reason that these other theoretical platforms could *mutatis mutandis* say something to transhumanism. Let us take deontology, for instance. Remedios and Dusek assert that the ethics of duty would be inadequate for a transhumanist scenario. They state that '[d]eontology is… inappropriate for Humanity 2.0 because [it] is based on duty. Enhancement can hardly be considered a duty in itself though this is what Fuller seems to claim' (Remedios and Dusek 2018, 110). In other words, the rightness of human enhancements may not be a universalizable belief—a Kantian condition for a maxim to become a categorical imperative.

Contra Remedios and Dusek, however, the literature on human enhancement as a duty—even if not specifically Kantian and *even if not specifically applied to transhumanism*—abounds (Harris 2010; Savulescu and Bostrom 2009; Persson and Savulescu 2012). Indeed, there is another way of seeing a transhumanist rapprochement to deontology. Fuller emphasizes Kant's exaltation of human autonomy as compatible with the transhumanist agenda, and as such, morphological freedom would be, in the end, nothing other than an instance of our exercise of free will. This rational and autonomous act could find formidable instantiation in volunteering for a rightful cause—i.e., for science research that would carry implications for the common good. In that context, a citizen's contribution to the advancement of knowledge (via volunteering in scientific experimentation) might qualify as a matter of universal duty—which is not to say that research that leads to human enhancements would be a matter of universal duty. However, it is conceivable that some forms of enhancement can be understood as dutiful (Sparrow 2011).

Immanuel Kant believed that the whole edifice of morality is constructed upon human reason itself—the former being mapped upon the latter, thus attaining consistency and universality. In terms of the *foundations* of morality, empirical concerns are deemed irrelevant, since this moral structure has to belong in an *a priori* fashion to what he called 'pure philosophy':

> if a law is to be morally valid, i.e., is to be valid as a ground of obligation, then it must carry with absolute necessity ... the ground of obligation here must therefore be sought not in the nature of man nor in the circumstance of the world in which man is placed, but must be sought a priori solely in the concepts of pure reason; ... Thus not only are moral laws together with their principles essentially different from any kind of practical cognition in which there is anything empirical, but all moral philosophy rests entirely on its pure part (Kant 1785, 389).

Within that framework, what gives an action moral worth would be its being performed for the sake of duty. The operative word here is, of course, 'duty'. What is this for Kant? It is nothing other than obedience to a categorical imperative. And how do we reach that? By universalizing a maxim and verifying that it does not incur in a logical contradiction, the just alluded moral edifice's consistency, relying upon the structure of reason, comes in handy. Pursuing 'enhancements', the conditions by which a person becomes cognitively and physically more powerful (understood as more adaptive to its environment) would perhaps not be considered a duty indeed—but so would not be moving a chair nor repairing a radio. However, a society constituted by many, perhaps a majority, of 'enhanced' persons (be it due to philosophical, theological, or moral reasons, or simply out of sheer survival from environmental or man-made threats) might elevate the bar in what regards the default standards of what could be considered a 'normal' quality of life. After all, it is not unlikely that in Kant's estimation, being literate, or pursuing higher education, could also be considered a matter of universal duty. On the opposite end, the social regard towards those who, for example, refuse to be enhanced (thereby affecting productivity for the individual, and if normalized, for a nation) will demand these maxims to be universalized and tested—not in a dissimilar manner in which we judge the action of those refusing to vaccinate their children. Thus, it is safe to assume that as new moral challenges emerge, new maxims will be forged,[6] and perhaps new categorical imperatives will be devised.

In what looks like an absolution of duty ethics by association, Fuller reminds us that utilitarianism has a deep correlation with deontology that has not been recognized enough in the literature—but that was noticed by Hannah Arendt, who observed that 'once Kant's political philosophy is taken into account, his overall normative position is, perhaps surprisingly, not so far from Bentham's' (Fuller 2016, 29). Both modern ethical theories, Bentham's and Kant's, reacted against a vision that normalized what humans were capable of until that time. In a *scala natura* ('Great Chain of Being') worldview, the person's purpose was to find its place in such a spectrum and be the best she can be within those limits. Fuller explains that this 'joint refusal to let the future of humanity be dictated by its past most clearly marked Kant and Bentham as philosophers who, unlike the virtue theorists and evolutionary ethicists, do not let "ought implies can" to slide into "is implies ought"' (Fuller 2016, 30). In other words, utilitarianism and duty ethics do not fall into the naturalistic fallacy associated with virtue ethics, where allegedly natural facts imply moral dicta or desirable features.

4 Virtue Ethics and Transhuman Flourishing: Problems

Another feature in common to both ethics of modernity (i.e., Bentham's utility and Kant's duty) is that they gradually 'superseded' an understanding of morality based upon the formation of 'character' and the practice of 'virtue', where action follows being—having originated in Aristotle and

[6] A mild (and for some, not so mild) taste of these new moral challenges already obtain while navigating through a social-media structured world—ignorance of which usually resulted in socially painful consequences. 'One shall not post on social media personal information that may come back to hunt you' could arguably be a sufficiently novel maxim (a virtual variation of 'discretion') that may one day qualify, after we know more about the nature and social future of these online sharing milieus, as a categorical imperative.

later developed by Thomas Aquinas. However, the rupture between these two ethics of modernity and virtue ethics arguably was based (at least in part) upon the displacement of the ultimate agent in the moral act within the 'divine corporation': from God to the human person (Schneewind 1984). It is against this backdrop that one should understand how '[a]fter all, as Kant made explicit in his famous essay that popularized the term "Enlightenment" in German (Aufklärung), the maturation of humanity as a species rests on the gradual removal of those features of social life that keep us in a childlike state longer than is biologically necessary' (Fuller 2016, 31). Thus, whereas for virtue ethics human nature determines the sphere of possibilities for the moral act, for utility and duty this nature is the springboard, or raw material, from which human agency reaches for new heights of moral and personal accomplishments.

Fuller laments that the scholarship has largely missed 'that science and technology might actually enhance native human capacities to render us capable of meeting the high ethical demands of Kant—or, for that matter, that other standard bearer of ethical modernism, Jeremy Bentham' (2016, 29). In contradistinction, contemporary resurgences of aretaic systems (e.g., Martha Nussbaum, Alasdair MacIntyre) had their basis for human nature relinquished to (or hijacked by) empirical findings of biology-oriented disciplines (e.g., evolutionary psychology). Fuller asserts that 'there is an unholy alliance between virtue theory and "evolutionary ethics" when it comes to downsizing human moral demands to fit the default capacities of natural born members of Homo sapiens' (2016, 29). Indeed, since newer virtue theory incorporates empirical discoveries to its core, it ends up locking human nature and its acts within a Darwinian predetermined picture of what is biologically feasible. The backbone of the virtues upon which Nussbaum identifies human 'capacities and capabilities' end up being swapped by what is genetically viable (Fuller 2016, 29). As a result, human nature gets 're-naturalized', undergoing a semantic exchange with biology, and rendering virtue as a version of 'gene expression':

> [V]irtue theorists such as Martha Nussbaum… trust our spontaneous moral sensibilities, which they then try to ground in 'human nature'. Unfortunately, they follow MacIntyre's lead in relying on Aristotle, sometimes as filtered through Aquinas, when in fact the biological basis for human nature has shifted to Darwin, with the emphasis now on the 'nature' rather than the 'human' part of human nature. As a result, virtue ethics, perhaps unwittingly,

> has contributed to the re-naturalisation of the human, with 'virtue' often nowadays functioning as a quaint word for gene expression (Fuller 2011, 89).

This 'unholy alliance' between virtue theory and evolutionary ethics, bridged via Neo-Darwinism, boxed in human nature within its current capacities bounded by its biological substratum—the opposite of what transhumanism stands for. Perhaps in an attempt to remain relevant in a world shaped by science, virtue ethics gets self-bound by empirical findings of fields like evolutionary biology and evolutionary psychology, blunting the possibilities for a radical rehaul of human nature (Fuller 2011, 86–95) and sliding into bioconservatism.

In 2002, president George W. Bush ordered the formation of the 'President's Council on Bioethics', led by biochemist and bioethicist Leon Kass.[7] This ethics committee, whose members were handpicked by the president, shared critical views against embryo and cloning research, effectively providing the basis for laws that would shut down federal funding for research on these subjects. This event fostered an odd alliance between conservative Christians and environmentalists, both endorsing a view of the human being as a 'finished entity'. Under this view, current *homo sapiens* are the culmination of an evolutionary process—the operative word being 'culmination'. Contemporary humans represent the end and goal of evolution in what regards this species. Thus, they shall stop evolving. The natural body (understanding as 'natural' what we now can see in its current form) is regarded as evolutionary complete, locked, and to be shielded from change.

Kass introduced an intuitively compelling (for some) pragmatic moral probe that would help us discern possible distortions to this finished product of natural evolution—humans as they look now. This probe is expressed as a 'yuck factor' (Schmidt 2008) and underpinned by a 'wisdom of repugnance' (Kass 1997): an instinctive feeling that gives a moral pointer regarding the wrongness of profound human alterations. In other words, our 'guts' would be telling us where to stop in what concerns the biological modification of humans—and the red flag would be raised once a treatment or process stops being strictly therapeutic.

[7] It was preceded by President Bill Clinton's National Bioethics Advisory Commission and succeded by President Barack Obama's Presidential Commission for the Study of Bioethical Issues.

Expectably, this view is inimical to the possibility of exerting substantial alterations on the human biological makeup. Endeavours such as genetic engineering, seen by its advocates as having the role of the midwife of natural processes—merely helping nature to get back on track (Dyson 2007)—would have to be firmly opposed. In this vein, Michael Sandel (2007), another member of Kass' council, did precisely that, condemning parents that would be willing to have their children enjoy desirable enhancing attributes. No wonder why Steve Fuller is not fond of virtue ethics: An Aristotelian morality would seemingly naturally bend towards a stance à la Kass, whose views might represent one of the most bioconservative positions in the last two decades against what we referred to above as 'morphological freedom'.

5 Aristotelian Ethics and Transhumanism: Problems

Remedios and Dusek seemingly side with Fuller in that transhumanism and Aristotle are not a great match, but from a very different standpoint. They imply that transhumanist enhancements would steer the subject away from a virtuous balance, towards either an excess or deficiency, violating the famous Aristotelian 'Golden Mean'. Remedios and Dusek (2018, 110) assert that '[v]irtue ethics would not be appropriate because it is about moderation, while enhancements for transhumanism are to go beyond the human that may be considered immoderate'. Even if their pitching of Aristotelian balance *against* the radical nature of the transhumanist proposals sounds reasonable, we still need to ask some questions. Is this moderation an essential feature of virtue ethics (i.e., a condition without which virtue theory would not survive)? Furthermore, is this moderation *the* feature that made virtue theory intrinsically intertwined with Christianity as we know it? Ultimately, is the sharing of this feature (moderation) the main cause behind the alleged incompatibility between Christianity and transhumanism? Let us see.

Aristotle uses in the *Nicomachean Ethics* two words when referring to the dispositions we need to *both* stay away from either vicious extreme *and* stay close to the hard-to-attain virtuous mean. Σωφροσύνη is used for referring to self-restraint (NE 3.10.1), and ἐγκράτεια, coined by his contemporaries (Jaeger 1943, 54), is used for referring to self-control or self-discipline (NE 7.8.4). Four centuries later, both St. Luke (Acts 26:25) and

St. Paul (1 Timothy 2:9,15) used σωφροσύνη to denote sobriety against excess. St. Luke (Acts 24:25), St. Peter (2 Peter 1:6) and St. Paul (Galatians 5:23) used ἐγκράτεια for referring to self-control against—once again—excess. This last Greek word, translated to the Latin Vulgate as *continentia*, made it to Church doctrine as one of the Fruits of the Holy Spirit.[8] Perhaps more importantly, St. Thomas Aquinas specifically indicated the virtuous mean as compatible with Christian morality, when he asserted that 'it is evident that the good of moral virtue consists in conformity with the rule of reason. Now it is clear that between excess and deficiency the mean is equality or conformity. Therefore it is evident that moral virtue observes the mean' (Aquinas 1485, I-II, q. 64, a.1). As we can see, the efforts for incorporating a moderation view of morality into the Christian doctrine are explicit.

There are problems, however. Becoming a saint is considered a feat of radical excellence, being humanly accomplishable only in cooperation with divine grace. But even with the help of God's grace, the amount of effort on our side can consume the totality of our lives. Sainthood is, after all, a universal call to straight-up *perfection*[9]—not 'moderation'. Certain dispositions necessary for the accomplishment of holiness call for unrestraint. One of those—but not the only one—is the act of love:

> [O]n the Christian worldview Aquinas accepts, there is no excess of love, and it is not good to love in moderation. And love is not the only relational good for which moderation is ill-suited. There are many others. Consider forgiveness, for example. On Aquinas's Christian worldview, forgiveness is meant to be unstinting, not moderate. (Stump 2012, 33)

To these acts, which should be performed without moderation, one could add that of self-sacrifice, seemingly resulting *not* from a 'balanced' attitude between excess and deficiency, but extreme in itself and yet virtuous (good). There seems to be something off regarding self-sacrifice—for an ideal or for love—as an internal movement towards a mathematically centred harmony so that the self remains virtuous.

[8] 'The fruits of the Spirit are perfections that the Holy Spirit forms in us as the first fruits of eternal glory. The tradition of the Church lists twelve of them: 'charity, joy, peace, patience, kindness, goodness, generosity, gentleness, faithfulness, modesty, self-control, chastity' (Catholic Church 1993).

[9] The Gospel of St. Mathew reads: 'So be perfect, just as your heavenly Father is perfect' (Mt 5:28 - NAB). The Second Vatican Council affirms that 'all the faithful of Christ of whatever rank or status, are called to the fullness of the Christian life and to the perfection of charity' (Lumen Gentium 1964, 40).

In this context, one could ask regarding the ancient Greek take on moderation: 'Why did they—or more particularly, why did Aristotle—fail to see the supreme significance of such actions of self-sacrifice and self-denial…or try to reduce them to such spurious arithmetic to cases of self-seeking?' (Stocks 1931, 176). The answer may be rooted in a Hellenic idiosyncrasy whose first representation could be traced back to a painting on the Oracle of Delphi's wall in 440 BC: μηδὲν ἄγαν (do nothing excessively). Despite being standardly presented as a key notion in virtue theory, one could question whether 'moderation' is *essential* to it. There have been calls for revisiting this feature as being inherent to Aristotle's ethics. Indeed, the Greek cultural overarching view of a balance *away* from two extremes seemingly may have stemmed from their views on nature—including their take on medicine. As Rosalind Hursthouse puts it, for the Greeks '[d]isease is not conceived of as an alien something getting in to the human body (as we now know it often is) but as its natural elements getting out of balance (or harmony or due measure or proportion or symmetry)' (Hurtshouse 2006, 110).

Peter Losin argues that this pre-scientific understanding of medicine compels us to take Aristotle's moderation aspect of his ethical doctrine with caution:

> Aristotle's father was a physician, and medical concepts and examples played an important and widely-recognized role in the philosophizing of Aristotle's day. Health was believed to lie in a balance of powers, in a mixture so constituted that none of its constituent elements eclipsed the others. The author of the Hippocratic treatise On Breaths writes that "opposites are cures for opposites. Medicine is in fact addition and subtraction, subtraction of what is in excess, addition of what is wanting."… Aristotle himself expresses this view… Proper balance or proportion makes for health, lack of it for disease… (Losin 1987, 330)

If the 'moderation' aspect of Aristotle's ethics is a deep remnant of ancient medicine (and socio-cultural circumstances), instead of an essential element of it, we could at least *conceive* of an aretaic narrative void of it. One could then ask: If it is possible to think of a version of virtue ethics whose emphasis is shifted away from the traditional requirement for the accomplishment of a behavioural middle sweet spot, could it then be applied to transhumanism (given the radical propositions of the latter)?

As it turns out, 'fixing' the virtue aspect of the theory would not be enough, since we still have that other major cornerstone of Aristotelian ethics to take care of. Indeed, a fundamental aspect of virtue ethics that would have to be dealt with is its reliance on the development (or change) of human 'character'—accomplishable once the subject, after persistent 'continence' via the practice of virtue, becomes 'temperate'—attaining 'excellence' (effectively, becoming 'good'). Under this view, a change of behaviour (i.e., for the better) follows a change of self. Aristotle uses clever examples of entities that accomplish the right functions once their 'structures' are set right.[10]

> [A]ll excellence makes what has it good, and also enables it to perform its function well. For instance, the excellence of an eye makes the eye good and enables it to function well as an eye; having good eyes means being able to see well. Likewise, the excellence of a horse makes it a good horse, and so good at galloping, carrying its rider, and facing the enemy. If this is true in all cases, then, the excellence of a human being will be that disposition which makes him a good human being and which enables him to perform his function well. (Aristotle 1908, 1106a16-25)

This notion of a subject's 'action' following its 'being' was picked up and developed by the Thomistic *agere sequitur esse* ('action follows being'), which later constituted the basis of Catholic natural law (Green 2014, 202). Thomas Aquinas advanced that humanity's 'first nature' is 'what we share universally with other humans', which could be understood in modern parlance as 'what metaphysically and biologically makes us members of the species Homo sapiens' (Green 2014, 203–204). Traditionally, whereas first nature is somewhat fixed and immutable (except for modifications after medical interventions), our 'second nature' (culture, education, and the rational aspect of us) is subject to change. Morality would exercise influence on the latter, but it cannot do much about the former—other than recognizing how its various forms of existence shape our behaviour (Green 2014). That is part of the point of finding our place within the *scala natura*.

[10] The issue of behavior versus structure as the source of an entity's identity is still very much present in philosophy and science—dismissal of which sometimes resulted in devastating consequences. For the contribution of this issue to the downfall of classical cybernetics, see Malapi-Nelson 2017, chs. 6 & 7.

This is where the problem of virtue theory with transhumanism begins to show: The ethics of modernity (duty and utility), as we saw, dramatically shifted the emphasis of the moral agency from nature/God to the human being. We no longer comply with nature's (or God's) designs, but instead forge new destinies and open new realms of moral existence. Since transhumanism advocates for (ideally) the complete restructuring of human nature, it would seem that only deontology and utilitarianism could be compatible with it—and in turn, incompatible with an ethical position where the subject's actions are subsumed to its (allegedly immutable) being.

This transhumanist objection to virtue ethics is solid, and it arguably lies at the core of Steve Fuller's rejection of it. After all, transhumanism advocates (and maintains the inevitability of) the radical transformation of, in Aquinas' nomenclature, both first and second human natures. Indeed, this inevitability would redefine the Aristotelian and Thomist approaches as archeological creatures, held to be irrelevant after humanity reached the point of maturity pointed out by Kant's views on the Enlightenment alluded above—now expanded by radically changing its biological nature.

And yet, it would be difficult not to perceive in Aristotle's above quote a call for a personal amelioration that could well result in both an amplified performance of our human functions and the improvement of the body politic—something openly promoted by transhumanism. Would painstakingly working toward 'flourishing', now aided by self-biohacking and genetic augmentations, not be what most transhumanists would say they are after? Moreover, could not this sentiment be what is behind—notwithstanding the damning above Enlightenment observation—the later transhumanist narratives of flourishing? Indeed, we might be witnessing an emergence of transhumanist-friendly positions that are nevertheless inspired by some sort of virtue theory (Vallor 2016; Helmer 2017; Green 2014, and the Vatican's humanity2-0.org). What are we to make of these? Are these ruminations to be short-lived after their supporters realize the utter contradiction of their positions? Or there are aspects of virtue ethics that deserve closer scrutiny after incorporating contemporary developments in technology and science (which might steer its theoretical core)?

6 How Would an Aristotelian Transhumanism Look Like?

The question of whether transhumanism can attain a *consistent* set of principles seems to remain open. Indeed, arguably most of Fuller's contemporary work is devoted to giving this position a coherent philosophical backbone, as evinced by his published book 'series' on transhumanism (Fuller 2011, 2012, Fuller & Lipinska 2014, and particularly, Fuller 2019). Thus, putting the 'consistency of transhumanism' question aside, the case could still be made for further studying ways in which virtue ethics could be modified to address transhumanist ethical concerns in ways that duty and utility ethics cannot. Indeed, without exhaustively laying out alternatives for an 'aretaic—or virtue-based—transhumanism' (if such a thing is feasible), one could still suggest possible avenues of exploration that may exploit points of connection between the ethics of human flourishing and a technologically driven philosophy of human thriving.

For instance, foreseeable new capacities that shall emerge from enhanced natures could arguably call for an ethical articulation that may be closer to an aretaic narrative than to one of utility or duty. As indicated above, Aristotle based his ethics on the understanding that a (moral) action follows a being. This sequential linearity was itself derived from his metaphysical understanding of how objects come to exist—expressed in his famous doctrine of the 'Four Causes'. To recall, the existence of each and every object (persons including) could be explained by referring to its matter (ὕλη), form (εἶδος), agent (κινοῦν), and finality (τέλος)—what constitutes it, what shapes it, the agency required for it to exist, and its aim or goal, respectively. The final cause, or the teleological explanation, encompassing the other three, suggests what the object is aimed at (i.e., a slim and muscular horse, bred by racing horses, is aimed at racing)—recall the above-quoted Aristotelian passage where perfection follows a well-fulfilled function.

In a similar fashion, the material, formal, and efficient characteristics of a human being already advance her final aim: self-fulfillment and flourishing. Alasdair MacIntyre was keenly aware of the role of this *telos* as the ultimate aim of a human being, to be reached via ethics:

> We thus have a threefold scheme in which human-nature-as-it-happens-to-be (human nature in its untutored state) is initially discrepant and discordant with the precepts of ethics and needs to be transformed by the

> instruction of practical reason and experience into human-nature-as-it-could-be-if-it-realized-its-telos. (MacIntyre 1981, 53)

For MacIntyre, the pursuing of this telos is precisely the safeguard we need against the radical indeterminism of a lifestyle solely based upon human will and the unforeseeable actions emanating from it. This teleological driving force would be the antidote against a purported relativism—even nihilism—to which we are bound to end up if we are to pursue the ethics of modernity (Collin 2013). And although after modernity referring to telos as an explanation or cause was deemed unscientific, contemporary scientific pursuits have nevertheless attempted to bring it back, prominently among them being classical cybernetics (Malapi-Nelson 2017, 24–33), itself a paradigmatic instantiation of transhumanist thought.

Once we have telos back in place, we find ourselves before two positions, transhumanism and virtue ethics, which share one common goal. To put it another way, if the ethics of modernity share with transhumanism the methodology for pursuing its aim (i.e., morphological freedom), the ethics of virtue share with transhumanism the *very goal*: the flourishing of the human. To be sure, transhumanism will allegedly modify the human condition so that the four causes outlined above, when applied to humans 2.0, will themselves account for the change of, in Aquinas' terms, the 'immutable' human first nature. Since the structure or being will be altered, the ethical lifestyle (actions) will be modified too. And thus, a virtue theory 2.0 may have to emerge.

For instance, newly acquired powers and capacities (resulting from genetic or prosthetic enhancements) will entail new moral assessments. Could we say that a person with enhanced abilities (cognitive sharpness, agility, health, longevity, strength, etc.) would carry the same moral responsibility as one who does not have them? Building upon Christ's teachings on how those with more positive 'attributes' will be judged,[11] the Medieval Jesuit exegete Cornelius a Lapide reminded us that according to St. Gregory, '[w]hen gifts are increased, the responsibilities are increased also' (a Lapide 1637, 317). Virtue theory will have something to say about this new human condition, given that the theory's core builds upon the (now modified) being and not the (moral) action. The better

[11] 'Much will be required of the person entrusted with much, and still more will be demanded of the person entrusted with more' (Lk 12:48 - NAB)

fulfillment of an entity's function will contribute in different ways to its flourishing, opening new ethical scenarios due to the novel responsibilities acquired.

On the other hand, it is not difficult to see how one could build a virtue ethics narrative upon the widely held transhumanist belief that a 'persistent humanity' will subsist despite (or on account of) radical alterations to our biological existence (Fuller 2011, 95). Indeed, a big part of the motivation for promoting emerging technologies is precisely to defend this humanity from the perils of evolution (i.e., extinction), the environment (i.e., an ecological catastrophe), or technological threats (i.e., 'grey goo'). And precisely this human aspect 'that we shall not lose' is in most transhumanist accounts a straight-up Cartesian one: What will prevail as identifiably human, regardless of the locus of its existence (i.e., in silico, in the cloud, etc.), is the mind. This identification of the mode of human existence as a rational one might call for virtue theory as a platform of ethical articulation in a way that it may not do with utility and duty—since these last two rely on actions, not being (in this case, a *res cogitans*).

As seen above, virtue ethics relies first on a subject's being, in order to guide its actions thereafter. Aristotle puts a strong emphasis on the right structure of a being, which allows it to fulfill its function. The reliance on the rational aspect of this being is extreme, since this aspect, if properly informed, would effectively govern its (ethical) life towards its telos (flourishing). It is difficult to see, due to the action-based core of their systems, how duty or utility can attain a similar level of emphasis on flourishing—arguably the main purpose of transhumanism at the end. What is not difficult to see is the need for a more thorough assessment of the possibility of articulating transhumanist ethical concerns from alternative ethical systems.

7 Concluding Remarks

The near equalization of transhumanism with utilitarianism is expectable. The self-determined well being of the individual towards the welfare of the many seems to be a shared goal of both perspectives—regardless of how we get there (i.e., technological improvement of our biology), as long as the desired consequences are reached (i.e., the 'uplifting' of the species as a whole). We can see, however, that grave ethical conundrums surface, and they seem to have no solution as long as we remain within the utilitarian framework (e.g., temporal superpredators seeking radical life-extension, thus jeopardizing the social integration of individuals not yet born).

When we turn to deontology, we can envisage novel categorical imperatives that may be compatible with enhancing technologies—just as higher education can be conceived as a form of holistic human improvement. In contrast, when we turn to virtue ethics, there seem to be more issues that need to be dealt with, due to the alleged inimicality of aretaic ethics with the proposed radical transformation of human beings. Upon closer inspection, however, we find that these rigid isomorphisms (i.e., what we currently are with what we can do) are later instantiations of virtue ethics combined with bioconservative stances, problematically using evolutionary ethics and psychology as a bridge.

Once we turn to the source, namely Aristotle's ethics, we find that certain interpretations of his moral philosophy can indeed be found to be incompatible with fundamental transhumanist tenets (i.e., the ancient Greek emphasis on moderation may be difficult to fit with a philosophy of radical human transformation). Once again, upon closer inspection, we can see that, beyond canonical but possibly insufficiently informed interpretations, there are fundamental aretaic tenets that seem to be in deep sync with the most important transhumanist pillars. These are the extreme emphasis on human flourishing, and perhaps more interestingly, the stress on the improvement of a *being*, before the *actions* that emanate from it—as opposed to the two ethics emerging from modernity, where the moral core is located on the actions that are triggered, regardless of the being. This latter feature, in fact, may put virtue ethics closer to transhumanism than utilitarianism itself.

Transhumanism ontologically asserts a persistent being that survives all possible instantiations, i.e., a persistent humanity deeply related to mind and rationality but superseding those, incarnated currently on homo sapiens but in principle possibly instantiated in other structures—or on no physical structure at all. This aretaic emphasis on being places Aristotelian virtue as a potential super-ally of transhumanism, given that it is concerned with the improvement of this very being, before the (improved) actions that may come thereafter. Further incorporating a Thomistic reading of the actions stemming from improved natures, we can envisage new responsibilities and novel virtues for a future flourishing society.

After reckoning the profound social consequences triggered by transhumanist undercurrents implemented via pervasive converging technologies, we may see that broadening our spectrum of ethical articulation for transhumanism, making it inclusive of divergent moral narratives, shall acquire an urgent character.

References

A Lapide, C. 1637. *The Great Commentary of Cornelius a Lapide: St Luke's Gospel*, ed., Translated by C.G. Ross. London, UK: John Hodges, 1887.

Aquinas, T. 1485. *Summa Theologica*, 2nd Rev. Ed. Translated by Fathers of the English Dominican Province. London, UK: Benzinger Brothers, 1920.

Aristotle. 1908. *Nicomachean Ethics*. Translated by W.D. Ross. Oxford, UK: Clarendon Press.

Catholic Church. 1993. *Catechism of the Catholic Church*. Citta del Vaticano: Libreria Editrice Vaticana.

Collin, F. 2013. Two Kinds of Social Epistemology. *Social Epistemology Review and Reply Collective* 2 (8): 79–104.

Dyson, F. 2007. *A Many-Colored Glass: Reflections on the Place of Life in the Universe*. Charlottesville, VA: University of Virginia Press.

Fuller, S. 2011. *Humanity 2.0: What It Means to Be Human Past, Present and Future*. Hampshire, UK: Palgrave Macmillan.

Fuller, S. 2012. *Preparing for Life in Humanity 2.0*. London, UK: Palgrave Pivot.

Fuller, S. 2016. *The Academic Caesar: University Leadership Is Hard*. London, UK: SAGE.

Fuller, S. 2019. *Nietzschean Meditations: Untimely Thoughts at the Dawn of the Transhuman Era*. Basel, Switzerland: Schwabe Verlag.

Fuller, S., & Lipinska, V. 2014. *The Proactionary Imperative: A Foundation for Transhumanism*. Hampshire, UK: Palgrave Macmillan.

Fuller, S., and S. Sorgner. 2019. The Humanities and Posthumanities: New Ways of Being Human. In *8th Winter School of the Estonian Graduate School of Culture Studies and Arts*. Tallinn University, 28 January–1 February. https://vimeo.com/315508439. Accessed 08 June 2021.

Green, B. P. 2014. Transhumanism and Catholic Natural Law: Changing Human Nature and Changing Moral Norms? In *Religion and Transhumanism: The Unknown Future of Human*, ed. C. Mercer and T. Trothen, 201–215. Santa Barbara, CA: Praeger.

Harris, J. 2010. *Enhancing Evolution: The Ethical Case for Making Better People*. Princeton, NJ: Princeton University Press.

Helmer, J. 2017. Enhancing Moral Goodness: Toward a Virtue Ethics of Moral Bioenhancement. In *Religion and Human Enhancement: Death, Values, and Morality*, ed. T. Trothen and C. Mercer, 229–243. Cham, Switzerland: Palgrave Macmillan.

Hughes, J. 1996. Embracing Change with All Four Arms: A Post-Humanist Defense of Genetic Engineering. *Eubios Journal of Asian and International Bioethics* 6 (4): 94–101.

Hurtshouse, R. 2006. The Central Doctrine of the Mean. In *The Blackwell Guide to Aristotle's Nicomachean Ethics*, ed. R. Kraut, 96–115. Malden, MA: Blackwell.

Jaeger, W. 1943. *Paideia: The Ideals of Greek Culture. Volume II: In Search of the Divine Centre.* Translated by G. Highet. New York, NY: Oxford University Press.

Kant, I. 1785. *Grounding for the Metaphysics of Morals: With on a Supposed Right to Lie Because of Philanthropic Concerns.* Translated by J.W. Ellington. Indianapolis, IN: Hackett (Third Edition: 1993).

Kass, Leon R. 1997. The Wisdom of Repugnance. *The New Republic* 216 (22): 17–26.

Kompridis, N. 2009. Technology's Challenge to Democracy: What of the Human? *Parrhesia* 8:20–33.

Levin, S. 2020. *Posthuman Bliss?: The Failed Promise of Transhumanism.* New York, NY: Oxford University Press.

Losin, P. 1987. Aristotle's Doctrine of the Mean. *History of Philosophy Quarterly* 4 (3): 329–341.

Lumen Gentium: Dogmatic Constitution on the Church, promulgated by His Holiness Pope Paul VI on 21 November, 1964. 1975. In *Vatican Council II: The Conciliar and Post Conciliar Documents*, ed. A. Flannery. Dublin, Ireland: Dominican Publications.

MacIntyre, A. 1981. *After Virtue: A Study in Moral Theory.* London: Bloomsbury.

Malapi-Nelson, A. 2017. *The Nature of the Machine and the Collapse of Cybernetics: A Transhumanist Lesson for Emerging Technologies.* Cham, Switzerland: Palgrave Macmillan.

More, M. 1993. Technological Self-Transformation: Expanding Personal Extropy. *Extropy 10* (4/2): 15–24.

More, M. 2013. The Philosophy of Transhumanism. In *The Transhumanist Reader: Classical and Contemporary Essays on the Science, Technology, and Philosophy of the Human Future*, ed. M. More and N. Vita-More, 3–17. West Sussex, UK: John Wiley & Sons.

Pearce, D. 1995. *The Hedonistic Imperative.* https://www.hedweb.com/welcome.htm. Accessed 08 June 2021.

Pedersen, D. B. 2013. Who Should Govern the Welfare State 2.0? A Comment on Fuller. *Social Epistemology Review and Reply Collective 2* (12): 51–59.

Persson, I., and J. Savulescu. 2012. *Unfit for the Future: The Need for Moral Enhancement.* Oxford, UK: Oxford University Press.

Rappert, B., and M. J. Selgelid, eds. 2013. *On the Dual Uses of Science and Ethics: Principles, Practices, and Prospects.* Canberra, Australia: Australian National University E Press.

Rawls, J. 1971. *A Theory of Justice.* Cambridge, MA: Harvard University Press.

Remedios, F., and V. Dusek. 2018. *Knowing Humanity in the Social World: The Path of Steve Fuller's Social Epistemology.* London, UK: Palgrave Macmillan.

Rousseau, J.J. 1762. *Emile, or, on Education: Includes Emile and Sophie, or, the Solitaries*, ed., Translated by C. Kelly and A. Bloom. Lebanon, NH: University Press of New England, 2010.

Sandel, M. 2007. *The Case Against Perfection: Ethics in the Age of Genetic Engineering*. Cambridge, MA: Belknap Press.

Savulescu, J., and N. Bostrom, eds. 2009. *Human Enhancement*. Oxford, UK: Oxford University Press.

Schmidt, C. W. 2008. The Yuck Factor When Disgust Meets Discovery. *Environmental Health Perspectives* 116:A524–A527.

Schneewind, J. 1984. The Divine Corporation and the History of Ethics. In *Philosophy in History*, ed. R. Rorty, J. Schneewind, and Q. Skinner, 173–192. Cambridge, UK: Cambridge University Press.

Singer, P. 2001. A Utilitarian Defense of Animal Liberation. In *Environmental Ethics*, ed. L. P. Pojman and L. Pojman. Boston, MA: Wadsworth.

Sorgner, S. 2010. Beyond Humanism: Reflections on Trans- and Posthumanism. *Journal of Evolution and Technology* 21 (2): 1–19.

Sparrow, R. 2011. A Not-So-New Eugenics: Harris and Savulescu on Human Enhancement. *The Hastings Center Report* 41 (1): 32–42.

Stocks, J. L. 1931. The Golden Mean. *The Monist* 41 (2): 161–179.

Stump, E. 2012. The Non-Aristotelian Character of Aquinas' Ethics: Aquinas on the Passions. *Tópicos, Revista de Filosofía* 42: 27–50. https://doi.org/10.1002/9781118321997.ch6.

Vallor, S. 2016. *Technology and the Virtues: A Philosophical Guide to a Future Worth Wanting*. New York, NY: Oxford University Press.

An Aristotelian Typology and Evaluation of Transhumanism: Concordance and Conflict

Brian Patrick Green

We ought not to follow the makers of proverbs and 'Think human, since you are human' or 'Think mortal, since you are mortal.' Rather, as far as we can, we ought to be pro-immortal, and go to all lengths to live a life in accord with our supreme element.[1]
Aristotle, *Nicomachean Ethics* (1999, 164–165)

Whatsoever is in conformity with nature is pleasant.
Aristotle, *History of Animals* (1984a, 923)

Nothing which is contrary to nature is good.
Aristotle, *Politics* (1941d, 1282)

[1] "Pro-immortal" is translated from *athanatizein*, meaning "supporter of immortality," "supporter of the immortals," or to "make oneself immortal" (Aristotle 1999, 310).

B. P. Green (✉)
Markkula Center for Applied Ethics and School of Engineering, Santa Clara University, Santa Clara, CA, USA
e-mail: bpgreen@scu.edu

G. Theiner, A. Malapi-Nelson (eds.), *Technologies of the Future Self*, Palgrave Studies in the Future of Humanity and its Successors,
https://doi.org/10.1007/978-3-031-99902-4_4

Philosopher Nick Bostrom and others have rightly noted that the ancient Greek philosopher Aristotle is a "perfectionist" philosopher, one who believes that humans should reach for the highest perfection possible (Bostrom n.d., 3; Campbell and Walker 2005; Walker 2002). But Aristotle was also mindful of fulfilling and cultivating human nature rather than opposing or reconstructing it, as well as the importance of moderation and avoiding extremes (Malapi-Nelson 2021, 8–16). Within the works of Aristotle, there is a tension between the normativity of nature and the normativity of transcending nature. This tension can be placed in fruitful dialogue with contemporary debates on the normativity of current human nature versus the normativity of human enhancement.[2]

To investigate this issue of the relationship between transhumanism and Aristotelian thought, we must first know what transhumanism is, what perfection it is reaching for, what means are to be used to attain this end, and to what extent transhumanism is "natural" in an Aristotelian sense. My goal in this paper is to create an Aristotelian typology of transhumanism, categorizing strains of transhumanism into three major types. This typology first divides transhumanism into two strands: embodied transhumanism, which maintains a human body, and disembodied transhumanism, which does not. Embodied transhumanism will be further subdivided into transhumanism which cultivates and fulfills human nature and transhumanism which replaces or reconstructs human nature. In light of this typology, I will then investigate the tension between the normativity of the natural and the perfect in Aristotle's works, and morally evaluate each strand of transhumanist thought through an Aristotelian lens.

My thesis is that embodied transhumanism which cultivates nature could be morally acceptable from an Aristotelian perspective, disembodied transhumanism would not be acceptable, and that embodied transhumanism which replaces or reconstructs nature requires more specific investigation at the time of actual decision; in Aristotelian terms, prudential discernment of the particulars (Aristotle 1999, 1141b15-24). I will conclude that some aspects of transhumanism are in concordance with Aristotelianism, while others are in conflict with it, and that both Aristotelianism and transhumanism could benefit from further intellectual interaction so as to stretch and improve the ideas in both systems. With

[2] This paper is derived from "Would Aristotle Be a Transhumanist?" presented to the Transhumanism and Religion Consultation of the American Academy of Religion, Montreal, Canada, November 9, 2009 (Green 2009).

this knowledge we will be better able to understand what is at stake in potential transhumanist manipulations and how better to respond to these ideas. The goal is not advocacy, but explanation.[3]

1 Aristotelian Background

Aristotelian thought is full of ideas such as soul, form, matter, finality, essence, nature, four cause theory, and act and potency. The age of these ideas does not make them any less useful; indeed Aristotelian ideas are important not only to philosophy but also to science, where they provide a foundation for empiricism and inductive logic. But because these ideas are somewhat obscure, before applying Aristotle to transhumanism we must first investigate what some of these major relevant ideas are.

To begin, act and potency are important aspects of the Aristotelian system. To explain the existence of change, Aristotle asserted a distinction between actuality and potentiality (1941b, 1019a15-1020a6, 1042b9-1043a28, 1045b28-1052a12, 1065b5-1066a34; 1941c, 200b12-202b29). Actual things currently exist, while potential things do not yet exist but have the power to come into existence in the future. Typically, potency lies in actual things in an unrealized way, waiting for the proper conditions for their actualization. These two concepts are very important for the relationship between Aristotelian thought and transhumanism because implicit in transhumanist theory is the idea that transhumanity lies in potential in standard humanity as it is currently actualized (Green 2015a, 206–207, 211). For Aristotle, in human nature there are natural potencies that should be actualized; for example, because we are talking/rational animals, the potencies for speech and reasoning ought to be actualized for a human to be considered normal (Aristotle 1941d, 1253a10, 1332b5; 1999, 1098a3-8, 1102a5-24). While natural potencies should be actualized, unnatural potencies (which are not aligned with natural entelechies, such as mental incapacitation) should not. So for Aristotle and transhumanism, the question of which potencies

[3] I write this chapter as a Catholic neo-Thomist who is not a transhumanist and has grave reservations about the transhumanist project. However, here I desire to be as impartial as possible in order to find a clear understanding of how these two disparate systems can interact. I have explored related topics like this before with Aristotle and cognitive science (Green 2012) and transhumanism and the Catholic Church (Green 2015b).

transhumanists ought to actualize, and whether this actualization is "natural" to humanity or not, becomes a key question.

Aristotelian four cause theory is also extremely relevant to the relationship of Aristotelianism and transhumanism. Four cause theory states that everything that changes can be said to change due to four causes: material, formal, efficient, and final. The material cause of something is the matter of which it is made. Specifically, the material cause is *hyle prote*, "first matter" or "prime matter," which is pure potency, and which therefore can never exist in nature except as the underlying substance which endures changes in form (Aristotle 1941b, 1028b33-1029a33, 1042b9-11; 1941c, 189b30-191a22, 193a1-29). In contemporary terms it would be like the mass-energy substrate underlying the universe; it can exist in one form or the other, and convert between forms, but never exist as simple mass-energy without form (Wallace 1996, 7–9). Prime matter is actualized by giving it a form, *forming* it, which creates a *hylomorph*—a matter-form. Thus, the formal cause is that which actualizes prime matter into something.

Matter plus form is the essence or nature of an entity, and for living entities, the form is its soul. The soul is what gives life to its matter, and without a soul "in-form-ing" living matter the hylomorph immediately begins the process of disintegration, departing from actuality back into potency, from a whole into parts—raw materials for something new. It should be mentioned that this is not substance dualism because matter and form can never exist in a free state without the other; rather it is properly thought of as a monism with dual aspects.

The efficient cause is that which causes change in form, for example, a sculptor is the efficient cause of a statue, changing free-form stone into sculpted stone. Lastly, the final cause is the cause for the sake of which the change occurred. For Aristotle, this included everything from the reason why a sculpture was commissioned (e.g. to honor a hero) to the reason a rock fell (e.g. down is the place for rocks to go).

Because living hylomorphs contain final causes towards which they naturally incline, these organisms have a normativity built into them, in their souls. The final cause of most living creatures is simply to survive and reproduce (Aristotle 1984b, 717a13-16, 731a25-731b6; 1991, 415a23-b7; Johnson, 173–178), but for humans there is something more besides just this inclination basic to all life. Humans are rational animals and therefore our final cause is also to ponder the deepest questions of the universe—the questions of philosophy, and especially theoretical, speculative philosophy (Aristotle 1999, 1177a12-1178a9; 1941b, 980a22-983a23). To perform this highest task—which, in all of existence, only humans can perform—we must cultivate the virtues and particularly the intellectual virtues of science (knowledge), understanding (the relationships between pieces of knowledge), and wisdom (the purpose towards which understanding should be put) (Aristotle 1999, 1139b15-1141b23).

Herein lays the origin of Aristotle's tension in human nature. It is our nature to have normatively human potencies that must be actualized. However, the most important potency for human actualization—transcendent wisdom—is one towards which we are, seemingly, *inadequately disposed* (Aristotle 1941b, 982b28-983a11). The full actualization of this potency may not be within the realm of possibility for human nature, and so human nature would seem to be deficient and therefore need some help. And transhumanism might seem to be able to offer that help. However, there are additional problems.

Aristotle's assertion in *Politics* VII.3 that "nothing which is contrary to nature is good" reveals his opposition to the unnatural, and his general belief that nature is teleological and normative. Nature is telic and *telos* is good, and whatever detracts from the *telos* is therefore not good at best, or downright bad at worst (Aristotle 1999, 1094a1-1094b12). In the context of transhumanism, the question then becomes which "nature" are we talking about—our limited one which we share with lower life forms or our higher one which makes us distinct from them? (The physicalist vs. rationalist split in the related but distinct field of Catholic natural law ethics; see Green 2015a, 201–214). This is a metaphysical question, not empirical, and thus remains open to investigation only through more theoretical means such as via Ian Barbour's four criteria for assessing metaphysical theories: coherence (makes sense within the context of other theoretical commitments), correspondence with data, comprehensiveness (scope), and consequences (fruitfulness or fertility) (Barbour 1990, 34–39). In the words of Alasdair MacIntyre, Aristotle has a "metaphysical biology" which roots his ethics (1999, x). Natural kinds have their own internal purposes, the fulfillment of which is good for that creature and the thwarting of which is bad; in other words, a creature's nature has a final cause, an entelechy or inbuilt goal which is its end and good (Aristotle 1941a, 641b24-27; 1941b, 1013a24-1013b3; 1941c, 198b10-199b33; Rollin 2005, 318; Johnson 2005).

When the question of technological intervention is applied to Aristotle's theory of nature, it creates an interesting division. Following the philosopher of technology Carl Mitcham, in matters of technology, Aristotle differentiates between the arts of *cultivation* and *construction*. In Mitcham's words:

> Cultivation is making or using, which helps nature to produce either more quickly or more perfectly what it could produce by itself; examples are the arts of medicine, teaching, and farming. Construction constrains nature to produce things it would not otherwise produce, the primary example being architecture.
>
> In Aristotle's cultivation-construction distinction, design would appear to be foreign to the arts of cultivation. In the arts of construction, by contrast, design is central. (Mitcham 1980, 308, cf. 306–309, 322)

Cultivation draws a living entity towards its natural perfection; while construction constrains nature into a form it would not otherwise produce (cf. Aristotle 1941c, 192b8-193b22, 198b10-199b33; Durbin 1981, 272–273). The primary distinction between the two categories is the role of humans in relation to the system in question. In the case of construction, humans are working with their own design, *techne*, but in cultivation, we are interacting with—so to speak—"someone else's work," in Aristotle's case, *physis*: nature's. When working with natural objects, we are always hijacking another system (particularly living entities, but also materials like minerals and elements), and thus we must work with it to whatever extent necessary in order to bend it towards our ends, while artificial objects have no ends other than those we place on them; they are therefore much easier to work with because they typically do not try so hard to "escape" towards their own ends. While in our contemporary world biotechnology, geoengineering, and other mixings of the natural/biological and artificial have begun to make these categories blur, there is still great use in these old categories as we shall see in the next section.

Moving on, we must finally consider what exactly it is from an Aristotelian perspective that makes an act moral. In order to make moral judgments some parameters must first be set. As mentioned above, the most profound distinction is whether the action in question leads towards human natural good. Aristotle, of course, mentions in some detail what the human good is—the philosopher/intellectual having the best life of all, and the political/moral life being secondary to that (Aristotle 1999, 1177a12-1178a9). These are the lives that lead most profoundly towards *eudaimonia*—true happiness. But in order to achieve these heights of eudaimonia, certain human capacities must be cultivated towards virtue: our capacities for intellectual and moral pursuits in particular. The virtuous person becomes the paradigm from which goodness is judged. Furthermore, the virtuous person does not act purely for the sake of good consequences, but also for good reasons. It is not enough to make the right choice, one must also make the right choice for the right reason, or else the action does not cultivate one towards virtue, but rather is merely a blind imitation that lacks understanding and is therefore not useful for the purposes of effective development of prudence (which is both a moral and an intellectual virtue). While careful imitation can start to habituate courage, temperance, and other moral virtues, prudence—also known as practical wisdom, which relies on judgment in particular cases—requires understanding (Aristotle 1999, 1139b15-1141b23). Therefore in any judgment of the morality of any particular transhumanist manipulation of human nature, we must consider not only its feasibility (impossible pursuits waste effort) but also whether it achieves good consequences with respect to human nature (lower or higher being a prime question) and also whether the motivations and justifications are correct (because doing the right thing for the wrong reason does not generate growth in virtue).

The question now is "how do these Aristotelian ideas align with transhumanism?" For this, I will present a typology with two forks, and then evaluate these types using some of the above ideas.

2 The Typology: Embodied or Disembodied, Cultivated or Constructed?

Transhumanism is a movement with the general goal of enhancing human capacities. However, transhumanist means for attaining these ends are highly diverse. Using two Aristotelian distinctions, transhumanism can be separated into three major schools.

The first distinction is one of embodiment vs. disembodiment from the natural human body. Some transhumanists (though perhaps not all would claim the title) seek to maintain a human body and merely perfect that body into its highest state, thinkers such as Aubrey de Grey with his SENS "strategies for engineered negligible senescence" (de Grey et al. 2002, 453), Lee Silver who says "we can already see the powerful forces that will drive the development of reprogenetics—the desperate urge to have biological children and the incredible lengths to which people will go to give their children the best possible lives" (Silver 1998, 308), and Gregory Stock who states "The question is no longer whether we will manipulate embryos, but when, where, and how" (Stock 2003, 2). Nicholas Agar likewise sees a future "granting prospective parents a limited prerogative to use enhancement technologies to choose their children's characteristics" (Agar 2004, 175). These thinkers represent an embodied form of transhumanism in which humans still live in and reproduce through biological bodies.[4]

Other transhumanists seek to upload their consciousness out of the human body and into another artificial state, even seeking to exist as information in a network, with no localized home. The mind-uploading process described by Hans Moravec in the 1970s gives some sense of this transition from neural to circuitry:

> [The robotic surgeon's] attention is directed at a small clump of about 100 neurons somewhere near the surface. It examines, non-destructively, the three-dimensional structure and chemical makeup of that clump with neutron tomography, phased array radio encephalography, etc., and derives all the relevant parameters. It then writes a program which can simulate the behavior of the clump as a whole, and starts it running on a small portion of the computer next to you. It then carefully runs very fine wires from the computer to the edges of the neuron assembly, to provide the simulation with the same inputs the neurons are getting. (Moravec 1976/1978)

Ray Kurzweil also describes uploading in quite a bit of detail, though, at least at the start, he also thinks that humans may want to be embodied, even if perhaps only virtually or through nanotechnology (Kurzweil 2005, 199). Further into the future, most humans might not live localized existences at all (Kurzweil 1999, 241), in the words of Joel Garreau, "transcended" like "the Christian portrayal of angels" (Garreau 2005, 104). This is a disembodied form of transhumanism where "substrates" can come and go at will, and entities (not exactly "lives") are spread across networks. The Carboncopies Foundation seeks "To identify, facilitate, and conduct the research which will deliver whole brain emulation," which permits the reinstantiation of human mental processes in computer substrates. They seek this because "we believe that the possibilities and opportunities enabled by Whole Brain Emulation (WBE) represent the most promising technological path to overcoming our fundamental limitations as a species" (Carboncopies

[4] Of note, over the past two decades this more biologically focused transhumanism has been somewhat eclipsed by a more computer and AI focused one.

Foundation 2024), in other words, to break out of humanity's biological shackles. This embodied/disembodied distinction is important because from an Aristotelian perspective, as hylomorphic beings, our nature inheres in our body, and maintaining the body maintains the being, while separating the two creates an unnatural discontinuity.

The second distinction is one of means for achieving the embodied or disembodied state. This distinction is between cultivative and constructive forms of transhumanism as discussed above. Cultivative transhumanism seeks to enhance human capacities by drawing natural human capacities towards their intrinsic perfections; as such it has to be embodied in a biological body, even if that might involve genetic and medical enhancements, as described by Agar, de Grey, Silver, and Stock. In contrast, constructive transhumanism seeks to constrain nature to produce things it would not otherwise produce, like uploading and brain emulation as described by Moravec, Kurzweil, and the Carboncopies Foundation. This is relevant because it affects how "natural" or "artificial" the system is (artificial not necessarily being grounds for dismissal), and if moral evaluation is judged by achieving the highest good for the highest reason, the means used may be dictated by these judgments.

While the categories of embodied/disembodied and cultivative and constructive at first might seem to be a 1-to-1 overlap, there is another possibility, as seen in this chart:

```
                      Embodied or 2) Disembodied, Constructive
                        /         \
                       /           \
1) Embodied, Cultivative or 3) Embodied, Constructive
```

The three possible categories are (1) embodied cultivative transhumanism, (2) disembodied constructive transhumanism, and (3) embodied constructive transhumanism. (A fourth option—cultivating disembodied transhumanism—is incoherent because it involves cultivating—a term defined as being for natural embodied entities—an artifact—something which by definition is not a natural body.) While the transhumanists mentioned above tend to fall into categories (1) and (3), category (2) is also worth examining because it tests Aristotle's philosophy in unique ways, not to mention providing a different perspective on transhumanist manipulations themselves.

This is not the only way to apply Aristotle to transhumanist thought; many other approaches exist. But I do believe the typology I present here is useful because it both encompasses and parses most, if not all, of transhumanist thought, and it does so in a way that allows thought-provoking further analysis not only of transhumanism, but also of the Aristotelian philosophy upon which the typology is based.

In the next section, I will investigate more deeply each type of transhumanism described above and then give an Aristotelian moral evaluation of that type of transhumanism.

2.1 Embodied, Cultivating Transhumanism

2.1.1 Description

Embodied cultivating transhumanism includes those transhumanist options presented by Agar, de Grey, Stock, Silver, and others, though that is not to say that these thinkers might not accept other more dramatic "constructive" approaches if available. But in their publications, they seem to fit this category best as a first approximation.

Fundamentally, this is perhaps the most "traditional" type of transhumanism, going through the father of eugenics Francis Galton (1883, 1892), all the way back at least to Plato's and Aristotle's eugenic proposals. As David Galton (no relation to Francis Galton) notes (1998, 263–267), in ancient Greece, both Plato and Aristotle recommended eugenics in their own ways. Plato in his *Republic* advocated for state-controlled human selective breeding (1992, 133–136), and in his *Politics*, Aristotle also supported state-controlled eugenics, though he was relatively less invasive into the private lives of the citizens of his state; he advocated the avoidance of deformed offspring through means including state restrictions on who could reproduce, infanticide, and abortion (1941d, 1334b29-1336a3).

How, then, might Aristotelian thought relate to contemporary transhumanism ideas for genetic manipulation, cognitive enhancing drugs, or similar human enhancements? As is demonstrated by his advocacy of abortion and infanticide, Aristotle allowed for valuing human traits above human lives. Human genetic manipulation, whether through PGD embryo selection, the use of eugenic sperm and eggs, biotechnological manipulations such as germ line gene therapy, and so on, might be thought of as analogous to the means Aristotle states as acceptable, though of course the manipulations would need to be limited to cultivating nature (and some scholars such as Harold Baillie actually use Aristotle to critique genetic engineering (Baillie 2005, 226–231)).

Similarly, based on his willingness to phenotypically (and thus by extension genetically) parameterize state citizens, Aristotle could perhaps find an analogy with the use of cognitive-enhancing drugs, as advocated by Greely et al. (2008), provided they had no negative side-effects; which could be a major exception if denial of the drug would lead to loss of the excellence in the individual, for this vulnerability would remove self-sufficiency from the individual and make them dependent on the drug for their virtue. Virtue is both a means and an end, and so its loss is a loss indeed.

As for external human technological enhancements that render one a "fyborg," or functional cyborg (Chislenko 1995; Stock 2003, 25–31), instead of an actual cyborg (for example wearing high-tech devices, possessing smartphones, or advanced sensory devices), those would seem to be as acceptable as any tools of the past; they need not harm our natures, and do greatly increase our capacities and thus our potential for virtue. There is a question here as to the depth of our embeddedness in the technical world, and how Aristotle might interpret that; for example, a Vygotskyan approach takes material culture extremely seriously because of its intimate connection with our own cultivated/constructed nature (Theiner and Drain 2017; this will be examined more below). However one interprets the extent and depth of the intersection of cultivation and construction, on the more cultivative side the rationale is that with greater capacities and powers for action are greater potential for action and therefore greater potential to actualize these capacities into excellence. It is the difference between having a teacup or a teapot; a filled teacup actualizes excellence, but a filled teapot does the same even more so. In less metaphorical terms, a smart person might be able to generate the theory of relativity, but the same person with expanded cognitive powers might be able to create a theory uniting quantum mechanics with gravity.

These technological "extras" would count as *external goods* to Aristotle, and while external goods may be necessary to actualize certain virtues, such as generosity, which requires having something to give away (Aristotle 1999, 1099a28-1099b1), they are not constitutive of the virtues themselves. The issue would then be one of directing these powers towards the good ends of human life rather than towards the vices, but this is a matter of judgment of ends, not a question of the means themselves. Which brings us to moral evaluation.

2.1.2 *Evaluation*

Are the manipulations feasible? Can they have "good" effects if done for the right reasons? What are the right reasons for pursuing these ends? These manipulations are feasible, and the desired effects are good. Though not Aristotelian, even Pope Pius XI agreed that the ends of eugenics—helping people be healthier, smarter, etc.—were good, though the means bad (Pius XI 1930, 63–71). Bracketing the means (Aristotle was certainly open to more intrusive, and even destructive, means than the Pope), the question then is one of whether these manipulations would be done for the right reasons or not.

Aristotle argues that the ultimate reason for pursuing goods is happiness because if we fulfill our natural functions we will be doing what we are supposed to do. All other goods must be subordinate to this end of becoming excellent at our human function, which is rational activity. The right reason, then, to act upon these manipulations is to further the end of human excellence in rational activity, and indeed some transhumanists agree (Bostrom n.d.; Bostrom 2014, 26–62; Kurzweil 2005, 316–317; Young 2006, 37).

Therefore, transhumanism which maintained the human body and sought to perfect its natural capacities might be acceptable to an Aristotelian approach to ethics, under certain conditions. If transhumanism cultivated natural human capacities and virtues towards perfection, whether intellectual, moral, or bodily, then it could be good. Of course, other ethical systems (and interpreters of Aristotle) are free to disagree.

2.2 *Disembodied, Constructive Transhumanism*

2.2.1 *Description*

On the opposite end of the spectrum from embodied cultivating transhumanism is transhumanism which seeks to disembody and construct human existence outside of a biological substrate. For an example of this, I will give an Aristotelian evaluation of mind-uploading.

In his book *The Singularity is Near*, Ray Kurzweil discusses mind-uploading in terms of "backing ourselves up" and reinstantiating brain patterns into a different, more "powerful" substrate (Kurzweil 2005, 323–324). Kurzweil considers himself a "patternist" as opposed to a materialist, so in Kurzweil's thinking, copying the person's pattern ought to copy the person (Kurzweil 2005, 5, 383–388). He acknowledges that there will be questions as to whether this reinstantiation is the same entity as the original in any meaningful way (Kurzweil 2005, 198–202, 382–386). Indeed, a hylomorphic understanding of life would indicate that even an identical form instantiated into different matter is indeed a different entity, thus making mind-uploaded immortality the immortality of a copy, not an original.

The removal of the human mind from the human body would seem to be extremely unnatural from an Aristotelian perspective and raises the question of its acceptability.

2.2.2 *Evaluation*

The Aristotelian perspective on this form of transhumanism is that it would be extremely unnatural, and therefore unsuitable for humans to participate in. From Aristotle's perspective, it is not only unnatural, it is an impossible, unfeasible course of action for true immortality, creating only some sort of electronic copy of a person, and is therefore irrational and should not be pursued. It may have other valid uses, but immortality is not one of them.

This is Aristotle's argument. In *De Anima*, Aristotle discusses the nature of the soul, analogous to a "pattern" in Kurzweil's thinking, and the soul's role as the form of its matter (Aristotle 1991, 412a1-414a30). Superficially this would seem to support Kurzweil's position that the soul, as a pattern, ought to be copiable and transferable. However, a more thorough understanding of Aristotle would make clear that the copied "soul" would not be the same soul as the original because it would be instantiated in different matter. This would be a new *hylomorph* (or "matter-form"), a new entity not equivalent to the original. If the original creature were maintained, there would now be two entities, one the biological original, and the second a computerized copy. (Additional resources for thinking about hylomorphic theory of mind and the metaphysics of uploading include Jaworski (2011, 269–355), Corabi and Schneider (2012), and others.)

What kind of entity would the disembodied copy be, from an Aristotelian perspective? According to Aristotle, there are three levels of soul: the nutritive level, natural to all life; the sensitive level, natural to all animal life; and the rational level, natural to human life (Aristotle 1991, 414b1-415a15; 1999, 1102a5-1103a10). Instantiated into a computer, the created entity would not clearly have a nutritive soul (it would not be growing, metabolizing, and maintaining "life" in a normal sense, but rather in an electronic sense). It also might not have a sensitive soul, though possibly data inputs and movement of some sort would count as this type of soul. It might have a rational soul, however, even this aspect of the entity would be unclear, since it is not yet known if artifacts can be rational, or even what the concept of "rationality" would really mean in this context, with a human mind re-instantiated into electronic matter, which would most likely not function like organic brain tissue, but rather would simulate or emulate the function of that tissue (obviously there is a question as to whether there is a significant distinction between simulating and emulating brain function and actually *living* brain function).

If the entity no longer has any of the three "souls" left in it, then it is not, in fact, a duplicated living thing, rather it is a dead artifact. This is not a good aim for one seeking the highest things. On the other hand, if the entity had no nutritive or sensitive soul, but did still have a rational soul, the resulting creature would be a rational, conscious "nonbiological entity," as Kurzweil notes (2005, 385). This is a type of entity that Nobel Prize-winning neuroscientist Gerald Edelman once claimed to be creating, in fact, not by upload, but from scratch, at the Neurosciences Institute in La Jolla, California, though he called it a "conscious artifact." In the late 2000s, Edelman claimed he had a device equivalent to a cat brain which had "intrinsic activity," that is, activity even with no sensory input (Kruglinski 2009).

Given these very strange characteristics, whether an uploaded entity could be considered "human" or "life" would be debatable. Of course, transhumanism already knows this: that is why it is called *trans*-humanism, and not simply *humanism*. Given the dramatic differences of an uploaded existence, it is an open question as to whether this would be a superhuman or a subhuman existence. What would thought or experience be like in an electronic substrate? Can consciousness or thought even inhere in such a substrate or are they inherently a property of, perhaps, the element carbon, or water, or of certain organic compounds or formats?

At an even deeper level, the entity created would not seem to be alive and therefore would be more like an artifact than a living being. Artifacts by definition do not contain their own principle of growth, metabolism, and so on, they have no internal teleological nature which drives them forward in life; rather, they are created by external forces and remain subject to these forces. Choosing death over life is irrational for one seeking excellence. Should the artificer become the artifact? Even if granting it were feasible (which from an Aristotelian perspective it is not), Aristotle would not approve.

Of course, the deeper question is whether humans could actually construct artifacts with internal "natures" that is, new life forms with their own "souls" and *telei*. Indeed, the field of synthetic biology is attempting to do precisely this. At this point, the distinction between construction and cultivation would be lost, and humans would have accomplished a task in some ways analogous to the original genesis of life on Earth. There are two conceivable ways to do this—by reworking extant living material, or from scratch. While the latter is the work of the disembodied constructive path of transhumanism (with the problems noted above), the former can be the work of embodied constructive transhumanism, a bridging category between life and artifact, more akin to synthetic biology or other forms of bioengineering. The next question is, then, exactly how this could be done and at what ontological level—atomic, molecular, organelle, cellular, tissue, organ, organ system, etc.

And so, as a final Aristotelian note on disembodied transhumanism, it must be stated that the ontological level of the uploading may affect whether the upload is successful in maintaining "soul-continuity" or not (Wiley 2014, 31–63). Certainly, there has been active research in this area regarding uploading particularly at the cellular level, with nematode models (OpenWorm 2024), mouse models (HumanBrainProject 2015; EBRAINS 2023), and more. From an Aristotelian perspective, the *matter* and *form* take a particularly important place, however. Arguments for "substrate independence" and "multiple realizability" do not work in an Aristotelian system, because it is not only the *form* that matters, but also the *matter* that matters.

The neuronal tissues of the brain change form over time (Shors et al. 2012), but overall, after the early stages of brain development, the actual matter of the neurons—specifically the Carbon-14—stays largely the same during a lifetime, as demonstrated through nuclear bomb radiocarbon pulse research (Grimm 2008, 1434–1437), Bhardwaj et al. saying "our results demonstrate that, whereas nonneuronal cells turn over, neurons in the human cerebral neocortex are not generated in adulthood at detectable levels but are generated perinatally" (Bhardwaj et al. 2006, 12564–12568). This is in contrast with Kurzweil's statements in *The Singularity Is Near*, where he claims that "My body is temporary. Its particles turn over completely almost every month. Only the pattern of my body and brain have continuity" (2005, 371). In other words, when it comes to the brain, Kurzweil is wrong, and our identity might well have something to do with our neural matter regardless of pattern; it is not a certainty, but it is a distinct possibility.

How do matter and form operate to maintain identity? Do we feel a continuity of identity because the majority of our neural connections stay the same over time, or because the matter of our neurons stays the same? Or something else? If, like Kurzweil, we believe it is the form that counts and not the matter, and if the transition from protein-based brain to computer-based brain were done at a low enough level, for example at the synaptic or molecular level, then there is at least the theoretical possibility for continuity of form, and thus of soul. At this point, the uploaded entity could either remain in its original body (thus keeping its lower levels of soul) or be fully computerized (thus possibly losing those levels). Bracketing the question of identity (the uploaded hylomorph is certainly a new entity), while the latter (full computerization) would not be acceptable to Aristotle, the former might be, which brings us to the interesting middle case, between full embodiment and full disembodiment.

2.3 *Embodied, Constructive Transhumanism*

2.3.1 *Description*

Embodied transhumanism which replaced nature, such as replacing organs or body parts with artificial organs or parts (e.g., heart, lung, spine, limbs, and in the extreme case, the brain), would not cultivate the natural parts of the body, but rather constrain and reconstruct them. Examples might include extreme body modification, implanted, attached, or remotely operated biomedical devices, and artificial organs. This is not mere fiction: the case of the "cyborg rat" with a partially artificial cerebellum presents clear possibilities for this type of transhumanism (Geddes 2011), as do experiments on spiny lobsters (Szucs et al. 2000); indeed, both of these experiments, and experiments on humans (Gupta et al. 2023), have demonstrated neural tissue does seem to be interfaceable and replaceable by electronics, at least at a small scale.

How much more extreme can these reconstructions of the body be made? In addition to nerves, we have artificial hearts, artificial limbs, artificial eyes, artificial tracheas, and more. So we can foresee, perhaps, a future not disembodied, but dismembered, disheartened, disemboweled, "disembrained." And what if all these reconstructions were brought together? While each individual part might be justifiably replaced for the sake of the cultivation of the whole, what if all the parts were replaced into a fully artificial whole? Where does cultivation become construction?

2.3.2 Evaluation

While Aristotle might at first face strongly oppose this as unnatural (an embodied whole, but "dis-body-parted" ("disorganed"?) existence), Aristotle might have trouble determining exactly why, because several Aristotelian category distinctions would come into interference with each other. The levels of the soul might remain intact, as might identity, at least in theory. Because "the whole is of necessity prior to the part," if the whole is retained, then the nature and *telos* of the creature would most likely remain intact as well (Aristotle 1941d, 1129, 1253a20). Could a being have artificial parts and a biological whole, which retained a human nature, yet with enhanced powers? Do the three levels of the soul count as wholes in themselves with their own ends, or are they all one whole together, with the rational soul being the uppermost in the hierarchy? Aristotle himself seems ambiguous regarding this issue, knowing that some animals and plants can be divided and still live as pieces that then can become new wholes, thus meaning that they contain potentially several souls (Aristotle 1991, 413b1-30, 432a10-b5); but whether the logical distinctions in the human soul are in fact separable is another matter entirely. Aristotle suggests that the rational faculty may in fact be separable from the other parts, "as that which is eternal from that which is perishable" (Aristotle 1991, 42, 413b25-6). While Aristotle saw this as a possible means for the immortality of the soul in an afterlife, for a transhumanist it might be the means for the immortality of the soul in this material world.

For Aristotle, soul and nature are determined and defined primarily by form and function. If an ax were a natural object, and its "axeity," or form, were removed, it would cease to be an ax even if it somehow retained its matter (Aristotle 1991, 412b11-5). Human nature is to be a nutritive, sensing, and rational creature, with rationality being our highest function and *telos*: our defining characteristic (Aristotle 1999, 1098a1-21). Matter, while necessary, is much less important than form and function, as Aristotle says "For the formal nature is of greater importance than the material nature" (Aristotle 1941a, 647, 640b29). While matter is a substrate for parts, the essence of a thing is in its whole (Aristotle 1941b, 1013b23). If humans could better fulfill their rational nature through dramatic changes in our matter or parts, this could, from this perspective, be grounds for choosing to do so.

At this point, it might be important to pause and note that there are other ways to interpret human nature and Aristotle's approach to it. As noted above, more "cyborg anthropology" which sees a more blurred boundary between nature and culture, *physis* and *techne*, cultivation and construction—or perhaps no boundary at all—causes the categorical separation to itself blur. This approach disagrees with Aristotle in a fundamental way, and we might, given current anthropology and philosophy of mind, decide that Aristotle simply got his dichotomization of cultivation and construction wrong. Authors such as Theiner and Drain have offered a Vygotskian perspective as an alternative; one which appreciates the co-evolved "nature" of human biology and culture (see Drain, this volume, as well as Theiner and Drain 2017; Theiner 2021a; Theiner 2021b). In this perspective, we might say that becoming (fully) human is the emergent, cultivated-constructed outcome of a cultivated-constructed spiral of mutual entanglements between biological bodies, material culture, and social and institutional realities. Language is a prime example of this: biologically we are disposed to acquire a language that exists as a social construction, an artifact of technology, outside of us before we are born; yet once acquired, the language-speaking animal is thus enabled to inhabit cognitive and experiential niches to which it would otherwise be existentially closed off. Authors who advocate this more expansive approach including "extended mind" or "cyborg anthropology" include Clark (2003, 2010), Deacon (1997), Dehaene (2010, 2011), Dennett (2008), Donald (1993), Ihde and Malafouris (2019), Malafouris (2013), Taylor (2016), Sinha (2017), Sorgner (2022), and many more.

But returning to Aristotle, this contemporary perspective is not clearly present in his text. Whether, given the opportunity to update or rephrase his ideas, he would be disposed toward it or not, is a question we unfortunately cannot answer. What we *can* do is utilize Aristotle's ideas as he presents them insofar as they are useful—and not, insofar as they are not.

3 Concordance and Conflict Between Aristotelian Thought and Transhumanism

3.1 Concordance

Regarding points of concordance, first, for Aristotle, reaching for the best is good and natural. Our *telos* is excellence, and the greatest excellence is that which is hardest to achieve and most unique to humanity: our rational excellences of, first, speculative wisdom, and, second, practical wisdom (Aristotle 1999, 1141a18-b9; 1941b, 981b13-983a23). Because intelligence augmentation is a long-running goal of transhumanists, this is a clear place of concordance—with the caveats noted above regarding means. First, both pursue the highest things, and insofar as these coincide from their perspectives the systems may be of some interest to each other.

Second, both Aristotle and transhumanists agree that the whole is prior to the parts. For transhumanists, the individual's survival—a "whole" in this case, perhaps a "pattern" or another sort of entity—even at the cost of all of their body parts, is acceptable (e.g., Kurzweil's infinitely disposable "bodies" (1999, 240–243)). In Aristotle, there is likewise an emphasis on the priority of the whole over the parts (1941d, 1253a20), though never explicitly to the extreme of some contemporary transhumanists. Aristotle simply never had the occasion to wonder about mind uploading and whole brain emulation, and were he alive today it is not at all clear how he would view them, other than as "most unnatural." Nevertheless, the prioritization of whole over part is a point of concordance.

Third, Aristotle and transhumanists agree on the legitimate manipulability of the human form. Aristotle allows for eugenics, and transhumanists allow for the same and much more. There is no "hands-off" human nature in Aristotle except for the rejection of the "unnatural." Still, Aristotle lacked important information and power that we now have; he had no way of conceptualizing how humankind could be manipulated or enhanced other than by selective breeding (as Greeks well knew they could do with animals) or by education, and particularly relevant for Aristotle, philosophical education. Divine knowledge is a clear target of human pursuit, even if we are not truly able to attain it: we should try (1941b, 982b28-983a11; 1999, 1177b32-1178a1). Just exactly to what extent it would be acceptable or "natural" to try is the precise question. Should humankind accept the pursuit of constructive approaches to enhancement as Moravec, Kurzweil, Bostrom, and others argue? In Aristotle, the theoretical possibility is maintained for certain types of manipulations that are constructive and not cultivative, but again, there are many unknowns because Aristotle can only "answer" these questions from the perspective of a non-transhumanist context, millennia ago.

3.2 Conflict

Regarding points of conflict, first, for Aristotle, in contrast with the beliefs of Kurzweil, Moravec, and others, copying the formal cause ("pattern") and placing it into new matter will not work to preserve a hylomorphic entity. The copy is a copy, not the original. Aristotle is very clear that the same form instantiated into different matter is a different hylomorph. Therefore, Aristotelianism can only be extremely skeptical about transhumanist mind-uploading.

Second, to Aristotle, goodness is about having the right feelings "at the right times, about the right things, towards the right people, for the right end and in the right way, is… the best condition, and this is proper to virtue" (1999, 24, 1106b21-24). Goodness requires all of the above. If any of those are corrupt then the moral nature of the act and end are corrupted too. Transhumanism seems, in some cases, to forget the importance of maintaining goodness throughout every moral decision and action, instead lending itself towards consequentialism/utilitarianism, as Malapi-Nelson notes (2021), citing Pearce (1995) and Hughes (1996). Even if utilitarianism is not a necessary aspect of transhumanism, it often permeates it (Malapi-Nelson 2021; More 2013, 13; Sorgner 2010, 9). Aristotle would never accept utilitarianism, on its own, as a proper ethic for anything, though perhaps it might be useful as one component of the virtue of prudence.

Third, transhumanism's lack of respect for nature is a source of conflict with Aristotelian philosophy. Aristotle was not only a lover of nature, e.g., spending many books seeking to understand animals, but also understood nature to have normative qualities that were worthy of respect. Transhumanism tends to treat nature irreverently, as mere matter to manipulate at will, to serve human ends (reminiscent of Francis Bacon, and as described by Horkheimer and Adorno (2002), yet with significant exceptions, such as the work of Ilia Delio (2020)). This is a significant theoretical difference, one which can create significant practical, behavioral differences as well.

Lastly, transhumanism's goals themselves, and the reasons, morals, and virtues behind those goals, are not always so "high" as Aristotle would like. The *telos* and final cause of transhumanism as a movement allow us to discern its nature or essence (from an Aristotelian perspective), and thus evaluate it on the widest scale, regardless of the particulars. Do the goals of transhumanism coincide with the goals of human nature? The precise goals of transhumanism are in many respects unclear; it is a diverse and libertarian movement, with far more varied understandings of human flourishing than Aristotle could imagine (see, e.g., Sorgner's work (Schussler and Balistreri 2024)). But there are certainly a few goals set forth, such as immortality, enhanced cognition, enhanced moral virtue, and enhanced power in general (Bostrom n.d.). But what is the point of immortality? What is the purpose of cognitive enhancement? Why enhanced moral virtue? What is the meaning in pursuing power? For the sake of what are all these enhancements?

If the goal of immortality is for the sake of seeking wisdom, then it is a good goal (Aristotle 1941b, 980a22-982a2); however, if it is simply a response to fear, then transhumanism is a pursuit of cowards, not the virtuous (Green 2015b, 196). If the goal is to understand the deepest meanings of the universe it is a good goal, but if it is aimed at any subordinate science (or arbitrary or immoderate desires) then it is an inferior goal. If the goal is to make oneself and the world more virtuous and morally good, then that is a good goal; if the purpose is to enslave the masses into complacent obedience to vice, then it is an evil end.

And if the pursuit of power is for the sake of helping others, ruling justly, and contemplating the highest things then it is good, but if it is for the sake of tyranny or pleasure or other lower goods then it is a harmful and destructive pursuit. As merely one example of this, the transhumanist Zoltan Istvan claimed through his fictional character Jethro Knights that "Extraordinary aims require extraordinary expedience," and proposed the genocidal "Humanicide Formula" to determine which humans to kill first: "Any who are not going along with the program to a sufficient degree, in this hypothetical scenario, will be eliminated. It's as simple as that" (Armstrong 2014). These are the words of a vicious tyrant, and in Aristotle's words "great is the honor bestowed... on him who kills a tyrant" (1941d, 1160, 1267a15-6).

From an Aristotelian perspective, the ends of transhumanism are not clearly enough focused on good pursuits, and so the movement, regardless of its means, needs to find its function and a truly good goal (Green 2018). Otherwise, no matter what means it chooses to achieve its ends, it will be morally corrupting, destructive, and the world would be better off without it. Transhumanism should be a pursuit of the wise, the just, and the brave, not of the foolish, the unjust, the vicious, or the fearful. And while some transhumanists have described themselves, at least implicitly, in those positive terms (e.g., Young 2006; Kurzweil 2005; Bostrom n.d.; Armstrong

2014), other thinkers have warned of the presence of the opposite. For example, virtue ethicist Shannon Vallor warns of the importance of humility (Vallor 2016, 237–246), Alasdair MacIntyre recommends such virtues as "the ability to live with uncertainty," "nonmanipulative relations," "accepting one's death," and "hope" and key virtue when considering our future (MacIntyre 1979). Lastly, scholars of religion such as Brandon Gallaher (2022, 161–175), Hava Tirosh-Samuelson (2022, 183–209), and others (Green 2022, 147, 150–154) have warned of the Satanic and idolatrous aspects of transhumanism—such phrasing is not Aristotelian *per se*, but even from the view of Aristotle, the words describe clear misprioritizations of goods, substituting the lower for the higher.

All of this includes the caveat that Aristotle had values in some ways very different from ours as well as a much more limited range of facts and experiences upon which to draw. After all, during his life he assumed that civilization was already far advanced, and that things that we consider horrific, like slavery and female servitude, were perfectly normal and "natural" (Aristotle 1941d, 1252a33-1255b40), while such seemingly normal activities to us, e.g., lending money at interest, were "most unnatural" (Aristotle 1941d, 1258b1-8).

4 Conclusion

Aristotelian thought, then, ought to be of some interest to transhumanism.

Could there, then, be an authentically "Aristotelian cyborg"; a state which would permit humans to pursue the divine and immortal by embodied constructive manipulations while maintaining some semblance of naturality? Possibly. Our functional nature is rational activity seeking the most divine objects of wisdom, Aristotle accepts cultivating manipulations to help us achieve this function, but radical constructive manipulations may constrain or even destroy the lower parts of the human soul for the sake of the higher part. There remains a genuine tension between our natural rational *telos* of divine knowledge and our identity as human beings, made of organic matter, requiring nutrition and having sensation, and existing in a human bodily form with our various lower functions. Human nature exists on the spectrum from animal to divine; our nature is not only to grow and to sense, but also to seek the highest things. Would Aristotle agree to trade all the lower goods for the highest good and goal of human life? Perhaps. In reality, we do not yet have enough information about how these transformations would take place, or even if they could take place, nor can we ever truly know Aristotle's perspective on the matter—we cannot invite him to speak with us here today, and an Aristotelian AI chatbot simply will not suffice. Practical wisdom and judgment require experience and discernment of particulars (Aristotle 1999, 1141b24-1142a32), and we do not yet have experience to tell us whether this path would be right or wrong from an Aristotelian perspective, or yet know all the necessary particulars about many of these still theoretical ideas.

In conclusion, some kinds of transhumanism could be acceptable to Aristotelian philosophy and others not. While Aristotle certainly desired human perfection, he always considered perfection to be within limits. Transhumanism stretches Aristotelian philosophy in a way Aristotle could have never anticipated, perhaps in some ways near its breaking point. Aristotelianism likewise challenges transhumanism by claiming that only certain kinds will work, while others, like mind uploading, must fail.

Aristotle's system is extremely comprehensive and adaptable, and even in the face of extraordinary ideas such as those of transhumanism, Aristotle can still help us to more deeply understand the issues involved (whether we agree or not), and to consider our proper moral response.

References

Agar, Nicholas. 2004. *Liberal Eugenics: In Defence of Human Enhancement*. Malden, MA: Blackwell.

Aristotle. 1941a. De partibus animalium. Translated by W. Ogle. In *The Basic Works of Aristotle*, ed. R. McKeon. New York: Random House.

Aristotle. 1941b. Metaphysics. Translated by W.D. Ross. In *The Basic Works of Aristotle*, ed. R. McKeon. New York: Random House.

Aristotle. 1941c. Physics. Translated by R.P. Hardie and R.K. Gaye. In *The Basic Works of Aristotle*, ed. R. McKeon. New York: Random House.

Aristotle. 1941d. Politics. Translated by B. Jowett. In *The Basic Works of Aristotle*, ed. R. McKeon. New York: Random House.

Aristotle. 1984a. History of Animals. Translated by d'A.W. Thompson. In *The Complete Works of Aristotle*, ed. J. Barnes. Princeton: Princeton University Press.

Aristotle. 1984b. Generation of Animals. Translated by A. Platt. In *The Complete Works of Aristotle*, ed. J. Barnes. Princeton: Princeton University Press.

Aristotle. 1991. *De anima*. Translated by R.D. Hicks. Buffalo, New York: Prometheus Books.

Aristotle. 1999. *Nicomachean Ethics.* Translated by T. Irwin. Indianapolis, Ind.: Hackett Publishing Company.

Armstrong, Christopher T. 2014. *At Any Cost: A Guide to The Transhumanist Wager and the Ideas of Zoltan Istvan*, archived website. https://web.archive.org/web/20140624020224/; http://transhumanistwagerguide.com/essays/.

Baillie, Harold W. 2005. Aristotle and Genetic Engineering: The Uncertainty of Excellence. In *Is Human Nature Obsolete?* ed. H. W. Baillie and T. K. Casey, 209–232. Cambridge, Mass: The MIT Press.

Barbour, Ian. 1990. *Religion in an Age of Science: The Gifford Lectures.* Vol. 1. New York: HarperSanFrancisco.

Bhardwaj, Ratan D., et al. 2006. Neocortical Neurogenesis in Humans Is Restricted to Development. *Proceedings of the National Academy of Sciences* 103:12564–12568. https://www.pnas.org/doi/pdf/10.1073/pnas.0605177103.

Bostrom, Nick. 2014. *Superintelligence: Paths, Dangers, Strategies.* Oxford: Oxford University Press.

Bostrom, Nick. n.d. Transhumanist Values. http://www.nickbostrom.com/ethics/values.html.

Campbell, Heidi, and Mark Walker. 2005. Religion and Transhumanism: Introducing a Conversation. *Journal of Evolution and Technology* 14 (2). http://jetpress.org/volume14/specialissueintro.html.

Carboncopies Foundation. 2024. Carboncopies website. https://carboncopies.org/

Chislenko, Alexander. 1995. Legacy Systems and Functional Cyborgization of Humans. http://www.lucifer.com/~sasha/articles/Cyborgs.html.

Clark, Andy. 2003. *Natural-born Cyborgs: Minds, Technologies, and the Future of Human Intelligence.* Oxford: Oxford University Press.

Clark, Andy. 2010. *Supersizing the Mind: Embodiment, Action, and Cognitive Extension.* Oxford: Oxford University Press.

Corabi, Joseph, and Susan Schneider. 2012. The Metaphysics of Uploading. *Journal of Consciousness Studies* 19 (7–8): 26–44.

de Grey, Aubrey D. N. J., et al. 2002. Time to Talk SENS: Critiquing the Immutability of Human Aging. In *Increasing Healthy Human Life Span: Conventional Measures and Slowing the Innate Aging Process*, ed. Denham Harman, vol. 959, 452–462. New York: New York Academy of Sciences.

Deacon, Terrence W. 1997. *The Symbolic Species: The Co-evolution of Language and the Brain.* New York: WW Norton & Company.

Dehaene, Stanislas. 2010. *Reading in the Brain: The New Science of How We Read.* New York: Penguin.

Dehaene, Stanislas. 2011. *The Number Sense: How the Mind Creates Mathematics, Revised and Updated Edition.* New York: Oxford University Press.

Delio, Ilia. 2020. *Re-Enchanting the Earth: Why AI Needs Religion*. Maryknoll, NY: Orbis Books.

Dennett, Daniel C. 2008. *Kinds of Minds: Toward an Understanding of Consciousness*. New York: Basic Books.

Donald, Merlin. 1993. *Origins of the Modern Mind: Three Stages in the Evolution of Culture and Cognition*. Cambridge, Mass.: Harvard University Press.

Durbin, Paul T. 1981. Aquinas, Art as an Intellectual Virtue, and Technology. *The New Scholasticism* 55 (3): 272–273.

EBRAINS. 2023. Mouse Brain Atlas. EBRAINS website. https://www.ebrains.eu/tools/mouse-brain.

Gallaher, Brandon. 2022. Technological Theosis? An Eastern Orthodox Critique of Religious Transhumanism. In *Religious Transhumanism and Its Critics*, ed. Arvin Gouw, Brian Patrick Green, and Ted Peters, 161–181. Lanham, MD: Lexington Books.

Galton, Francis. 1883. *Inquiries into Human Faculty and Its Development*. London: Macmillan.

Galton, Francis. 1892. *Hereditary Genius: An Inquiry into Its Laws and Consequences*. London: Macmillan.

Galton, David J. 1998. Greek Theories on Eugenics. *Journal of Medical Ethics* 24:263–267.

Garreau, Joel. 2005. *Radical Evolution: The Promise and Peril of Enhancing Our Minds, Our Bodies—And What It Means to Be Human*. New York: Doubleday.

Geddes, Linda. 2011. Rat Cyborg Gets Digital Cerebellum. *New Scientist*. https://www.newscientist.com/article/mg21128315-700-rat-cyborg-gets-digital-cerebellum/.

Greely, Henry, B. Sahakian, J. Harris, et al. 2008. Towards Responsible Use of Cognitive-Enhancing Drugs by the Healthy. *Nature* 456:702–705. https://doi.org/10.1038/456702a.

Green, Brian Patrick. 2009. Would Aristotle Be a Transhumanist? Paper presented to the Transhumanism and Religion Consultation of the American Academy of Religion. Montreal, Canada, November 9.

Green, Brian Patrick. 2012. Teleology and Theology: The Cognitive Science of Teleology and the Aristotelian Virtues of *Techne* and Wisdom. *Theology and Science* 10 (3 (13 August)): 291–311. http://www.tandfonline.com/doi/abs/10.1080/14746700.2012.695247.

Green, Brian Patrick. 2015a. Transhumanism and Catholic Natural Law: Changing Human Nature and Changing Moral Norms. In *Religion and Transhumanism: The Unknown Future of Human*, ed. C. Mercer and T. Trothen, 201–215. Santa Barbara, CA: Praeger.

Green, Brian Patrick. 2015b. Transhumanism and Roman Catholicism: Imagined and Real Tensions. *Theology and Science* 13 (2 (May)): 187–201. https://doi.org/10.1080/14746700.2015.1023528.

Green, Brian Patrick. 2018. The Technology of Holiness: A Response to Hava Tirosh-Samuelson. *Theology and Science* 16 (2): 223–228. https://doi.org/10.1080/14746700.2018.1455271.

Green, Brian Patrick. 2022. A Roman Catholic View: Technological Progress? Yes. Transhumanism? No. In *Religious Transhumanism and Its Critics*, ed. Arvin Gouw, Brian Patrick Green, and Ted Peters, 143–160. Lanham, MD: Lexington Books.

Grimm, David. 2008. The Mushroom Cloud's Silver Lining. *Science* 321:1434–1437. https://www.science.org/doi/full/10.1126/science.321.5895.1434.

Gupta, A., N. Vardalakis, and F. B. Wagner. 2023. Neuroprosthetics: From Sensorimotor to Cognitive Disorders. *Communications Biology* 6:14. https://doi.org/10.1038/s42003-022-04390-w.

Horkheimer, Max, and Theodor W. Adorno. 2002. *Dialectic of Enlightenment: Philosophical Fragments.* Stanford, CA: Stanford University Press.

Hughes, James. 1996. Embracing Change with All Four Arms: A Post-Humanist Defense of Genetic Engineering. *Eubios Journal of Asian and International Bioethics* 6 (4): 94–101.

HumanBrainProject. 2015. A Simulated Mouse Brain in a Virtual Mouse Body. *HumanBrainProject YouTube Channel.* https://www.youtube.com/watch?v=ldXEuUVkDuw.

Ihde, Don, and Lambros Malafouris. 2019. Homo Faber Revisited: Postpheno menology and Material Engagement Theory. *Philosophy & Technology* 32:195–214.

Jaworksi, William. 2011. *Philosophy of Mind: A Comprehensive Introduction.* Malden, Mass.: Wiley & Sons.

Johnson, Monte R. 2005. *Aristotle on Teleology.* Oxford: Oxford University Press.

Kruglinski, Susan. 2009. What Makes You Uniquely 'you'? interview with Gerald Edelman. *Discover*, January 16. https://www.discovermagazine.com/mind/what-makes-you-uniquely-you.

Kurzweil, Ray. 1999. *The Age of Spiritual Machines: When Computers Exceed Human Intelligence.* New York: Penguin Books.

Kurzweil, Ray. 2005. *The Singularity Is Near: When Humans Transcend Biology.* New York: Viking.

MacIntyre, Alasdair. 1979. Seven Traits for the Future. *The Hastings Center Report* 9 (1 (Feb.)): 5–7. http://www.jstor.com/stable/3561692.

MacIntyre, Alasdair. 1999. *Dependent Rational Animals.* Chicago: Open Court.

Malafouris, Lambros. 2013. *How Things Shape the Mind: A Theory of Material Engagement.* Cambridge, MA: MIT Press.

Malapi-Nelson, Alcibiades. 2021. Is Transhumanism Necessarily Utilitarian? Recasting Alternative Ethical Systems Towards a Future Human Flourishing. *Postdigital Science and Education.* https://doi.org/10.1007/s42438-021-00246-4.

Mitcham, Carl. 1980. Philosophy of Technology. In *A Guide to the Culture of Science, Technology, and Medicine*, ed. Paul T. Durbin, 282–363. New York: The Free Press.

Moravec, Hans. 1976/1978. Today's Computers, Intelligent Machines and Our Future. A speech at Stanford University, July 21, 1976, published 1978. https://web.archive.org/web/20221219111328/https://www.frc.ri.cmu.edu/~hpm/project.archive/general.articles/1978/analog.1978.html.

More, Max. 2013. The Philosophy of Transhumanism. In *The Transhumanist Reader: Classical and Contemporary Essays on the Science, Technology, and Philosophy of the Human Future*, ed. M. More and N. Vita-More, 3–17. West Sussex, UK: John Wiley & Sons.

OpenWorm. 2024. OpenWorm website. https://openworm.org/.

Pearce, David. 1995. *The Hedonistic Imperative*. https://www.hedweb.com/welcome.htm. Accessed 08 June 2021.

Pius XI. 1930. *Casti Connubii*. Vatican. https://www.vatican.va/content/pius-xi/en/encyclicals/documents/hf_p-xi_enc_19301231_casti-connubii.html

Plato. 1992. *Republic*. Translated by G.M.A. Grube, revised by C.D.C. Reeve. Indianapolis: Hackett Publishing Company.

Rollin, Bernard E. 2005. Telos, Value, and Genetic Engineering. In *Is Human Nature Obsolete?* ed. Harold W. Baillie and Timothy K. Casey, 317–336. Cambridge, Mass.: The MIT Press.

Schussler, Aura Elena, and Maurizio Balistreri, eds. 2024. *Metahumanism, Euro-Transhumanism and Sorgner's Philosophy: Technology, Ethics, Art*. Budapest, Hungary: Trivent Publishing.

Shors, T. J., L. M. Anderson, D. M. Curlik, and S. M. Nokia. 2012. Use It or Lose It: How Neurogenesis Keeps the Brain Fit for Learning. *Behavioural Brain Research* 227 (2 (Feb 14)): 450–458. https://pmc.ncbi.nlm.nih.gov/articles/PMC3191246/pdf/nihms301551.pdf.

Silver, Lee M. 1998. *Remaking Eden: How Genetic Engineering and Cloning Will Transform the American Family*. New York: Perennial/HarperCollins.

Sinha, Chris. 2017. Language as a Biocultural Niche and Social Institution. In *Ten Lectures on Language, Culture and Mind*, 138–154. Leiden, Neth: Brill.

Sorgner, Stefan Lorenz. 2010. Beyond Humanism: Reflections on Trans- and Posthumanism. *Journal of Evolution and Technology 21* (2): 1–19.

Sorgner, Stefan Lorenz. 2022. *We Have Always Been Cyborgs: Digital Data, Gene Technologies, and an Ethics of Transhumanism*. Bristol: Bristol University Press.

Stock, Gregory. 2003. *Redesigning Humans: Choosing Our Genes, Changing Our Future*. Boston: Mariner Books.

Szucs, Attila, Pablo Varona, Alexander R. Volkovskii, Henry D. I. Abarbanel, Mikhail I. Rabinovich, and Allen I. Selverston. 2000. Interacting Biological and Electronic Neurons Generate Realistic Oscillatory Rhythms. *Computational Neuroscience* 11 (3, (28 February)): 1–7. https://web.archive.org/web/20030419143905id_/http:/inls.ucsd.edu:80/~hdia/neurorpt.pdf.

Taylor, Charles. 2016. *The Language Animal: The Full Shape of the Human Linguistic Capacity*. Cambridge, MA: Harvard University Press.

Theiner, Georg. 2021a. Uniquely Human – Without Exception. *Culture e Fede* 29 (3 (Supplement)): 21–25.

Theiner, Georg. 2021b. The Extended Mind: A Chapter in the History of Transhumanism. In *The Mind-Technology Problem: Investigating Minds, Selves and 21st Century Artefacts*, 275–321.

Theiner, Georg, and Chris Drain. 2017. What's the *Matter* with Cognition? A 'Vygotskian' Perspective on Material Engagement Theory. *Phenomenology and the Cognitive Sciences* 16:837–862. https://doi.org/10.1007/s11097-016-9482-y.

Tirosh-Samuelson, Hava. 2022. The Transhumanist Pied Pipers: A Jewish Caution Against False Messianism. In *Religious Transhumanism and Its Critics*, ed. Arvin Gouw, Brian Patrick Green, and Ted Peters, 183–214. Lanham, MD: Lexington Books.

Vallor, Shannon. 2016. *Technology and the Virtues: A Philosophical Guide to a Future Worth Wanting*. New York: Oxford University Press.

Walker, Mark Alan. 2002. Prolegomena to Any Future Philosophy. *Journal of Evolution and Technology* 10 (March). http://jetpress.org/volume10/prolegomena.html.

Wallace, William A. 1996. *The Modeling of Nature: Philosophy of Science and Philosophy of Nature in Synthesis*. Washington, D.C.: The Catholic University of America Press.

Wiley, Keith. 2014. *A Taxonomy and Metaphysics of Mind-Uploading*. Seattle, WA: Humanity+ Press and Alautun Press.

Young, Simon. 2006. *Designer Evolution: A Transhumanist Manifesto*. Amherst, NY: Prometheus Books.

Revolution and Transformation: Vygotsky, *Eudaimonia*, and the Social Essence of (Trans)human Development

Chris Drain

1 Introduction: Cultural-Historical Psychology and Vygotsky's "higher type"

As the tumult of the October Revolution settled, psychologists in the nascent Soviet academy were eager to develop a system worthy of the newly minted socialist state. This meant rebuilding a psychology that could live up to the Marxian aspirations of the fledgling republic by breaking free of the systems of pre-Revolutionary bourgeois metaphysicians, whose Neo-Kantian, Christian, Subjectivist, or otherwise mystical commitments conflicted with the tenets of dialectical materialism (Kozulin 1984, 5–39). By the early 1920s, several camps were in contention to be the "officially sanctioned" psychology of the Bolsheviks, and the ensuing decade saw several attempts to bridge psychology with the age-old Platonic

---

C. Drain (✉)
Institute of Writing and Rhetoric, Dartmouth College, Hanover, NH, USA
e-mail: christopher.p.drain@dartmouth.edu

G. Theiner, A. Malapi-Nelson (eds.), *Technologies of the Future Self*, Palgrave Studies in the Future of Humanity and its Successors,
https://doi.org/10.1007/978-3-031-99902-4_5

ideal of utopian social engineering. Here I focus on the particular project of L.S. Vygotsky, whose approach centered on explaining "higher mental functions" (e.g., semantic memory and conceptual cognition) with respect to their ontogenetic, phylogenetic, and cultural-historical development. According to Vygotsky, these distinctly human powers arise during the hominin transition away from environmentally driven reactions to the culturally and technically mediated activity of the becoming self-conscious agent. Such an agent is a *self-regulating* entity that having mastered its "reactive functions" can enter into new and varied goal-directed relations with its environment based on meaningful, rather than simply functional, affordances (Vygotsky 1999, 63).

In resituating Marx and Engels in the Soviet academy, Vygotsky's "cultural-historical" psychology centralized the idea that uniquely human consciousness is a social product inherently connected to the objectification of nature (Vygotsky 1999, 63–64; see also Marx and Engels 1978, 150). More primitive mental functions are, in other words, transformed into the higher cognitive capacities of the historical human through the internalization of sociocultural practices:

> We call this withdrawal of the operation inward, this interiorization [интериоризация (*interiorizatsiia*)] of higher mental functions connected with new changes in their structure *the processes of revolution* [вращивание (*vrashchivanie*) ingrowing, cultivation], having in mind mainly the following: the fact that higher mental functions are constructed initially as external forms of behavior and depend on an external sign is not to any extent accidental, but, on the contrary, is determined by the very psychological nature of the higher function, which... does not arise as a direct continuation of elementary processes, but is a social method of behavior applied to one's self. (Vygotsky 1999, 53)[1]

For Vygotsky, then, *Homo sapiens* is a self-transcending species, whose phylogenesis can only be explained by its concurrent cultural-historical genesis.

Yet Vygotsky was not simply concerned with the past—such a backward gaze may have belonged to the Freudian, whose timeless unconscious dictated a future beset with neurotic discontent (Freud 1961). Countering depth psychology, Vygotsky proposed a "peak" psychology that looked to

[1] See Yasnitsky (2018) for an analysis of translation and conceptual issues regarding Vygotsky's use of "internalization."

the future (Vygotsky 1997a, 137). Vygotsky was fascinated by what the human body (and therefore mind) may as yet be capable of, and he anticipated a socialist future with a materially new type of human, a "superman" (Vygotsky 1994, 182).[2] Such Nietzschean aspirations were not uncommon in the early days of the Soviet Republic (Rosenthal 1994). But for Vygotsky, Nietzsche "imagined that the development of this higher type of man was subject to the same law of biological evolution, the struggle for life and selection based on the survival of the fittest, which prevails in the animal world" (Vygotsky 1994, 182).[3] Vygotsky's Nietzsche, in other words, was committed to a vulgar Darwinism that "ignore[d] the fact that the laws of historical evolution of man differ fundamentally from the laws of biological evolution and that the basic difference between these two processes consists in the fact that a human evolves and develops as a historical, social being" (loc. cit.). Borrowing much from Trotsky to centralize the Marxian point of self-mastery (Yasnitsky 2014), Vygotsky highlights how the human of the future must emerge in the same manner as the human of the evolutionary past:

> In this sense Engels, who had examined the process of evolution from the ape to man, said that it is labour which created man. Proceeding from this, one could say that new forms of labour will create the new man and that this new man will resemble the old kind of man, 'the old Adam', in name only. (Vygotsky 1994, 183).[4]

This "new man," it is clear, is the product of a novel organization of the social and its concomitant tools and material scaffolds. In terms of morphology and function, it is a being with a novel organic composition—a "socialist cortex" (Wolfe 2010)—which correspondingly grants novel cognitive powers. For Vygotsky, then, the human is "always already" a

[2] See Wolfe (2010) and van der Veer (1984) for treatments of Vygotsky's Spinozist connections in this regard.

[3] Vygotsky's interpretation here is controversial. As Passmore (2000) points out, Nietzsche was critical of Darwin and presumed that the superman would be wrought not by natural selection but by a Pelagian "exercise of the will" (394). But on this note, we could say simply that Vygotsky reads Nietzsche as a Spencerian in considering the foundation of such an "exercise" to be a biological struggle (Vygotsky 1994, 182). For an alternate transhumanist reading of Nietzsche that centralizes his idea of limitations (or lack thereof) of powers and capacities in ontogenesis as well as his "values perspectivalism," see Sorgner (2009, 2021).

[4] See also Peters (this volume) for a discussion of the Christian conception of a "new Adam" and its transhumanist relevance.

project of biological transcendence and, as such, it is the proper goal of collectivist politics to secure the *transformation of human nature* in a way that accords with the Marxian goal of human self-determination.

Regarding the tradition of human perfectibility, which aims not at a God-like stasis of ultimate completion, but instead the "endless progressive improvement" (Owen 1836, as quoted in Passmore 2000, 242) of human capabilities, Vygotsky occupies a curious place.[5] Inheriting from Marx the ideas that capitalist relations obstruct human improvement and that social must precede individual transformation (Steinhoff 2014, 9), Vygotsky makes a compelling case for a particular brand of eudaimonist transhumanism where human excellence is understood as radically contingent on historical, cultural, and technological factors. Distilling this latter claim is the object of the next part of this paper. There we see that Vygotsky's "higher type" blends aspects of transhumanism as a project of *transcendence* and *enhancement*. The interpretative gains of this reading bear on how we understand the ethical dimensions of Vygotsky's project in the remaining sections.

2 Transcendence and Enhancement: Vygotsky and Developmental Transformation

2.1 *Humanity Transcended*

According to Fuller (2011), this original sense of transhumanism is found in Julian Huxley's eugenicist program, where the artificial selection of genetic traits is governmentally engineered: "Advances in knowledge of genetics allow these interventions to become increasingly upstream, which pre-empt traditional forms of social engineering, such that eventually we acquire Plato-like powers to legislate the desirable distribution of traits in a population" (103). If this is what we mean by *humanity transcended*, then Vygotsky's place here might be tenuous at best. For Vygotsky, humans have already been undergoing an "artificial selection"—it was

[5] Recently, Theiner (2021) and Maidansky and Biryukov (2021) discuss Vygotsky in their respective accounts of transhumanism. But Vygotsky is a relatively small part of either project, the former focusing on the transhumanist history of the idea of "the extended mind" (where Vygotsky is seen as a historical antecedent) and the latter on rethinking transhumanism more generally with Marx and Russian Marxism. By contrast, Vygotsky's transhumanism is the sole focus of the present work.

responsible for *Homo sapiens* in the first place, supplanting biological pressures on the course of evolutionary development (Vygotsky 1994, 175). Thus, privileging gene manipulation as the preeminent mode of "artificiality" that will steer humanity to new heights seems to be at odds with Vygotsky's greater conception of the place of artificiality in human becoming. Vygotsky does share the conviction that Darwinian natural selection must be overcome through artificial means to ensure the emergence of a "higher type." However, his account of human origins maintains that anthropogenesis takes the characteristic of an "epiphylogenesis" (Stiegler 1998, 177; see also Drain 2022), which, in keeping with Stiegler's neologism, holds socially distributed and culturally transmitted techniques as potentially biologically generative phenomena. For Vygotsky, then, the artificiality that will enact the emergence of a "higher type" will be found not in genetic engineering but in a more general cultural shift.[6]

Any concord with Huxleyan transcendence, moreover, seems to be precluded by Vygotsky's overall wariness of biological determinism, especially as it surfaces in biologistic psychologies where cognitive development is seen as the mere unfolding of innate natural processes or unavoidable recapitulation of primordial psychological stages (Vygotsky 1997b, 9). Biology may explain evolutionary aspects of human morphology, but higher mental functions—those products and producers of self-transcendence—can only be explained by the culturally and historically contingent development of "artificial organs" (Vygotsky 1997b, 16).

But Fuller's reading of Huxley is not the only one. While Huxley no doubt advocated state-sponsored genetic engineering, Passmore (2000) understands him as acknowledging the importance of history and culture to human development, a point familiar to Vygotskians. For instance, Huxley knows that humans are unique in the animal kingdom regarding their extended neoteny. Such a delay in development, in as much as it renders the human child extraordinarily dependent on adults (Gould 1977), is the precondition for human-specific cultural transmission:

> The adults on whom the child is dependent pass on to him general concepts, tradition, a stock of information.... So it comes about... that by the gradual

[6] This is not to say that there was no support for eugenics among the Marxian left in the early twentieth century. See Paul (1984) and Lucassen (2010) for general treatments of leftist eugenics in Britain, the United States, and Europe; see Adams (1989) for a survey of the controversies surrounding Soviet eugenics in particular.

> increase in his scientific knowledge, man can substitute new methods of 'effective, progressive change' for natural selection. (Passmore 2000, 390)

Huxley seems poised to recognize the profound effects of cultural and cognitive artifacts (Norman 1991; Hutchins 1995; Malafouris 2013) on human development, and such would put him close to the Marxist position that so inspired Vygotsky. Nonetheless, he still takes transcendence to be attained by manipulating heritable gene expressions, i.e., "biogenetic engineering." The progress and transmission of knowledge and culture, and the engineering of an environment suited to such, are ultimately in service of the deliberate pursuit of biogenetic perfection (Passmore 2000, 392). Thus Huxley comes close to recognizing the transformative potential of neoteny that is so crucial to the Vygotskian project, but it's a near miss.

Vygotsky's commitment to explaining higher cognition by recourse to both onto- and phylogenesis adds a double valence to his idea of transcendence that distances him from Huxley's by now traditional position. Vygotsky indeed admits that proactive human intervention will affect the course of evolution and produce a biologically novel species: "This change in human behavior… must inevitably lead to further evolution of man and alteration of the *biological type of man*" (Vygotsky 1994, 182). But this is an obvious point for Vygotsky—we are animals whose biological makeup is already the product of culture and labor (Vygotsky and Luria 2016; see also Engels 1946). Further speciation events could of course be traced to novel cultural-technical interventions, so it's not a stretch to assume that artificial selection would produce a "new type" since it already produced the "old type." However, this Huxley is not ready to apprehend, for it would put undue pressure on culture rather than genetics as the key to human development (Passmore 2000, 392). Cultural configurations, by contrast, are of the utmost importance to Vygotsky's transhumanism. Any transformation of the biological organization of man must result from "*the social liberation of man*" (Vygotsky 1994, 183), i.e., a revolutionary transformation in political organization that concurrently spurs novel forms of technological and intersocial relationships (ibid., 181). And such transformations necessarily bear on the *ontogenetic* realization of a "higher type."

Vygotsky, and those who currently find inspiration in Vygotsky, thus takes not biogenetic but ontogenetic engineering as the key to human transcendence, where any biological outcomes can be viewed from proximate and distal levels. At the distal level is what we have just gone over—the new biological type, a speciation event. But at the proximate level, there are biological alterations that consist in plastic reconfigurations of the cortex. An early advocate of neuroplasticity (see Drain 2022), Vygotsky takes ontogenesis not simply as a period of maturation, but the ground of neural reorganization via the internalization of cultural forms of activity. Given such, the outcome of cultural mediation in ontogenesis is de facto biological:

> The historical development of the human type is based upon the functions of the cortex. All that development, mediated from outside, as all psychological and cultural development of the child likewise determined from outside by social environment, is accomplished in its main in the cortex and via the cortex. "The growing cortical mediateness of human physiology in the end is of great importance as a progenerative factor" (Zalkind 1927). History, changing the human type, depends upon the cortex; the new socialist man will be created through the cortex.... [I]n Marxist paedology, the primacy of social mediation of personality development includes the role of the cortex in development as the main question. (Vygotsky, as quoted in van der Veer and Valsiner 1991, 320)

Thus the transhuman for Vygotsky, the "new socialist man" (sic) will be effected not through the manipulation of DNA but through social mediation and its resultant effects on the plastic brain. If we can escape the "Modern Synthesis" with its myopic focus on gene selection and follow Vygotskians Packer and Cole (2019), who advocate for an "Extended Evolutionary Synthesis" that conceives of organism and environment as co-constitutive and takes ontogenesis to be an extended system of obligatory collaboration in the dynamic niche of human caregiving (202), then it's clear that the force of the Vygotskian challenge to Huxley's biogenetic engineering rests on how one conceives "genetic" alteration. If we take "genetic" to mean, as most anglophones do, that which concerns natural selection, with alleles, DNA, RNA, and chromosomes all playing their respective parts, then casting Vygotsky's transhumanism as Huxleyian makes little sense. But if we take "genetic" [генетическая (*geneticheskaya*)] in Vygotsky's sense as "developmental," which includes both the phylogenetic and ontogenetic in its intension, then it is clear that the transhuman is the result of intentional modifications of the cultural environment.

2.2 *Humanity Enhanced*

"Enhancement" captures what many in the popular press (e.g., Bennett 2021) think of when conjuring the "human of the future"—a human "cyborg," availing itself of nanotechnology and prostheses to improve native capacities. While there are radical and moderate or weak versions of "enhancement transhumanism," placing Vygotsky squarely in either camp has its difficulties. Moderate enhancement can be framed negatively according to Brennan (2023, 243–245) as avoiding

1. a new or extended capacity/ability that is an impossibility for any unenhanced human
2. a change or high likelihood of a change to moral status
3. a significant change in physical and/or cognitive vulnerability
4. an inability for the enhanced and the unenhanced to relate to one another

As we'll see, Vygotsky's transhumanism avoids the most extreme variants of these qualifications. But it's not a perfect match. The third presents a problem insofar as Vygotsky maintains that aging and disease are biological constraints that ought to be overcome (Vygotsky 1994, 182–183). While it is unclear whether this will occur through onto- or phylogenetic engineering, it's a radical aspiration in either case. Also problematic is the fourth qualification, not least because Vygotsky's reconstruction of Nietzsche's *Übermensch* poses a "higher type" that resembles current humans "in name only." But figuring out the details here is far too speculative. (Would there even be an older type left in Vygotsky's world?) Fuller's explanation of weak enhancement, which explains "the prospect of humans acquiring improved versions of their current powers without the more extreme implications of prolonging life indefinitely or upgrading us to a superior species" (2011, 103), is inadequate for the same reasons. Given Brennan and Fuller's descriptions, it seems imprudent to drop the "radical" qualifier from Vygotsky's transhumanism altogether. As we'll see, enhancement is at the *root* of Vygotsky's theory of cognitive transformation, and it is this notion of transformation that separates Vygotsky from more standard treatments found in transhumanist literature (e.g., Bostrom and Savulescu 2009a).

In this regard, it may be tempting to see Vygotsky as presaging Clark's "extended mind theory" (EMT), which holds that humans often find themselves in "coupled systems," where cognitive processes involve both internal neuronal and external environmental states (Clark 2008). Indeed, for Clark, these systems are so ubiquitous that he dubs humans "natural born cyborgs" (2003), furtively promoting a variety of transhumanism in augmenting the classic *Homo faber* motif (Bergson 1911). For Clark, the human is indeed the tool maker, but the tool is also the human (Clark 2003, 6–7). As Theiner eloquently puts it,

> From Clark's cyborgian perspective, most if not all of what travels under the banner of "advanced human cognition" must be seen as "always already" technologically enhanced rather than as accomplishments of our "naked," i.e., unaugmented biological brains.... Drawing an explicit analogy between the future-oriented agenda of using science and technology to enhance the human condition, and our inborn appetite for using mind-expanding tricks and tools, the EMT thus effectively "normalizes" transhumanist aspirations as the continuation of something we have been doing all along. (Theiner 2021, 281)

Vygotsky is close to such a vision, but his focus on transformation critically differentiates his transhumanist project from the more enhancement-minded EMT.

Much of this stems from how Vygotsky subsumes externally *and* internally directed tools under the general concept of mediation. Instrumental tools are understood in the typical sense as externally directed, mediating between the subject and an object; psychological tools, however, are internally directed, effecting a change in the subject herself (Vygotsky 1997b, 85). A certain tool, then, may enhance aspects of the human body and improve the efficiency of a task in the objective world (cf. Kapp 2018). But if it were to also take on a semiotic character, it would effect a transformation in the subject herself (Vygotsky and Luria 2016, 138). To further this point, and to see just how radical Vygotsky is regarding enhancement, it is illustrative to consider his account of the cognitive effects of language. No doubt Vygotsky would take it as uncontroversial that a language-agent coupled system has enhanced capacities compared to a nonlinguistic agent (Clark 1998). However, Vygotsky would add that the inward effects of linguistic mediation are always more than the mere enhancement of cognitive capacities—in linguistic activity, lower mental functions also undergo a *radical reconstruction*. In ontogenesis, even perceptual and motor systems are transformed after (repeat) engagements with signs and other (cultural) psychological tools (Vygotsky 1999, 31). As such, mediation leads not only to the augmentation of an aspect of a system's external "behavior" (cf. Norman 1991) but also to a radical internal reorganization of the agent herself. "Enhancement" thus captures a part of Vygotsky's transhumanism insofar as it describes an aspect of a mediated activity. Instrumental tools no doubt enhance an agent's capacity to act on the world. Psychological tools, however, not only enhance the systemic competency of the agent-tool hybrid but also transform the nature of cognition itself.

This latter point is not lost on some transhumanists. Bostrom and Savulescu note that from a certain point of view, *all* technology amplifies native human capacities and cognitive artifacts and the learned skills associated with them "are a kind of enhancement of our mental software" (2009b, 2). Despite the attachment to such a dualistic metaphor (Searle 1984), the authors go on to affirm the *transformative* nature of such inward enhancement:

> When we learn to calculate and read we are literally reprogramming the micro-structure of our nervous system, with physiological effects just as real as those resulting from the ingestion of a psychoactive drug, and often more durable and with more profound consequences for our lives. At the limit of this line of reasoning, *all* learning could be viewed as physiological enhancement. (Bostrom and Savulescu 2009b, 3)

Vygotsky is at the radical limit of such reasoning, taking enhancement to only be developmentally relevant as concerns its transformative effects. While at the phylogenetic level mediated transformations (pre)historically amounted to a speciation event (anthropogenesis qua epiphlyogenesis), and may in the future amount to the biological emergence of an "other" type (a meta-anthropogenesis), the ontogenetic relevance of transformation is more pertinent to contemporary ethical concerns and any practical realization of a "higher" type. Vygotsky, that is, sees technical-cultural activity as the key to "social liberation" and the transformation of humanity in the here and now (Zavershneva 2012, 56).

3 Virtue Ethics, Essence, and Capabilities

The ethical and political import of Vygotsky's insistence that "social liberation" must accompany the genesis of the transhuman is self-evident. Like many of his generation, Vygotsky saw in Soviet communism the opportunity to fulfill Marx's emancipatory project of securing an association where the individual is free to develop as a "species being" [*Gattungswesen*]. In addressing the alterations of productive processes that necessitate "the creation of a new type of human being" (1994, 179), Vygotsky specifically recalls Marx's developmental perfectionism: "the individual who has been turned into a fraction, the simple bearer of a fractional social function, would be replaced by a fully developed individual for whom the various social functions represent alternating forms of his activities" (Marx 1990, as quoted in Vygotsky 1994, 179). Such an imminent transformation, we just saw, is conceptually tied to an understanding of anthropogenesis as already an "epiphylogenesis," where human cognitive and bodily capacities—"sensuous capacities" (*sinnliche kraft*)—are "cultivated" over the course of historical development: "all these come into being only through the existence of *their* objects, through *humanized* nature" (Marx 1992, 352–353). The human, in other words, produces and objectifies nature and in confronting her product and the products of others is gradually objectified herself, her organs and cognitive capacities becoming *more* human, potentiating new inter- and intra-mental relations and activities. According to such a Marxian understanding, human transformation is a process contingent on the social and historical practices and products of the "species life." Vygotsky, in highlighting Marx's "fully developed individual," thus looks to a future individual whose capacities and capabilities are free to develop in novel configurations based on the activity of a liberated "species being."

Such talk of capabilities prima facie seems to position Vygotsky in the eudaimonistic tradition, which takes self-actualization as the ultimate human virtue. Marx's engagement with Aristotle on these matters has been well documented (McCarthy 1986). But it is still an open question whether Marx advocated a virtue ethics per se, and thus it is unclear how Vygotsky bears on contemporary eudaimonic concerns over the political securing of "capabilities" (e.g., Nussbaum 2011). At a surface level, Vygotsky shares with Marx the eudaimonic position that capitalism "systemically denies the majority of people the requirements of a 'good life', as identified by Aristotle" (Dean 2016, 509; see also Miller 1992,

277–278). But as we'll see, Marx's treatment of Aristotle is varied, and much of his relevance for Vygotsky's transhumanism falls on how one understands the "primary being" (*to tie en enai*) of the human.

3.1 Aristotle, Techne, and Political Being

For Aristotle, the flourishing of an individual is the highest good (Aristotle 1984a, I.7, 1097b19-21) and the telos of the polis is a state of self-sufficiency for the general purpose of living well (1984b, III.9, 1280b31-34; 1984c, I.6, 1362b10-12). More specifically, humans have a natural *impulse* [φύσει ἡ ὁρμὴ (*Trieb*, instinct)] towards political association (1984b, I.2, 1253a29-30), and thus one's *ergon* is to be self-sufficient by means of the polis. Since the *hexis* or active state of being human is to activate the capacity [δύναμις] of being political, the individual without a polis is not fulfilling her potential and is human in name only (ibid., 1253a30-31).

Aristotle is clear, then, that the human is "by nature a political animal" [ἄνθρωπος φύσει πολιτικὸν ζῷον] (ibid., 1253a1-2). But such should not be misconstrued as concretely differentiating humans from the animal kingdom or the natural world at large. Aristotle's description of the polis as the result of a natural inclination demonstrates this. The polis "belongs among the things that exist by nature" (loc. cit.), and there are many types of animals belonging to polis-like associations: "[T]he human, the bee, the wasp, the ant, and the crane" are all political animals insofar as they take part of a common *ergon* (1984d, I.1, 488a9-10). The impulse towards political association, then, is not unique to human beings.

Unlike later political thinkers, political association for Aristotle is not a contra-natural disposition. While for Hobbes (1994) the genesis of the polis rests on a specific utilization of a linguistic-technical apparatus,[7] namely, the collective linguistic action of transferring rights to an artificial person, for Aristotle no such renunciation of nature is necessary. Rather

[7]For Hobbes, words are but profitable inventions (Hobbes 1994, 16), the product of an arbitrary imposition of voice "to bring to … mind some conception concerning the thing on which it was imposed" (Hobbes 1928, 14). As vocal "marks," these artifacts are among "the first contra-natural objects" (Nerney 1991, 59), i.e. techno-logical innovations through which the human is allowed to stand in contrast to nature. The emergence from the "state of nature" is thus doubly marked for Hobbes: first comes renouncing "vocal anarchy" with the invention of speech; second comes utilizing speech to renounce "political anarchy" (see Pettit 2008).

than actualizing itself as a technical act that separates humanity from the natural world, the innate drive towards political association is such that it, when brought about towards its end, completes what nature cannot do itself: "art [τέχνη] in some cases completes what nature cannot bring to a finish" (Aristotle 1984e, II.8, 199a16-17). The polis, therefore, is a "natural-technical" extension of the natural world. While the polis does not have a principle of motion like that of a natural substance (cf. Aristotle 1984e, II.1, 192b8-20), its coming-into-being and preservation are still rooted in the natural human impulse towards politicality. The constituting act that brings forth the polis into being thus works *with* nature, allowing an already inherent and biologically grounded goal to realize itself.

3.2 *Marx and Aristotle: Capacities, Essence, and "species being"*

Marx seems at least congruent with Aristotle here. His introduction to the *Grundrisse* famously rebukes social contract theories and so-called Robinsonades where the individual, abstracted from her social essence, appears "[n]ot as a historic result but as history's point of departure.... [n]ot arising historically, but posited by nature" (Marx 1993, 83). With Aristotle, the foundation of the polis is a historical act that allows for a natural capacity to realize itself: "Human beings historically *discover* their highest capacities for political and ethical organization" (Gilbert 1992, 313). Marx takes this point to heart, invoking Aristotle's classic formulation of *anthropos* as the *zoon logikon*:

> The human being is in the most literal sense a ζῷον πολιτικὸν, not merely a gregarious animal, but an animal which can individuate itself only in the midst of society. Production by an isolated individual outside society... is as much an absurdity as is the development of language without individuals living *together* and talking to each other. (Marx 1993, 84)

But despite Marx's Aristotelian inflection, here we can't take him as signing off wholesale to a peripatetic understanding of the human. While some commentators see no problem in identifying Marx's "species being" with Aristotle's idea that the telos of the polis is to tend to human completion (McCarthy 1986; Ahmed 2018), there is still a glaring difference in how Marx and Aristotle conceptualize human capabilities and, indeed, human essence. True, Marx affirms Aristotle's characterization of *anthropos* as naturally political, and even more subtly affirms that political organization itself historically emerges as a "natural-technical" activity. But Marx's "productivist" anthropology maintains that human capacities are self-cultivated. Aristotle, on the other hand, recognizes no such thing; though he acknowledges the history of political association, this crucially does not entail the historicity of human potentials.

Recall that the well-functioning polis, while still realizing a naturally occurring goal (that of living well), is not simply a natural *product*. Rather, for Aristotle there is an added element to human political life, one which separates it from natural products insofar as political life is a collection of practices brought about through *reasoned* cultivation. Thus Aristotle's *zoon polikon* is first and foremost a *zoon logikon*. Aristotle's aprioristic commitment to positing human intelligence (and rationality is that human "capacity" *par excellence*) as biologically (and metaphysically) given betrays a stark difference with Marx's account of the social essence of the human. As Margolis (1992) aptly phrases it,

> Gattungswesen is a regulative idea that [Marx] associates with *what it is that distinguishes man from the animals*, whereas Aristotle's notion of man's essence is an idea of *what actually constitutes man's timeless nature*. There is, accordingly, no recognition of the historicity of human existence in Aristotle. (334–335)

For Marx, then, "species being" dictates that whatever the human *is* is a product of its objectifying activities; human "essence," and indeed our capacities and capabilities, is malleable and contingent upon particular historical configurations of production. To again quote Margolis:

> Marx maintains that the Wesen of man may change because it is nothing but the object of his "life activity;" and this activity becomes his Gattungsleben when he acts in accord with his unfettered awareness of what distinguishes him from the animals, namely, that he individuates himself "only in the midst of society" and that, with this understanding, he *could* act in a way that invoked his solidarity with *all* such beings... Aristotle holds, on the contrary, that—since man's nature is actually fixed, and theoretically discernable as such—man's "life activity" (his praxis) can vary only in the historical details of his particular work, not in the changeless *archai* by which it is rationally guided (the virtues). For Marx, the real world *is* the exteriority, the externalization of man's Wesen—hence, the product of praxis. For Aristotle, the thesis would be utterly incoherent. (ibid., 334)

In light of such opposing views of human nature, it's not unwarranted to ask whether Marx is simply incompatible with Aristotle. If so, there could be drastic effects for any interchange between Vygotsky's transhumanism and virtue ethics, the latter traditionally incorporating into the claim that human flourishing is the ultimate end the additional proposition that "there is a distinct human nature in terms of which this *telos* is defined" (Rasmussen 2012, 401).

For Marx and Vygotsky, it is clear that human nature is not definable in any fixed set of properties. It is a historical product that varies according to the sophistication and complexity of the social and technical assemblages of a given culture. If by this we mean that species being "*has* no determinable content *of its own*" (Margolis 1992, 340), then "functioning as a Gattungswesen" to distinguish itself from the animals still "does not yield *any* determinate essence the specific *potencies* of which *can* be actualized—and it cannot be made to yield a morality of virtues, rights, or principles," which under a traditional reading require an invariant human nature (loc. cit.). For Marx, there are only contingent "natures" that a particular individual at a given time and place may realize—"To actualize what man is as a class creature is to act in accord with the norms internal to, and determined by, the practice of a given epoch" (ibid., 341). But in a revolutionary sense, in overcoming capitalist alienation in accordance with one's species being, "actualization" refers to the "ridding" of such internalized values, such that "there are no longer any similar such potencies to consider *as determinately essential to his nature*" (loc cit.).

3.3 Marx, Vygotsky, and the "good life"

Given this, how can we square Vygotsky with a eudaimonistic account of the good life? Here we can take the influential Aristotelian "human capabilities argument" put forth notably by Nussbaum (1992a, 1992b) as our exemplar. At its core, Nussbaum's view is that the telos of political planning is human flourishing, which consists in the attainment and functional realization of certain basic human capabilities. Thus if, e.g., a certain level of cognitive ability is essential to being a human, the role of the polis is to ensure not only that this basic capability is functionalized (for if not, then the life in question would hardly be a human one) but that it can function beyond such a threshold so that we may judge such a life not merely as a human one, but a good one (Nussbaum 1992b, 221). The problem, noted by Margolis (1992), is that there is "no recognition of the very historicity of such capabilities in this approach" (335). For Marx, "man's Wesen is a historicized artifact of man's *Leben*" (loc. cit.). And prior norms, needs, and indeed *capabilities* that were previously contoured by external historical constraints are, in revolutionary praxis, "ridded" insofar as "acting individuals themselves [become] capable of controlling the historical consequences of their own social activity" (Márkus 1978, 55).

Even without endorsing the capabilities argument, however, there is still a moral valence to Marx's project of self-directed human development. Marx held the communist revolution not only as a historical and practical (and indeed necessary) solution for the "contradictions of capitalism" but also as a "*moral contrast*" to all previous antagonistic epochs of human history; it is a "morally affirmed form of social development… on the ground that in it men will realize their metabolism with nature 'under conditions most favourable to, and worthy of, their human nature'" (Márkus 1978, 2; see also Marx 1991, 959). Such a moral imperative seems to affirm, in an Aristotelian manner, that some social art (*techne*) ought to bring forth a more complete "human nature." But we have just seen that for Marx such a "nature" is entirely variable. What, if anything, remains a constant in such a radically historicized human being?

Here we can turn to Margolis (1992) and Márkus (1978). Margolis points out that though Marx's human "naturally" has no "fixed nature" and *is* "what contingently his historical praxis forms him as," nevertheless,

> his native powers distinguish him from the animals. He can act to make more efficient (or more "rational") the very mode of production that forms him and his "worlds"; he can transform himself thereby….And, unlike the animals, he can (through all that) then distinguish himself *as* that creature… (Margolis 1992, 351).

Márkus, too, notes certain invariable "native powers" in his reconstruction of a Marxian philosophical anthropology. For Márkus, Marx's "human essence" consists in the following, related, modes of being:

1. Work: "Work is first of all an activity aimed not directly, but only *through mediations* at the fulfillment of needs. Work makes its object suitable for human use by changing, forming it with the help of some other—nature-given or (usually) also man-made—object as instrument" (Márkus 1978, 5).

2. Consciousness: Human work begets *consciousness* insofar as it, as mediated activity, *indirectly* satisfies a need, and in doing so, "develops and presupposes the separation of need and its object, of subject and object, i.e. the emergence of consciousness and self-consciousness" (ibid., 25–26).
3. Sociality: Work is not "reducible to technical action, but designates the material activity of human self-transformation always in some social form" (ibid., 13). Even if consciousness *appears* as an isolated subjective fact, such results from fetishizing "certain characteristics of it by assigning to them a superhistorical relevance and making them independent of their real subjects, the socially concrete individuals" (ibid., 27). Thus, the self-conscious individual "is always a 'social being,' i.e., a being who can live only in and through society, and whose existence and nature are determined by the subsisting social relations" (ibid., 38).
4. Historical development: The three prior "native powers" are not to be understood as "empirically constant and shared characteristics of every human individual" (ibid., 39), since certain historical configurations can directly alienate them from humanity. But these traits nonetheless prove essential to humans as a historical species: "[T]his never-ceasing formation of 'human nature' is itself a unified process which can be comprehended and characterized in its unity" (loc. cit.).

Such a unified process *is* the historical process, one of continual "self-creation" (ibid., 40) where new forms of labor, sociality, and consciousness historically determine the abilities, needs, and wants of the species [*Gattung*] itself. Overcoming the discrepancy between the progress of the species as a whole and the development of singular individuals who are excluded from the "self-formative and creative aspects of human activity" (ibid., 46) *is* Marx's moral imperative. As Dean (2016) puts it, Marx looks "to a *eudaimonia* that [is] not… some latent potential of a *zoon politikon*" but is rather concerned with "human beings' continuing and ever-evolving struggle for a *better* life, for self-fulfillment, for moral progress" (514).

That humans work and socially develop sophisticated cognitive powers is thus the precondition for the eudaimonic realization of the species as generatively self-directing. As Márkus makes clear:

> The process of human genesis is... not completed with the formation of *Homo sapiens* as a biological species, a species to which organisms with definite, constant and identical biological and anthropophysical characteristics belong. Indeed, this is only the starting point and foundation from which man's *social-historical genesis*, the genesis of man as a 'generic being' [Gattungswesen] departs and begins. This social genesis is the process of 'prehistory', which at the end, gives rise to the human species as *mankind*, as a real and conscious unity of globally interacting and interconnected individuals, on the one hand, and to the concrete, many-sided and multidimensional human individuality which truly *represents* the historically achieved stage of development of the 'genus' [Gattung], on the other. And this twofold, but unified process can be completed only through the revolutionary communist transformation of the existing alienated system of social relations. (1978, 48)

Any venture to construct a list of invariant capacities in Marx that ought to be maximalized to achieve "the good life" would fall short if it misses this last point regarding historical self-direction. Work, sociality, and consciousness only "function well" *if* they affect humans' historical appropriation as developmentally self-directing beings. Thus, even if we transpose such "native powers" into a capabilities argument, where each represents a basic capacity of human being whereby the privation of which would render one "less than" human and the good execution of which would contribute to a "good life" (which politics is in the business of securing through distributive means)—such would miss that the real moral imperative of Marx is not simply to improve upon each element for its "own" sake (even if that means a local and historically discrete positive change in one's material conditions and vital powers). Such an empirical understanding would rob each trait of its philosophical substance and revolutionary import. The "good" functioning of the native capacities of work, sociality, and consciousness is only relevant insofar as it lays the foundation for the revolutionary attainment of historical self-realization, which consists in (a) the "conscious unity of globally interacting and interconnected individuals" and (b) the concrete emergence of a "many-sided and multidimensional human individuality" (Márkus 1978, 48).

Vygotsky accepts such a proposition as his starting place. Vygotsky's position on (a) is implied by his embrace of Marx's idea that nature is the human's "inorganic body" (Marx 1992, 328), to be socially worked over via historically concrete forms of labor that mediate not only external objective but also internal subjective relations. Human distinctiveness thus comes about (epi)phylogenetically through cultural-cognitive "engineering" (cf. Sterelny 2018)—the human environment has always been a "smart" environment for Vygotsky, with culturally and historically determined "psychological tools" affecting the development of human needs, capacities, and competencies. A "higher type" would emerge on the basis of such an interconnected environment. Regarding (b), such an enhanced sociality is ontogenetically relevant if it transforms the capacities of the *present* human. Thus for Vygotsky as with Marx, (a) the "unity of globally interacting individuals" must come about with (b) the emergence of a "multidimensional human individuality"—both are aspects of the same phenomenon.

Where does this leave us with other recent proposals and critiques of transhumanist ethics? We can see that the Marxian backdrop to Vygotsky's transhumanism might put him at odds with Wood's "social-democratic" project (this volume) that, to mitigate existential risk, calls for proactive "initiatives" concerning the optimization of various modalities of human being. Such might hazard an empirical understanding of the "native powers," to which the Marxian rejoinder would be what we have just put forth regarding transposing such powers into a "capabilities argument." It may be possible to secure such initiatives, despite taking place in a system that rewards certain capacities over others; however, it is only after a revolutionary embrace of the human as a historical project that such capacities can be reconfigured from the ground up. On the other hand, Green's "Aristotelian transhumanism" (this volume) offers some interesting categories to consider. That we develop tools and techniques to "cultivate" our capacities is a given for Vygotsky. But the radicality of his theory of (epi)phylogenesis and his transhumanist vision concerns how our capacities are themselves constructed and transformed in social-technical activity. As such, Vygotsky is in principle not opposed to disembodied forms of enhancement, though it might be too much to speculate as to how he would imagine, e.g., a disembodied ontogenesis. Finally, Sorgner's "Euro-Transhumanism" (this volume) might be amenable in some ways to Vygotsky's. It may be that Vygotsky was short-sighted in rejecting Nietzsche's *Übermensch* in that the latter's "values relativism" could be congruent with aspects of the Marxian ideal of historical self-direction. Nevertheless, while Nietzsche's will-to-power metaphysics shows the mutability of human capacities, his centralization of relations of dominancy and overall disdain for the "herd" frame his project in an individualistic register antithetical to Vygotsky's.[8] However, a closer inspection of all these issues is needed.

4 Conclusion

Vygotsky acknowledges that the revolutionary "dictatorship of the proletariat" is at least the initial impetus for the functional realization of the "capabilities" of work, consciousness, and sociality in the service of onto- and phylogenetic self-transcendence and eudaimonic historical self-direction. But insofar as he is writing in the first decade of (to him) a post-Revolutionary age, he saw that once such a condition is met, the "active state" of fulfilling such a potential consists in polytechnical education, which abides by the formula that "*everyone should know something about everything, and everything about something*" (Vygotsky 1997c, 202). A collectivism so implemented thus realizes a higher form of human freedom:

> [T]he result of achieving this, the crowning glory of this whole process of transforming human nature, should be the appearance of this higher form of freedom which Marx describes in the following way: "Only in community [with others has each individual] the means of cultivating his gifts in all directions." (Vygotsky 1994, 182; Marx and Engels 1978).

This is not simply the negative liberty of liberalism. It is a psychological freedom premised on self-regulation, or the emergence of

> the consciousness of the new man, the superman, but not in the Nietzschean sense of the word, not a new biological breed, but a socially organized superman, enlightened through and through, in every cache of the most elemental forces of the body, freed from the most terrifying enslavement—enslavement to the self—and from the most bitter dependence—on one's own nerves and mind—by subordinating to himself the play of the body's inner forces as he does the outer forces of nature (Vygotsky, cited in Zavershneva 2012, 56).

[8] Complicating things is the role of the verbal command in Vygotsky's sociogenetic account of self-regulation (Jones 2019; Drain 2021a, 2021b), which might suggest that domination plays a bigger role in Vygotsky's "genetic" story than one would otherwise assume. However see Potapov (2021), who provides a convincing explanation of why such still does not imply a Nietzschean power dynamics for Vygotsky.

We saw earlier that self-regulation is central to Vygotsky's account of *anthropogenesis*, both onto- and phylogenetically understood. It may thus come as no surprise that it features so centrally in his *transhumanism*. Such regulative activity might be seen as highly congruent with Aristotelian ethics, where virtuous action consists in holding oneself in such a condition whereby one's actions "can be clearly envisioned and freely chosen without being determined" by excessive or deficient affections (Sachs 2002, 30). But is more likely that here Vygotsky has in mind Spinoza, for whom virtue, power, and freedom are all interrelated,[9] and the "emendated intellect" is one that is freed from passive affections that transition it to lower levels of perfection. For Vygotsky, social liberation bears out in the individual who actively comprehends, and thus controls, the inner workings of her own body and mind.

This point seems lost in the enhancement hype that surrounds much of transhumanist discourse. Futurists, e.g., might long for the realization of a "Neuralink," which when implanted in the brain would allow internet access with the mind alone (Bennett 2021). But considering how current internet platforms are in the business of *controlling* users—surveilling their activities to canalize engagement from a motive of *profit* (Zuboff 2015; Srnicek 2017)—how could such enhancement work towards Vygotsky's goal of free self-cultivation? Maidansky and Biryukov, who in critiquing the so-called "liberal" transhumanism of Nicholas Agar (2005, though also see Buchanon 2011) and invoking Vygotsky's individual who "has been turned into a fraction" (Vygotsky 1994, 179) and reduced to "l'idiotisme du métier"? (Marx 1847), ask: "*Are transhumanist technologies to immortalize a professional cretin*, incapable of passions and hence lacking compassion *and, with it, the totality of affects hitherto identified as 'human'*?" (Maidansky and Biryukov 2021, 46 *emphasis added*). We can adjust this question to better fit with Vygotsky's Spinozist emphasis on self-mastery. It is not a human *incapable* of passions who is the subject of much of current transhumanist speculation, but rather an individual who is crafted to be *passion-dependent*, a slave to that "perpetual and restless desire of power and power that ceaseth only in death" (Hobbes 1994, 58).[10] Such a "liberal" transhuman would be enhanced, true enough. But

[9] See the scholium to proposition 11 in Book III of Spinoza's *Ethics* (Spinoza 1985, 500–501).

[10] Adjusted to account for the transhumanist aspiration for immortality, Hobbes' grim summation only gets more depressing.

if any transformation were to take place, it would not be a eudaimonic one. One's cognitive powers would be heightened, but still subject to tendencies and affects that hinder any semblance of self-mastery—"it is naïve to suppose that these technologies [of digital incorporation] simply expand the powers of our organic bodies. They create *dependencies*" (Seymour 2019, 71). Work, consciousness, and sociality might indeed be transformed in such a future, but not in the service of self-direction, both individually and socially construed.

References

Adams, Mark B. 1989. The Politics of Human Heredity in the USSR, 1920–1940. *Genome* 31 (2): 879–884. https://doi.org/10.1139/g89-155.

Agar, Nicholas. 2005. *Liberal Eugenics: In Defense of Human Enhancement.* Oxford: Blackwell Publishing.

Ahmed, Sabeen. 2018. Communism as Eudaimonia: An Aristotelian Reading of Human Emancipation. *International Journal of Philosophy & Social Values* 1 (2): 31–48.

Aristotle. 1984a. Nichomachean Ethics. In *The Complete Works of Aristotle*, ed. Jonathan Barnes. Translated by W.D. Ross and J.O. Urmson, Vol. 2, 1729–1867. Princeton: Princeton University Press.

Aristotle. 1984b. *The Politics.* Translated by C. Lord. Chicago: The University of Chicago Press.

Aristotle. 1984c. Rhetoric. Translated by W.R. Roberts. In *The Complete Works of Aristotle*, ed. Jonathan Barnes, Vol. 2, 2152–2269. Princeton: Princeton University Press.

Aristotle. 1984d. History of Animals. Translated by d'A.W. Thompson. In *The Complete Works of Aristotle*, ed. Jonathan Barnes, Vol. 1, 774–993. Princeton: Princeton University Press.

Aristotle. 1984e. Physics. Translated by R.P. Hardie and R.K. Gaye. In *The Complete Works of Aristotle*, ed. Jonathan Barnes, Vol. 1, 315–446. Princeton: Princeton University Press.

Bennett, Hayley. 2021. Rise of the Cyborgs: Inside the Technology Transcending Humanity's Biological Limits. *BBC Science Focus.* https://www.sciencefocus.com/future-technology/cyborgs-transhumans. Accessed 8 Aug 2024.

Bergson, Henri. 1911. *Creative Evolution.* Translated by A. Mitchell. New York: Henry Holt and Company.

Bostrom, Nick, and Julian Savulescu, eds. 2009a. *Human Enhancement.* Oxford: Oxford University Press.

Bostrom, Nick, and Julian Savulescu. 2009b. Human Enhancement Ethics: The State of the Debate. In *Human Enhancement*, ed. Nick Bostrom and Julian Savulescu, 1–22. Oxford: Oxford University Press.

Brennan, Cian. 2023. Weak Transhumanism: Moderate Enhancement as a Non-radical Path to Radical Enhancement. *Theoretical Medicine and Bioethics* 44 (3): 229–248. https://doi.org/10.1007/s11017-023-09606-6.

Buchanon, Allen. 2011. *Better than Human*. Oxford: Oxford University Press.

Clark, Andy. 1998. Magic Words: How Language Augments Human Computation. In *Language and Thought: Interdisciplinary Themes*, ed. Peter Carruthers and Jill Boucher, 162–183. Cambridge: Cambridge University Press.

Clark, Andy. 2003. *Natural Born Cyborgs*. Oxford: Oxford University Press.

Clark, Andy. 2008. *Supersizing the Mind*. Oxford: Oxford University Press.

Dean, Hartley. 2016. Eudaimonia and 'species being': A Marxist Perspective. In *Handbook of Eudaimonic Well-Being*, ed. Joar Vittersø, 507–520. Cham: Springer Nature.

Drain, Chris. 2021a. Commands and Collaboration in the Origin of Human Thinking: A Response to Azeri's 'On reality of thinking'. *Social Epistemology Review and Reply Collective* 10 (3): 6–14.

Drain, Chris. 2021b. Tomasello, Vygotsky, and the Phylogenesis of Mind: A Reply to Potapov's 'Objectification and the labour of the negative in the origin of human thinking'. *Social Epistemology Review and Reply Collective* 10 (6): 26–35.

Drain, Chris. 2022. Technics and Signs: Anthropogenesis in Vygotsky, Leroi-Gourhan, and Stiegler. *HPLS* 44 (4): 53. https://doi.org/10.1007/s40656-022-00539-2.

Engels, Frederick. 1946. The Part Played by Labour in the Transition from Ape to Man. In *Dialectics of Nature*, ed. Clemens Dutt, 279–296. London: Lawrence & Wishart.

Freud, Sigmund. 1961. *Civilization and Its Discontents*. Translated by J. Strachey. New York and London: W.W. Norton & Company.

Fuller, Steve. 2011. *Humanity 2.0*. Basingstoke: Palgrave Macmillan.

Gilbert, Alan. 1992. Marx's Moral Realism: Eudaimonism and Moral Progress. In *Marx and Aristotle: Nineteenth-Century German Social Theory and Classical Antiquity*, ed. George E. McCarthy, 303–328. Savage, MD: Rowman & Littlefield.

Gould, Steven Jay. 1977. *Ontogeny and Phylogeny*. Cambridge, MA: Harvard University Press.

Green, Brian Patrick. 2025. Aristotle and Transhumanism: Concordance and Conflict. In *Technologies of the Future Self: An Ethics for Transhuman Flourishing*, ed. Georg Theiner and Alcibiades Malapi-Nelson. Springer.

Hobbes, Thomas. 1928. In *The Elements of Law: Natural and Politic*, ed. Ferdinand Tönnies. London: Cambridge University Press.

Hobbes, Thomas. 1994. In *Leviathan*, ed. Edwin Curley. Indiannapolis: Hackett Publishing Company.

Hutchins, Edwin. 1995. *Cognition in the Wild*. Cambridge, MA: MIT Press.

Jones, Peter E. 2019. Vygotsky and Marx: Re-setting the Relationship. In *Questioning Vygotsky's Legacy: Scientific Psychology or Heroic Cult*, ed. Anton Yasnitsky, 22–42. Abingdon and New York: Routledge.

Kapp, Ernst. 2018. *Elements of a Philosophy of Technology: On the Evolutionary History of Culture*, ed. Jeffrey West Kirkwood and Leif Weatherby. Translated by L.K. Wolfe. Minneapolis: University of Minnesota Press.

Kozulin, Alex. 1984. *Psychology in Utopia: Toward a Social History of Soviet Psychology*. Cambridge, MA: MIT Press.

Lucassen, Leo. 2010. A Brave New World: The Left, Social Engineering, and Eugenics in Twentieth-Century Europe. *International Review of Social History* 55 (2): 265–296.

Maidansky, Andrey, and Nikolai Biryukov. 2021. Remolding the Human: The Transhumanist Ideas in Marx and Russian Marxists. *New Proposals: Journal of Marxism and Interdisciplinary Inquiry* 12 (1): 41–48.

Malafouris, Lambros. 2013. *How Things Shape the Mind: A Theory of Material Engagement*. Cambridge, MA: MIT Press.

Margolis, Joseph. 1992. Praxis and Meaning: Marx's Species Being and Aristotle's Political Animal. In *Marx and Aristotle: Nineteenth-Century German Social Theory and Classical Antiquity*, ed. George E. McCarthy, 329–355. Savage, MD: Rowman & Littlefield.

Márkus, George. 1978. *Marxism and Anthropology: The Concept of 'human essence' in the Philosophy of Marx*. Translated by E. de Laczay and G. Márkus. Assen: Van Gorcum & Company.

Marx, Karl. 1847. *Misère de la philosophie: Réponse à la philosophie de la misère de M. Proudhon*. Paris: A. Frank.

Marx, Karl. 1990. *Capital: A Critique of Political Economy*. Translated by B. Fowkes, Vol. 1. London: Penguin Books.

Marx, Karl. 1991. *Capital: A Critique of Political Economy*. Translated by D. Fernbach, Vol. 3. London: Penguin Books.

Marx, Karl. 1992. Economic and Philosophical Manuscripts. Translated by R. Livingstone and G. Benton. In *Karl Marx: Early writings*, 279–400. London: Penguin Books.

Marx, Karl. 1993. *Grundrisse*. Translated by M. Nicolaus. London: Penguin Books.

Marx, Karl, and Friedrich Engels. 1978. The German Ideology: Part 1. In *The Marx-Engels Reader*, ed. Robert C. Tucker, 146–200. New York and London: W.W. Norton & Company.

McCarthy, George. 1986. German Social Ethics and the Return to Greek Philosophy: Marx and Aristotle. *Studies in Soviet Thought* 31 (1): 1–24.

Miller, Richard W. 1992. Marx and Aristotle: A Kind of Consequentialism. In *Marx and Aristotle: Nineteenth-Century German Social Theory and Classical Antiquity*, ed. George E. McCarthy, 275–302. Savage, MD: Rowman & Littlefield.

Nerney, Gayne. 1991. Homo Notans: Marks, Signs, and Imagination in Hobbes's Conception of Human Nature. *Hobbes Studies* 4 (1): 53–75.

Norman, Donald A. 1991. Cognitive Artifacts. In *Designing Interaction*, ed. John M. Carroll, 17–38. Cambridge: Cambridge University Press.

Nussbaum, Martha. 1992a. Nature, Function, and Capability: Aristotle on Political Distribution. In *Marx and Aristotle: Nineteenth-Century German Social Theory and Classical Antiquity*, ed. George E. McCarthy, 175–212. Savage, MD: Rowman & Littlefield.

Nussbaum, Martha. 1992b. Human Functioning and Social Justice: In Defense of Aristotelian Essentialism. *Political Theory* 20 (2): 202–246.

Nussbaum, Martha. 2011. *Creating Capabilities.* Cambridge, MA: Harvard University Press.

Owen, Robert. 1836. *The Book of the New Moral World.* London: E. Wilson.

Packer, Martin J., and Michael Cole. 2019. Evolution and Ontogenesis: The Deontic Niche of Human Development. *Human Development* 62 (4): 175–211. https://doi.org/10.1159/000500172.

Passmore, John. 2000. *The Perfectibility of Man.* Indiannapolis: Liberty Fund.

Paul, Diane. 1984. Eugenics and the Left. *Journal of the History of Ideas* 45 (4): 567–590.

Peters, Ted. 2025. Our Future Self: Cyber Enhancement and Spiritual Transformation. In *Technologies of the Future Self: An Ethics for Transhuman Flourishing*, ed. Georg Theiner and Alcibiades Malapi-Nelson. Springer.

Pettit, Philip. 2008. *Made with Words: Hobbes on Language, Mind, and Politics.* Princeton: Princeton University Press.

Potapov, Kyrill. 2021. Vygotsky's Janus-Faced Theory of Language: A Reply to Drain's 'Tomasello, Vygotsky, and the phylogenesis of mind'. *Social Epistemology Review and Reply Collective* 10 (8): 17–29.

Rasmussen, Douglas B. 2012. Perfectionism. In *Encyclopedia of Applied Ethics*, ed. Ruth Chadwick, 395–403. London: Elsevier.

Rosenthal, Bernice Glatzer. 1994. Introduction. In *Nietzsche and Soviet Culture*, ed. Bernice Glatzer Rosenthal, 1–32. Princeton: Princeton University Press.

Sachs, Joe, ed. 2002. *Aristotle: Nichomachean Ethics.* Newburyport, MA: Focus Publishing.

Searle, John R. 1984. *Minds, Brains and Science.* Cambridge, MA: Harvard University Press.

Seymour, Richard. 2019. *The Twittering Machine.* London: Indigo Press.

Sorgner, Stefan Lorenz. 2009. Nietzsche, the Overhuman, and Transhumanism. *Journal of Evolution and Technology* 20 (1): 29–42.

Sorgner, Stefan Lorenz. 2021. *We Have Always Been Cyborgs: Digital Data, Gene Technologies, and an Ethics of Transhumanism.* Bristol: Bristol University Press.

Sorgner, Stefan Lorenz. 2025. Euro-transhumanism. In *Technologies of the Future Self: An Ethics for Transhuman Flourishing*, ed. Georg Theiner and Alcibiades Malapi-Nelson. Springer.

Spinoza, Baruch. 1985. In *The Collected Works of Spinoza*, ed. Edwin Curley, vol. 1. Princeton: Princeton University Press.
Srnicek, Nick. 2017. *Platform Capitalism*. Cambridge: Polity Press.
Steinhoff, James. 2014. Transhumanism and Marxism: Philosophical Connections. *Journal of Evolution and Technology* 24 (2): 1–16.
Sterelny, Kim. 2018. Culture and the Extended Phenotype: Cognition and Material Culture in Deep Time. In *The Oxford handbook Of 4E Cognition*, ed. Albert Newen, Leon de Bruin, and Shaun Gallagher, 773–792. Oxford: Oxford University Press.
Stiegler, Bernard. 1998. *Technics and Time: The Fault of Epimetheus*. Translated by R. Beardsworth and G. Collins. Stanford: Standford University Press.
Theiner, Georg. 2021. The Extended Mind: A Chapter in the History of Transhumanism. In *The Mind-Technology Problem: Investigating Minds, Selves and 21st Century Artefacts*, ed. Robert W. Clowes, Klaus Gärtner, and Inês Hipólito, 275–321. Cham: Springer Nature.
van der Veer, René. 1984. Early periods in the Work of L.S. Vygotsky: The Influence of Spinoza. In *Learning and Teaching on a Scientific Basis*, ed. Mariance Hedegaard, Pentii Hakkarainen, and Yrjö Engeström, 87–98. Aarhus: Psykologisk Institut.
van der Veer, René, and Jaan Valsiner. 1991. *Understanding Vygotsky*. Oxford: Blackwell.
Vygotsky, Lev Semyonovich. 1994. The Socialist Alteration of Man. In *The Vygotsky Reader*, ed. René van der Veer and Jaan Valsiner, 175–184. Oxford: Blackwell.
Vygotsky, Lev Semyonovich. 1997a. The Problem of Consciousness. In *The Collected Works of L.S. Vygotsky*, ed. Robert W. Rieber and Jeffrey Wollock, vol. 3, 129–138. New York: Plenum.
Vygotsky, Lev Semyonovich. 1997b. The History of the Development of Higher Mental Functions. In *The Collected Works of L.S. Vygotsky*, ed. Robert W. Rieber, vol. 4, 1–259. New York: Plenum.
Vygotsky, Lev Semyonovich. 1997c. *Educational Psychology*. Translated by R. Silverman. Boca Raton: St. Lucie Press.
Vygotsky, Lev Semyonovich. 1999. Tool and Sign in the Development of the Child. In *The Collected Works of L.S. Vygotsky*, ed. Robert W. Rieber, vol. 6, 3–68. New York: Plenum.
Vygotsky, Lev Semyonovich, and Alexander Romanovich Luria. 2016. *Studies on the History of Behavior: Ape, Primitive, and Child*, ed. and translated by Victor I. Golod and Jane E. Knox. Abingdon and New York: Routledge.
Wolfe, Charles T. 2010. From Spinoza to the Socialist Cortex: The Social Brain. In *Cognitive Architecture: From Biopolitics to Noopolitics*, ed. Deborah Hauptmann and Warren Neidlich, 186–206. Rotterdam: 010 Publishers.

Wood, David W. 2025. Human Flourishing Requires Transhumanist Flourishing. In *Technologies of the Future Self: An Ethics for Transhuman Flourishing*, ed. Georg Theiner and Alcibiades Malapi-Nelson. Springer.

Yasnitsky, Anton. 2014. Higher Functions & "height" Psychology: Vygotsky (ab) uses Leon Trotsky and Friedrich Nietzsche. In *Paper presented at the 2nd Congresso Internacional sobre a Teoria Histórico-Cultural and at the 13th Jornada do Núcleo de Ensino de Marília*. Marília, Brazil. http://individual.utoronto.ca/yasnitsky/texts/Yasnitsky%20(2014).%20Nietzsche-Trotsky-Vygotsky.pdf. Accessed 2 Aug 2024.

Yasnitsky, Anton. 2018. Vygotsky's Science of Superman: From Utopia to Concrete Psychology. In *Questioning Vygotsky's Legacy: Scientific Psychology or Heroic Cult*, ed. Anton Yasnitsky, 1–21. Abingdon and New York: Routledge.

Zalkind, Aron Borissovich. 1927. *Osnovnye Voprosy Pedologii [Basic Issues of Paedology]*. Moscow: Rabotnik Prosveshchenija.

Zavershneva, Ekaterina. 2012. The Key to Human Psychology: Commentary on L.S. Vygotsky's Notebook from the Zakharino Hospital (1926). Translated by N. Favorov. *Journal of Russian & East European Psychology* 50 (4): 16–41.

Zuboff, Shoshana. 2015. Big Other: Surveillance Capitalism and the Prospects of an Information Civilization. *Journal of Information Technology* 30:75–89.

In Search of the Human Style: Or, Is *Homo Sapiens* a Crime Against the Concept of Humanity?

Steve Fuller

1 Setting the World-Historic Scene: The Human After Foucault

Social theory doesn't take the nature of the 'human' as seriously as it should (cf. Chernilo 2017; Fuller 2020b). No less than normativity itself is at stake (Turner 2010). Is value possible in a world where the meaning of 'human' is now as fluid as it was before Carolus Linnaeus coined *Homo Sapiens* in th2e mid-eighteenth century? To be sure, when 'posthumanists' nowadays call for decentering the human as the locus of value in the world, they usually mean the removal of privilege from *Homo Sapiens* rather than the cessation of the moral intuitions associated with being human. They simply wish to spread those sentiments among a wider range of species. That seems hopeful, but the material consequences are bound to be

S. Fuller (✉)
Department of Sociology, University of Warwick, Coventry, UK
e-mail: S.W.Fuller@warwick.ac.uk

G. Theiner, A. Malapi-Nelson (eds.), *Technologies of the Future Self*, Palgrave Studies in the Future of Humanity and its Successors,
https://doi.org/10.1007/978-3-031-99902-4_6

significant. Moreover, the present is also being decentered as the temporal polestar of our moral compass. Some would have us do justice to the extended past (e.g., 'reparations') and others to the extended future ('future generations'). The law normally specifies limits on how far into the past and the future its normative reach can legitimately extend. This explains the existence of 'statutes of limitations', on the one hand, and 'sunset clauses', on the other. Taken together, this explosion of the material character of humanity from its modern *Homo Sapiens* base prompts perhaps the widest of normative considerations, what in religious times was called 'theodicy', or God's sense of justice in a world where one person's sense of good looks to another like evil (and vice versa). In secular culture, this is often discussed as 'the meaning of life'. In any case, so I shall argue, we are living through what Nietzsche would recognize as a 'transvaluation' of the human, which reopens theodicy in a way that Kant had tried to close in *Critique of Pure Reason*.

An interest in normativity runs as a red thread in my work, and it has eventually led back to the human. It started with my doctoral work comparing the rationality of scientific and legal decision-making, followed by my early work in social epistemology, which was largely about redirecting the course of science and later about managing the theological and biological challenges to the social sciences, to my more recent explorations of transhumanism. Over the past forty years, the focus of my interest in normativity has shifted from the nature of knowledge to the nature of the knowers, which brings me to concept of humanity. Throughout this period, I have had a sustained fascination with theodicy as the branch of theology from which (so I believe) all the normative concerns of modern philosophy derive (Fuller 2015). Susan Neiman (2002) deserves credit for making this point with such articulate explicitness. Theodicy is ultimately about what to believe and how to act in a state of profound ignorance and uncertainty. (Both Neiman and I began our inquiries with the work of John Rawls, who routinely flagged the significance of theodicy in his moral philosophy lectures [Rawls 2000].) The original object of this ignorance and uncertainty was the 'Mind of God', which human subjects were driven to fathom. But in our secular world, the object has become more diffuse, encompassing a natural and human world that operates independently of our knowledge and will. Moreover, the 'human subject' is itself arguably unprecedented conceptual pressure, given the demands made by animals and machines for recognition; hence, the recent rise of post- and trans-humanism.

I put my own cards on the table a decade ago, as one of nearly two hundred thinkers asked by John Brockman's self-styled 'third culture' website *Edge*, 'What scientific idea is ready for retirement?' My answer was the equation, 'human being = *Homo Sapiens*' (Fuller 2014). It is an idea whose time has literally come and gone. In this respect, I was simply reiterating Michel Foucault's (1970) striking thesis in *Les Mots et Les Choses*, which over the years has become increasingly salient, serving as a polestar for my development of 'Humanity 2.0' (Fuller 2011: chap. 2). I agree with Foucault on the ontological precariousness of the equation 'human being = *Homo Sapiens*' and follow him in crediting Linnaeus and Kant—not Aristotle or Darwin—with first realizing the equation's significance. However, we differ about the equation's prehistory and emerging 'posthistory'. I will start by presenting Foucault's vision of the relevant history and then my own in the next section.

Foucault seems to have thought that there was nothing ontologically unique about humanity prior to the mid-eighteenth century. Even when religious and secular authorities had stressed the human capacity for language, he imagines that it was against the backdrop of an 'always already' semiotically rich world, full of spontaneously occurring modes of representation without the need for any distinctly linguistic mediation; hence Foucault (1970) begins famously with a chapter entitled 'The Prose of the World'. The tendency for distant objects to behave similarly, which has to this day fed into both the literary and scientific imaginations, suggested a world order in which the 'microcosm' reproduced elements of the 'macrocosm'.

However, for Foucault, the coinage of '*Homo Sapiens*' to name the human object and 'anthropology' to name its science marked only a temporary stoppage to the above activity. The emergence of humanity 'suddenly' (in his rhetoric of 'epistemic ruptures') became what Bruno Latour (1987) would call an 'obligatory passage point' for all thought. In other words, humanity became a necessary condition for any signification. Henceforth, only humans could make meaning because only they use language, through which meaning is made. Representation was reduced to 'reference', which implied 'intentionality', a very targeted form of representation involving a special relationship between the speaker and the spoken about. Getting that relationship right starts to acquire enormous normative significance in all spheres of life in the modern era (Turner 2010). After John Searle (1983), the relationship has been called 'aboutness', with a nod to Franz Brentano , who suggested that the relevant linguistic function is rooted in the psychology of humans understood as natural creatures. Searle (1984) has continued to develop this theme in his later career, notably without mention of—and sometimes with hostility to—the phenomenological tradition of Husserl and Heidegger that historically emanated from Brentano.

Foucault would regard Searle as the death rattle of the sort of 'human' whose autopsy he was trying to perform in *Les Mots et Les Choses.* (This may help to explain Searle's signature role in the 1980s as the scourge of Jacques Derrida and other 'deconstructionists' during the first phase of the ongoing 'culture wars'.) However, Brentano *malgré lui* also paved Foucault's way, through his influence not only on Heidegger but perhaps more importantly on Freud, who worked on Brentano's project to get the works of John Stuart Mill translated into German. For Foucault, the archaeology of the unconscious—a concept shared by Brentano and Freud—serves alongside Marx's ideology critique and Nietzsche's genealogy of morals as harbingers of the ultimate undoing of *Homo Sapiens*, understood as the rational ape. At one level, the history that Foucault tells here is familiar, but the sort of future into which it opens is not. Normally, Marx, Nietzsche, and Freud are positioned as part of an ironic or self-subverting turn in nineteenth century thought, whereby reason comes to undermine reason. In that respect, the rational ape becomes the author of its own demise. But Foucault believes that it is a mistake to assign *Homo Sapiens* the role of protagonist in this narrative; hence, the much vaunted 'death of the author'. Rather, the anthropic object is more like a passing phase, perhaps like the sort of recurrent dream that Freud would diagnose as signifying neurosis in patients. In the end, however, the object dissolves, having served as a bottleneck, if not outright blockage, to the free flow of signification that existed prior to its emergence.

This sensibility perhaps explains Foucault's (2008) identification of neoliberalism with the rise of 'biopolitics', a theme that dominated his later work. Common to Marx and the great theorist of liberal political economy David Ricardo was a protean view of capital that makes its indefinite concentration the ultimate enemy—one that can appear in many guises, including inherited wealth, protectionist trade policies, and corporate monopolies. All these prevent the freedom needed for the realization of human potential, be it cast as 'innovation' in the sense that Joseph Schumpeter would later associate with entrepreneurship (Ricardo) or more broadly focused on a species-level sense of human emancipation (Marx). I have recently argued that this commonality has profound consequences for the prospects of academic knowledge as a public good (Fuller 2023a: chap. 5).

To be sure, Ricardo's abstract presentation of political economy, which Marx largely copies, creates an ambiguity about the exact source of the problem. Should capital be subject to maximally free exchange due to the nature of capital itself or to the fact that such an arrangement most benefits the humans who are involved in the exchange? This, in a nutshell, is the problematic of neoliberalism. Although both Ricardo and Marx officially subscribed to a 'labor theory of value', which seems to privilege the 'human' as a factor of production, they also believed—in their different ways—that the human is uniquely able to transcend its circumstances. In the transhumanist jargon, for both, humanity is 'morphologically free' (Fuller 2019a: chap. 2). And so, if exchange relations are regarded as the material equivalent of relations of signification, then arguably *Homo Sapiens* is tantamount to a semiotic version of illegitimately concentrated capital. And while such a conclusion might make Foucault appear too misanthropic, it would certainly conform to his self-image as a 'happy positivist'.

2 Recognizing the Transvalued Human in a New Theodicy

As suggested above, my view of the concept of humanity both before and after the emergence of *Homo Sapiens* differs from Foucault's in both substance and spirit. The biggest difference between us is that I regard humanity primarily as a concept—not an object. Whereas for me *Homo Sapiens* marks the moment when the concept of humanity coalesced around a concrete object, for Foucault that is the only time in which humanity has ever existed.

A good way into my point of view is to recall that when Nietzsche spoke of the 'transvaluation of all values' (*Umwertung aller Werte*), he was referring to a never-ending process of evaluating values, the result of which would always be that something else might serve the purported value better than the value's current standard-bearer. In this 'transvaluative' frame of mind, we might think of *Homo Sapiens* as a prototype or 'proof of concept' for *humanity*, but not the ultimate standard-bearer of the value—or set of values—represented by 'humanity'. In terms of modern logic, to be human is to satisfy the open predicate, 'X is human'. This is certainly the position of today's 'transhumanists', who envisage a 'Humanity 2.0' that will exemplify humanity better than *Homo Sapiens.* To be sure, there are many variants of transhumanism, depending on what needs to be retained/ amplified and eliminated/diminished in the constitution of *Homo Sapiens.* The gamut runs from humans remaining in their *Homo Sapiens* bodies indefinitely (say, through regenerative medicine), through some kind of 'cyborganization' (whereby humans become hybrid carbon-silicon beings), to the uplifting of some distinctly human value (generically called 'consciousness') into a silicon virtualization through which its capacities might be extended indefinitely.

The above narrative of transvaluation is predicated on *Homo Sapiens* as the starting point of the story of humanity. Here I agree with Foucault, but only on this point. Unlike Foucault, I believe that both before and after the Linnaean coinage of *Homo Sapiens*, 'humanity' was a free-floating signifier, not necessarily encompassing all *Homo Sapiens* and possibly including some non-*Homo Sapiens.* In that respect, before Linnaeus, to be 'human' was to be 'morphologically free' in the sense that transhumanists nowadays aspire to. Thus, while we regard Aesop's fables as 'anthropomorphizing' animals, 'anthropomorphism' is itself a late nineteenth century neologism of George Romanes, who worried that as people came to accept Darwin's view that humans are just another animal species, they might conclude that humanity can be found in other animal species. In other words, Darwin might increase rather than decrease the Aesopian tendency. For example, such a charge was lodged by early twentieth century American animal behaviorists against Wolfgang Koehler's claim to have detected 'insight' in ape problem-solving (Fuller 2020a).

Here it is worth recalling that Jeremy Bentham—who lived halfway in time between Linnaeus and Darwin—lobbied extensively for what he dubbed the 'humane' treatment of animals. He probably did not see Aesop as anthropomorphizing animals, but rather (perhaps like Francis of Assisi) as finding humanity in animals (Fuller 2019b; cf. Bourke 2011: Part Three). Bentham's position has been revived with gusto in recent times by Peter Singer, who grants that 'sentience' as a quality that is distributed across the animal kingdom is morally more relevant than whatever form of consciousness might be unique to *Homo Sapiens.* Nevertheless, he seems to value animals according to their capacity for 'suffering', a rather anthropomorphic way of interpreting the sensation of pain. In practice, it suggests a more hierarchical view of the relative view of animal lives than advertised by Singer's self-styled 'species egalitarianism'.

This opens the door to a rather different transvaluation of the human: posthumanism. Here, *Homo Sapiens* is taken as simply a touchstone for qualities that are already found across nature. If transhumanism is about the temporal extension of the human, then posthumanism is about its spatial extension, which Singer (1981) popularized as 'expanding the moral circle'. Put in economic terms, transhumanism is about *investing* in the distinctive value of being human, whereas posthumanism is about *distributing* that value as widely as possible. Implied here are two rather different attitudes towards *risk*, with transhumanists wishing to embrace it—and its potential benefits—and posthumanists to avoid it—and its potential harms: 'proactionary' *versus* 'precautionary', as I have called it (Fuller and Lipinska 2014: chap. 1). Thus, animal rights and environmental activists have appropriated Socialist rhetoric, notwithstanding its anthropocentrism, to call for *Homo Sapiens* to abstain from reproduction to 'make room' for other species and thereby maintain biodiversity. Sometimes it is even argued that if *Homo Sapiens* is the primary—and growing—source of suffering for all life on the planet (including themselves as well as other species), then perhaps the world would be a better place without *Homo Sapiens* altogether.

This line of reasoning—and the discourse surrounding it—is a secular technoscientific version of the problem of theodicy, the justification of God's ways to humans, especially in the context of the *prima facie* tendency for bad things to happen to good people (and vice versa). In its heyday in seventeenth- and eighteenth-century Europe, theodicy invited much controversial speculation, as intellectuals inside and outside the clergy felt emboldened by *both* faith and reason to effectively second-guess why God would allow Creation to unfold in such a strange way (Nadler 2008). It marked the moment when such now familiar explanatory concepts of the human sciences as 'invisible hand', 'cunning of reason', 'law of unintended consequences', 'moral luck', and 'moral hazard' were forged. Nowadays these concepts are discussed not in relation to an inscrutable transcendent deity but to *Homo Sapiens* as the *de facto* replacement deity—for better or worse; hence, the moral resonance of the term 'Anthropocene', which characterizes the recent period in which the Earth's ecology has been remade in our image and likeness (Fuller 2024).

From this standpoint, the classic polar positions in theodicy epitomized by, say, Leibniz ('this is the best of all possible worlds') and Schopenhauer ('it would be better not to have been born') correspond nowadays to, on the one hand, those who are strongly pro-human survival because the good that even a few *Homo Sapiens* have done (and might continue to do) outweighs whatever harms have been (and will be) committed by the rest of the species and, on the other hand, those who believe that human extinction (by whatever means) would end the tyranny of the Anthropocene, which would be the most effective way to restore an environment that is secure for the widest range of life to flourish (Torres 2024). Of course, many intermediate religious and secular positions are possible. But in all cases, a judgement is made on the handiwork of the 'Creator', be it divine or human.

Notwithstanding the considerable intellectual and affective stakes, no solution to the problem of theodicy has gained widespread acceptance. Judgements of good and evil suffer from the same variability as judgements of true and false, not least because the two are intertwined. Lurking here as a 'meta-problem' is a lack of agreement over the range of beings entitled to be treated 'humanely'. The force of this point is most immediately felt in the question: Who is to judge whether the world would be a better place with more or fewer (or no) *Homo Sapiens*—or perhaps only *Homo Sapiens* of a certain sort? Classical theodicy appeared to provide an obvious answer: God. But even this says less than one might think, since there are powerful arguments in Abrahamic theology for both the deity who would save (nearly) everyone and the one who would damn (nearly) everyone. The matter is further complicated by whether the divine judgement has been 'always already' delivered or is not delivered until 'the end of time'. And the complexity only increases once God is out of the picture, as the human decision-makers must now function as both Creator and creature (Fuller 2011: chap. 5; 2012: chap. 4).

The history of the concept of humanity provides some guidance in dealing with theodicy's conundrums, since prior to the Linnaean equation 'human being = *Homo Sapiens*', 'humanity' was understood as a complex of qualities that might be discovered in an indefinitely wide range of beings, not only *Homo Sapiens.* Indeed, a striking feature of the conceptual history of humanity has been the role played by *recognition*, as defined by two complementary questions: How do you recognize a human? How do you become recognized as a human? Education has always been central to answering these questions, even if the 'human' one learns to be does not enjoy the same standing as other so-called 'humans'. In any case, the existing self-recognizing human community is itself reproduced and evaluated in the face of each new candidate human.

Here perhaps a reflexive application of the Turing Test comes most readily to mind, whereby the burden of proof is shouldered by those who would *refuse* to recognize a non-*Homo Sapiens*—be it animal or machine or extraterrestrial—as human, even after the candidate had met the relevant test criteria that would be expected, if not required, of *Homo Sapiens*. Moreover, beyond the historical concerns about potentially racist, sexist, classist, and culturalist bias in the application of such criteria, our emerging 'transvaluative' moral world reveals biases based on *presentism* (the preemptive discounting of future generations), *speciesism* (the preemptive discounting of non-*Homo Sapiens* animals), and even *substratism* (the preemptive discounting of non-carbon-based entities). These new biases threaten to undermine the chances of humanity constituting what Kant, following the Stoics, called a 'cosmopolis', whereby the capacity for recognizing ourselves in the other expands indefinitely (Fuller 2022a, 2022b, 2023b).

In recent times, Derek Parfit (1984) interestingly reformulated the problem of recognition as one of *identification*. Inspired by John Locke, Parfit argued for a radically anti-essentialist view of personal identity that rests primarily on narrative continuity, which Locke himself associated with individual memory but which could equally involve 'collective' memory of some sort. Examples of the latter would include forms of legal personhood based on family membership, corporate role, or national citizenship. In these cases, the meaning of an individual's life (both in terms of ascribed intentions and overarching purpose) is judged in relation to their performance in the legally prescribed capacity. The individuals are effectively treated as 'tokens' of a certain 'type', as Charles Sanders Peirce would say. In this context, a certain amount of self-sacrifice is expected, as individual intentions blend into a larger, transhistorical sense of collective agency.

Moreover, the boundaries of this collective sense of identity may range still more widely in time and space than the law normally permits. With the advance of Darwin's theory of evolution, the idea of human stewardship of nature has been increasingly linked to our common ancestry with all living things. As opposed to the line of thinkers from Francis Bacon onwards (not least Marx), who saw nature as the ultimate enemy against which humans must organize to conquer, post-Darwinian humans would be positioned as repaying their considerable debt to the rest of nature. To be sure, some evolutionists, notably Julian Huxley, have tried to leverage this sentiment into a transhumanist investment in a 'eugenically' steered future (Fuller 2019a: chap. 1). But more recently, this sensibility has taken a somewhat different turn, largely in response to W.D. Hamilton's evolutionary explanation of altruism, which is based on the idea that organisms sacrifice themselves to promote their genes, what Richard Dawkins popularized as the 'selfish gene'. Instead of trying to directly engineer a desirable future, the emphasis is placed on trying to sustain what is necessary for a desirable future. Thus, the 'good society' morphs into the 'good gene pool' (Fuller 2006: chap. 10). In this context, 'effective altruism' has emerged as a movement specifically aimed at providing future generations with at least a comparable scope for flourishing as we and other living things have enjoyed (McAskill 2022).

It is worth parsing this approach, since it is usually promoted in the context of an impending planetary catastrophe, inspired by an argument that appears at the end of Parfit (1984). Here Parfit challenges the utilitarian intuition that 99% of the world's population perishing would be a greater loss of value than the final one percent perishing, simply because the former case involves many more people dead. However, Parfit sees the situation differently. He thinks that loss of the final one percent would mean that humans would not be able to regenerate as a species, which matters if you believe that humanity is the primary source of value in the world. Now, of course, you may believe that a more generic idea like 'life' or 'intelligence' is the source of value, and humanity is simply one form it might take. In that case, you might imagine a replacement species. But in the absence of God, humans are arguably the only beings who recognize value as value, so they are positioned to promote it in a way no other being can. To be sure, this begs many questions, not least it presumes that 'human being = *Homo Sapiens*'. Nevertheless, it marks a significant shift in the direction of thinking about the future of humanity. Specifically, it suggests that humanity's potential for value creation need not decrease, even with a radically diminished population of *Homo Sapiens*.

This development may be regarded as a reconciliation of two radically opposed demographic worldviews that have coexisted since the end of the eighteenth century. They are epitomized by, on the one hand, the Marquis de Condorcet's cornucopian vision of an expanding number of human brains working to push forward the frontiers of knowledge to exert greater control over nature and, on the other, the Reverend Thomas Malthus' vision of successive generations of mass human misery and mortality due to absolute limits in the Earth's carrying capacity, which he regarded as the legacy of Original Sin but which Darwin secularized as 'natural selection' (Fuller 2006: chap. 13). Evolutionists, including self-declared Darwinians, have been split between the two worldviews. The original codifiers of the modern evolutionary synthesis, Julian Huxley and Theodosius Dobzhansky, were clearly in Condorcet's camp, both supportive of the heretical Jesuit paleontologist, Pierre Teilhard de Chardin, who treated humanity as the vehicle of cosmic apotheosis. Elon Musk's dream of interstellar human settlement is the most recent manifestation of this tendency. Huxley and Dobzhansky looked to eugenics (even post-1945) as a strategy to *overcome* natural selection by giving direction to evolution. Indeed, on this view, humanity's purpose is to provide purpose to an otherwise purposeless world (Dobzhansky 1967). This broadly 'transhumanist' sensibility continues the revolt against Aristotle that in the Renaissance had revived the vision of an infinite universe promoted by Stoic cosmopolitanism, which then acquired mathematical expression during the Scientific Revolution (Koyré 1957).

In contrast, the Parfit-driven 'effective altruism' crowd dance to the beat of a Malthusian drummer. To be sure, they accept the uniqueness of humanity as the source of cosmic value, but they deny the Condorcetian assumption that a greater *Homo Sapiens* population necessarily makes for a greater humanity. Instead, what matters is the perpetual existence of the human potential, what August Weismann originally called our 'germ plasm' and nowadays is most closely associated with our genome. (Weismann, himself ignorant of Mendelian genetics and molecular biology, heroically assumed that the germ plasm was immortal.) For the sake of argument, let us ignore the fact that we have yet to discover any uniquely 'human' genetic markers. Nevertheless, the influence of this way of seeing humanity should not be underestimated. As Maurizio Meloni (2016) has observed, Emile Durkheim modelled his definition of 'the social' as *sui generis* on Weismann's germ plasm. Like a species, a society transcends the set of its current members. And whereas species members are defined by their common germ plasm, a society is defined by its scripts and other rules that serve to identify and structure the activities of its members—their common 'culture', in the sense coined by Kant's student Herder and developed in modern anthropology. In both the biological and sociological cases, the focus is primarily on maintaining the capacity for realizing the species or culture. By the 1950s, under the influence of mathematical information theory, both were using 'code' to refer to this commonality (Kay 1999). It amounts to what Karl Popper (1972) called 'objective knowledge', his so-called 'World Three', which he illustrated in terms of the surviving libraries and databanks after a neutron bomb blast enabling a civilization to be resurrected long after its original members had perished. A biologically based version of this idea has been floated in recent years by the Harvard geneticist George Church under the rubric of 'de-extinction', whereby the preserved DNA of extinct species is used to bring them back to life—for better or worse (Church and Regis 2012).

3 Transcending the Relativity of Judgment in a Transvalued Humanity: The New Theodicy as Research Program

Regardless of which transvaluations of the human are regarded as morally appropriate and conceptually intelligible, the spatiotemporal relativity of judgement potentially confounds any 'fair' assessment of whether various decisions, policies, and actions have been 'good' or 'evil'. How the past appears to the future and the future to the past—even if agreement could be reached by both sides of the spatiotemporal divide on the facts about the past and the future—is bound to depend at least somewhat on the observer's standpoint. For example, effective altruists generally suppose that future generations will thank us, if we now enact long-term policies to safeguard life on Earth. They assume not only that such policies would have the desired effect, but also that the interests of those future generations will be sufficiently like ours to appreciate whatever we might have done for their benefit. Note that this entire way of thinking presupposes that long-term collective memory is a moral imperative: We *should* remember our ancestors as they would have us remember them.

However, if one considers the radical changes that *Homo Sapiens* has undergone during its time on Earth, humanity's moral compass has been reset many times, albeit perhaps less at the level of general concepts and principles than of their specific interpretation and application. The truism that the world into which one is born sets the baseline for what counts as a 'normal' existence explains the recurrent intergenerational normative struggles in history (Fuller 2019a: chap. 3). Finally, in the more extreme transhumanist and posthumanist transvaluations, which would have *Homo Sapiens* identify mainly with such non-*Homo Sapiens* as machines and animals, one wonders whether the recipients would even be in a cognitive position to recognize any prior actions that we might have taken on their behalf. If not, then does the sort of cosmic justice promised by theodicy—and still pursued by its secular followers—even begin to make sense without a transcendent deity who delivers the final verdict? But rather than surrender the future of social theory to Divine Providence, I will end by providing a menu of four sets of research questions that follow in the wake of the considerations raised in this article about the concept of humanity that should be of interest—and perhaps even urgency—to those concerned with the future of social theory.

1. If the transvaluation of the human involves an expansion of the circle of moral concern —a reaching for 'cosmopolis', as it were—can a set of criteria be developed to identify those deserving of 'humane' treatment that avoids various biases against new candidates? Or is this the wrong way to approach the definition of the human in a transvalued world?
2. How does one arrive at a theory of justice that is fair to the wider range of beings that will accompany the expansion of the circle of moral concern? Is there a way of harmonizing the 'rights' and 'duties' of both *Homo Sapiens* and non-*Homo Sapiens*? Or is this the wrong way to approach inclusiveness in a transvalued world?
3. Are there still lessons to be learned from theodicy in terms of its various proposals for how benefits and harms are distributed over time and space in the name of an overall moral judgement about the world? If so, who or what counts as a secular proxy for the deliverer of divine justice? Is it possible to have an open-minded view about the forms that the 'human' might take, while remaining concerned with the ongoing distribution of harms and benefits in the beings that materially pass as 'human'? Or is that the wrong way to approach specific moral judgements in a transvalued world?
4. How should the claims of the past, the present, and the future be related in a transvalued world? Should accountability among these temporal moments be treated symmetrically or asymmetrically? What if any role should time-travel thought experiments play in these considerations? Should we imagine that we might somehow enter a dialogue with other moments in time? Or is this the wrong way to approach intertemporal justice in a transvalued world? After all, perhaps silence provides the most room for other times to flourish.

4 Postscript: Taking Seriously the Idea of Humanity as a Style of Being in the World

I began by agreeing with Foucault's provocative thesis that 'human' as the name of a distinct object began in the mid-eighteenth century and is gradually disappearing in salience. It corresponds to the Linnaean equation, 'human being = *Homo Sapiens*'. However, I differ from Foucault in claiming that before Linnaeus—and perhaps again in the future—'human' referred to a quality that was distributed across a wide range of beings who need not be members of *Homo Sapiens* (and perhaps might not even include all *Homo Sapiens*). These non-*Homo Sapiens* might include animals, intelligent machines, and even extraterrestrials. Moreover, their 'humanity' would be revealed through collective engagement involving mutual recognition, what Kant called *Gemüt*, which he famously used to characterize both the order in the 'starry heavens' and the internal moral order. Indeed, Kant's *Gemüt* may simply be an update of the Stoic *pneuma* that literally holds together the *kosmopolis*, a kind of psychic cosmology.

This understanding of humanity acquires a renewed urgency in today's world, as epitomized by the question: 'What is the added value of being human?' On the one hand, artificial intelligence increasingly matches if not replaces historically human-based value inputs to the economy; on the other hand, humans are held responsible for value degradation on a planetary scale as their activities impact adversely on the Earth's climate and biodiversity. Implied in this pincer attack is that 'intelligence' and 'life' are valued as much if not more than ever, but at the expense of the 'human', whose normative standing has been put into doubt. In this respect, humanity has become a 'stranded asset': Its true value has been masked by its all-too-frequent suboptimal realizations, which reflects the fact that humanity tends to be associated more with its people and products than its process. However, humanity is a second-order concept that refers to a style of being in world (what Stoics after Cicero called *humanitas*), not a first-order concept that is exhausted by a unique set of objects. Thus, it is only an initial guess (i.e., a 'prior probability') that a *Homo Sapiens* will display greater humanity than, say, a canine, but the proof lies in how either makes their way in the world. The idea that *Homo Sapiens should* be human led the ancient Greeks to develop *paideia*, the precursor to liberal arts education, but they had no illusions that every species member was up to the task (Jaeger 1939). In contrast, the Abrahamic religions made the possession of humanity a species aspiration, whose secular legacy is the centrality of education to the modern 'human rights' agenda (Fuller 2020a).

So, what is this 'humanity' that needs to be possessed? If it is indeed a stranded asset, as I have asserted, then it requires a rebranding. In that case, how does 'human' add value to the world? There are two general and somewhat opposed ways that 'human' might add value to the world, which our bipolar attitude towards artificial intelligence already demonstrates. The first involves treating the 'human' as the original, the standard by which all copies are judged. This is how both art and research produced by artificial intelligence are popularly treated—that is, as inferior if not fraudulent versions of whatever humans might produce. In this context, it is often said that artificial intelligence cannot be better than its human programmers. While this view remains serviceable for purposes of assigning intellectual property rights and liabilities (i.e., to the programmer not the machine), it becomes harder to maintain as artificial intelligence becomes more 'generative', in the sense of being capable of reproducing brain-like processing functions. Here the 'human' becomes whatever escapes attempts at its replacement, which is the second way 'human' might add value to the world. This more sporting approach to humanity recognizes artificial intelligence's increasing capacity to improve itself beyond its human programmers' expectations. In effect, 'humans' must continually raise their game to avoid replacement by their creations (Fuller 2019c). Thus, the human is *Homo Absconditus*, as modelled on *Deus Absconditus* of the eighteenth-century Deists—but with a twist. *Deus Absconditus* is, as Voltaire joked, the cosmic culprit who flees the scene of its crime, in this case Creation itself. Thomas Jefferson offered a tamer version, whereby the deity is depicted as the departed ancestor whose earthly estate has been inherited by its human progeny. In contrast, *Homo Absconditus* lives in a perpetually fugitive state, a trickster who makes artificial intelligence researchers wonder at the next human task that their machines will need to match.

The idea that humanity might possess a fugitive essence is closely tied to the moral value of free will, a point brought out clearly by Kant, who observed that one acts morally not by conforming to a code of ethics (or God's will, for that matter) but by a free creative act, the 'spontaneous expression of the will', which one would have everyone else do (i.e., the categorical imperative). It may be that, as a matter of fact, this free act corresponds to some prescribed ethical norm, but the morality of the action lies in the freedom of its commission, not its correspondence to the norm.

To be sure, from a behavioral standpoint, it is difficult to tell whether a course of action that happens to correspond to a prescribed norm has been self-legislated and hence truly moral. Unsurprisingly, under the circumstances, actions that deviate from the norm start to look morally luminous. Aspiring revolutionaries since the nineteenth century have exploited this observation to justify their own disruptive actions. A precedent closer to Kant himself is the Protestant demand for people to openly declare their faith, rather than simply inherit it or conform to its strictures.

In the modern period, this line of thought was first secularized and then naturalized, culminating in Thomas Henry Huxley's hypothesis that *Homo Sapiens* may be the only species whose proper specimen is an individual organism, rather than a collective, understood as the locus of biological autonomy. An interesting debate then ensued on whether such individuality marked our species as the pinnacle of maturity in the evolution of life (Huxley's view) or on the contrary, as August Weismann thought, it signaled the immaturity of our species, which over time will acquire the more collective and functionally differentiated character of our evolutionary ancestors, such as the ants and bees but also the wolves and apes (Minot 1884). Seen through a political lens, the disagreement tracked whether the vision of humanity under liberalism or socialism, respectively, was in the vanguard of history. It was also mirrored in Ferdinand Toennies' and Emile Durkheim's contrasting attitudes towards the division of labor as a defining moment in modern society: the former saw it as a path to greater human individualization, the latter to our greater functionalization—yet each regarded their own vision as 'progressive' (Aldous 1972).

In terms of the above discussion, the 'stranded' character of humanity as an asset amounts to the individual trying to break away from the mass. After all, Weismann's association of species maturation with the functional differentiation of individual organisms was ultimately about individuals coming to realize their dependency on the mass from which they came and hence the need to maintain for continued propagation. In the generation after Weismann, the Harvard entomologist William Morton Wheeler promoted this sensibility under the rubric of 'superorganism', which inspired E.O. Wilson (1975) to develop 'sociobiology'. In this context, a sense of humanity too closely associated with individuality may threaten the species survival of *Homo Sapiens.* However, to appreciate the argument pursued here, think about humanity as, in terms that Marshall McLuhan might appreciate, a message that transcends its medium. Two examples come to mind, one literal and the other more symbolic. First is the historical tendency for greater meaning to be assigned to the formal arrangement of words (syntax) as the media for their potential expression multiplies. The Code of Hammurabi, one of the earliest extended pieces of writing, commanded authority not simply (or perhaps even primarily) for what it said but for the solidity of the stone on which it was inscribed, which itself signaled permanence. However, as writing shifted to more accessible, portable, and even ephemeral media, especially paper, the value of the medium shrank in relation to the value of the message conveyed (Innis 1951). A more symbolic version of the same process relates to Jesus as a being whose divine message—in both word and deed—exceeded the *Homo Sapiens* medium of its conveyance, which helps to explain Biblical remarks about Jesus' tolerance in the face of severe criticism and other indignities. Thus, the history of Christianity has witnessed several ingenious attempts—from the Arian heresy to Karl Rahner's 'Transcendental Christology'—to understand the human as ultimately a finite realization of the divine, whose alternative realizations may even include extraterrestrial beings (Rahner 1965).

Put in more contemporary terms, the suggestion is that humanity is possessed by any being that serves as an interface to the divine, which most evidently involves expanding our circle of concern to incorporate the alien, or 'other'. The Turing Test can be repurposed in this direction by accepting as human those who unexpectedly pass the test. Even if that moment means that the criteria for passing require revision, the aim would not be to prevent such an individual from passing in the future, but rather to correct whatever biases might have prevented others like that individual from passing in the past. In that sense, the Turing Test tests itself. In statistical terms, it amounts to treating what might have seemed to be a 'false positive' as a 'false negative'. In other words, it signals not that we have humanized non-humans but that we have dehumanized humans. It recalls the famous late nineteenth century debate between William Clifford and William James over whether belief should be proportional to evidence, with Clifford saying yes and James no. James regarded even scientific hypotheses as beliefs that exceed their evidential content, but ones that are forever subject to severe tests, comparable to trials of faith, perseverance in the face of which requires a 'will to believe' (Fuller 2020c: chap. 10).

A useful way to understand the human as a kind of interface is to imagine the complementary sides of the communication process, understood in its most general form: *the conveyance of intention* and *the reception of new information*. Now think about the Turing Test as just such a communicative setting where these two sides are present. The former is about self-presentation in the face of a Turing-style investigation, the aim of which is to transmit signs of *humanitas* by whatever means are at the communicator's disposal. Quintilian's adage, 'The clothes make the man', popularized in the Renaissance by Erasmus, captured this insight for the Classical world, which in the twentieth century could be characterized in terms of our leading lives as 'method actors', filling our clothes as if they were made for us. The twenty-first century twist is that we can design entire 'digital wardrobes', networks of avatars, which increasingly secure our ontological standing in both the online and offline worlds. The sociologist Nobert Elias (1994) captured this general movement in the expansion of humanity as the 'civilizing process', the hallmark of which is that previously excluded agents and actions come to be included a common world through some 'artful' transformation of both the observer and the observed. In this way, the original Stoic second-order conception of *humanitas* is given a modern facelift as a *style of being in the world* (Simmel 1991).

The morphologically fluid character of 'style' immediately invites questions of limits: Granted that the human style may be displayed under a wide variety of material conditions and transformations, are there patterns of relations that must remain invariant across these differences? Drawing on his mathematical background, the Gestalt social psychologist Kurt Lewin (1936) famously proposed a 'topological psychology' based on 'field' properties that serve as dimensions along which patterns of relations among a range of objects can be gauged. Nowadays we might most easily associate this approach with the sorts of pattern recognition methodologies used in the search for extraterrestrial intelligence (SETI). In their own day, Lewin and his fellow Gestaltists were naturally understood as flipping Husserl's Descartes-inspired strategy of 'eidetic variation', whereby the essence of a known object is abstracted by removing its various phenomenal properties to see if it retains its ontological integrity. The Gestaltists flipped the procedure by starting with an ambiguous object and then studying its dynamics to see whether it manifests the requisite relational properties. Of course, the 'flip' is not perfect, since the 'requisite relational properties' of, say, 'being human', are not themselves fully known *a priori*. Indeed, the ambiguous object is equally meant to prompt the observer's intuition of those properties—a reminder of what he or she is trying to find. A landmark experiment in this vein is the so-called 'Heider-Simmel Illusion', whereby a simple animation consisting of two triangles and a circle moving around a partially open square elicited a wide range of elaborate narrative content from observers, who clearly detected human-like relationships on the screen (Heider and Simmel 1944).

However, the ability to 'see humanity' in the world is not limitless, and the limits seem to arise in metaphysically curious and morally awkward places. As mentioned earlier, there is a long and distinguished tradition of 'anthropomorphism' as an appropriate attitude to many—though not necessarily all—animals, which only came to an end once the Linnaean equation of 'human being = *Homo Sapiens*' became culturally ensconced in the late nineteenth century. In most traditional cases of anthropomorphism, the animals only loosely physically resemble the humans who recognize them. A contrasting point of reference is the 'uncanny valley', a phenomenon observed by researchers studying the assimilation of robots into the human lifeworld (Mori 2012). In essence, once a robot starts to physically resemble *Homo Sapiens* 'too much', humans find it increasingly difficult to attribute humanity to them, perhaps because it seems to be merely 'impersonating' surface qualities of being human without fully inhabiting the human form, as in the Quintilian adage. But the problematic character of the relationship between a physical resemblance to *Homo Sapiens* and genuine humanity runs deeper. Development aid workers have long noted the difficulty in maintaining global public interest in human mass suffering in the Global South, once images of the suffering recede from the forefront of collective attention and discussion. A striking point of comparison is the money raised for the rebuilding of Notre Dame Cathedral in Paris after it burned down in 2019 versus the money raised for Ethiopian famine relief in the two widely publicized 'Live Aid' concerts in 1985. Adjusted for inflation, the former raised three times the amount of the latter, resulting in French President Macron having to return most of it as surplus to needs. As for famine and strife in sub-Saharan Africa, it continues and has even increased over the past four decades, with only intermittent publicity shone on it (Fuller 2020a).

The morally difficult observation that lurks in these cases of liminal humanity is that the metamorphic character of the human as a style of being in the world applies not only to the range of beings potentially detected as being human (e.g. they need not be *Homo Sapiens*) but also to the range of relatability that is expected of the beings so detected. Put bluntly, humans must be relatable under a wide variety of conditions. Indeed, when the sociologist Georg Simmel (1991) proposed humanity as a style of being in the world, he closely linked it to the propensity for *fashion*, which evolutionists at the time were touting as a species marker for *Homo Sapiens*—namely, our seemingly endless capacity for self-reinvention under new conditions, both given and made (Simmel 1957). Such self-reinvention typically amounted to the reconfiguration of already existing elements 'in a new light' through the sorts of transvaluations that Gestalt psychology would later study systematically. Considered in this context, Notre Dame Cathedral is arguably more 'human' than millions of suffering sub-Saharans. When UNESCO designated the periodically burned down twelfth century church as one of the first 'World Heritage' sites, it was formally recognizing the artefact's relatability to a vast range of humans over time and space, an increasing number of whom have had no contact with its original cultural roots. In that respect, Notre Dame is properly 'cishuman', in that its humanity is recognized even without knowing how it became human (Fuller 2019b). In contrast, while the sub-Saharan Africans are clearly *Homo Sapiens*, they are perceived monolithically—as a mass. The basis was set by early philosophical and social scientific views that the largely hereditary nature of 'traditional' cultures made them static, restricting their behavioral range in the manner of animal cultures. These views were increasingly viewed through a capitalist lens that strongly associated humanity with the capacity for productive labor under changing market conditions. (Here socialists called for the unleashing of that hidden productivity through various interventions into traditional cultures.) The long-term effect has been that without direct sustained public exposure, sub-Saharans remain insufficiently relatable along the number of dimensions required to be fully recognized as humans.

This account of humanity as a style of being in the world may amount to a moral indictment of those who would uphold such an account. I leave that for the reader to judge. My main point here—and throughout this essay—has been to highlight the normative unsustainability of the Linnaean equation of 'human being = *Homo Sapiens*', in ways that go beyond Michel Foucault's treatment of the equation as a passing artefact in the archaeology of knowledge. At the very least, I have drawn attention to the broadly and deeply *aesthetic* dimension of judgements of humanity. It suggests that the Turing Test and kindred procedures need to be administered in the spirit of *connoisseurship*, in which the overall aim is to both widen and deepen the search for humans. Moreover, the exercise need not happen only once; indeed, one may wish or need to 'recertify' their humanity by engaging in the relevant forms of interaction prescribed by the Turing Test and its successors. In this context, given its human-based inputs and neural-based architecture, generative artificial intelligence may prove a useful trainer for 'certified' humans to enhance their humanity. Lea Peersman-Pujol, ex-personal assistant to the pioneer French cybernetic philosopher Edgar Morin, has developed an advanced consultancy in human resource management in this spirit for corporate purposes (www.lign.ai). However, there is no reason why the idea could not be applied more broadly to secure the brand 'human' for future generations.

References

Aldous, J. 1972. An Exchange Between Durkheim and Tönnies on the Nature of Social Relations, with an Introduction by Joan Aldous. *American Journal of Sociology* 77:1191–1200.

Bourke, J. 2011. *What It Means to Be Human: Reflections from 1791 to the Present.* London: Virago.

Chernilo, D. 2017. *Debating Humanity: Towards a Philosophical Sociology.* Cambridge, UK: Cambridge University Press.

Church, G., and E. Regis. 2012. *Regenesis: How Synthetic Biology Will Reinvent Nature and Ourselves.* New York: Basic Books.

Dobzhansky, T. 1967. *The Biology of Ultimate Concern.* New York: New American Library.

Elias, N. 1994. *The Civilizing Process (Orig. 1939).* Oxford: Basil Blackwell.

Foucault, M. 1970. *The Order of Things. (Orig. 1966).* New York: Random House.

Foucault, M. 2008. *The Birth of Biopolitics. (Orig. 1978–79).* London: Palgrave.

Fuller, S. 2006. *The New Sociological Imagination.* London: Sage.

Fuller, S. 2011. *Humanity 2.0: What It Means to Be Human: Past, Present and Future.* London: Palgrave.

Fuller, S. 2012. *Preparing for Life in Humanity 2.0.* London: Palgrave.

Fuller, S. 2014. Human Being = *Homo Sapiens.* (Response to the Question: What scientific idea is ready for retirement?) Edge.org. https://www.edge.org/response-detail/25396.

Fuller, S. 2015. *Knowledge: The Philosophical Quest in History.* London: Routledge.

Fuller, S. 2019a. *Nietzschean Meditations: Untimely Thoughts at the Dawn of the Transhuman Era.* Berne, SZ: Schwabe.

Fuller, S. 2019b. The Metaphysical Standing of the Human: A Future for the History of the Human Sciences. *History of the Human Sciences* 32:23–40.

Fuller, S. 2019c. Technological Unemployment as a Test of the Added Value of Being Human. In *Education and Technological Unemployment*, ed. M. Peters et al., 115–128. Berlin: Springer.

Fuller, S. 2020a. The Unity of Humanity. In *The Bloomsbury Handbook of Posthumanism*, ed. M. Thomsen and J. Wamberg, 171–182. London: Bloomsbury.
Fuller, S. 2020b. 'Our love-hate relationship with humanity'. Review of Chernilo (2017). *Distinktion* 21 (1): 67–73.
Fuller, S. 2020c. *A Player's Guide to the Post-Truth Condition: The Name of the Game*. London: Anthem.
Fuller, S. 2022a. Kant After Kant: Towards a History of the Human Sciences from a Cosmopolitan Standpoint. In *The Palgrave Handbook of the History of the Human Sciences*, ed. D. McCallum, 29–55. London: Palgrave.
Fuller, S. 2022b. The Mind-Technology Problem. *Postdigital Science and Education* 4:247–252.
Fuller, S. 2023a. *Back to the University's Future: The Second Coming of Humboldt*. Berlin: Springer.
Fuller, S. 2023b. Humanity's In-Betweenness: Towards a Prehistory of Cyborg Life. In *Humanity In-Between and Beyond*, ed. M. Michalowska, 63–80. Berlin: Springer.
Fuller, S. 2024. 'Catastrophe's Justice' (Review of Turner [2023]). *International Journal of Politics, Culture, and Society* 37:159–167.
Fuller, S., and V. Lipinska. 2014. *The Proactionary Imperative: A Foundation for Transhumanism*. London: Palgrave.
Heider, F., and M. Simmel. 1944. An Experimental Study of Apparent Behavior. *American Journal of Psychology* 57 (2): 243–259.
Innis, H. 1951. *The Bias of Communication*. Toronto: University of Toronto Press.
Jaeger, W. 1939. *Paideia: The Ideals of Greek Culture*. Oxford: Blackwell.
Kay, L. 1999. *Who Wrote the Book of Life? A History of the Genetic Code*. Palo Alto, CA: Stanford University Press.
Koyré, A. 1957. *From the Closed World to the Infinite Universe*. Baltimore: Johns Hopkins University Press.
Latour, B. 1987. *Science in Action*. Milton Keynes, UK: Open University Press.
Lewin, K. 1936. *Principles of Topological Psychology*. New York: McGraw-Hill.
McAskill, W. 2022. *What We Owe the Future*. New York: Basic Books.
Meloni, M. 2016. The Transcendence of the Social: Durkheim, Weismann and the Purification of Sociology. *Frontiers in Sociology* 1:11. https://www.frontiersin.org/journals/sociology/articles/10.3389/fsoc.2016.00011.
Minot, C. S. 1884. Death and Individuality. *Science* 4 (90): 398–400.
Mori, M. 2012. The Uncanny Valley. *IEEE Robotics and Automation* 19 (2): 98–100.
Nadler, S. 2008. *The Best of All Possible Worlds: A Story of God, Philosophers and Evil*. New York: Farrar Straus and Giroux.
Neiman, S. 2002. *Evil in Modern Thought: An Alternative History of Philosophy*. Princeton: Princeton University Press.

Parfit, D. 1984. *Reasons and Persons.* Oxford: Oxford University Press.
Popper, K. 1972. *Objective Knowledge.* Oxford: Oxford University Press.
Rahner, K. 1965. *Hominisation: The Evolutionary Origin of Man as a Theological Problem.* Freiburg: Herder and Herder.
Rawls, J. 2000. *Lectures on the History of Moral Philosophy.* Cambridge, MA: Harvard University Press.
Searle, J. 1983. *Intentionality.* Cambridge, UK: Cambridge University Press.
Searle, J. 1984. *Minds, Brains and Science. (Reith Lectures).* London: British Broadcasting Corporation.
Simmel, G. 1957. 'Fashion'. (Orig. 1904). *American Journal of Sociology* 62:541–558.
Simmel, G. 1991. The Problem of Style. (Orig. 1908). *Theory, Culture and Society* 8:63–71.
Singer, P. 1981. *The Expanding Circle: Ethics and Sociobiology.* Oxford: Clarendon Press.
Torres, E. 2024. *Human Extinction: A History of the Science and Ethics of Annihilation.* London: Routledge.
Turner, S. 2010. *Explaining the Normative.* Oxford: Polity.
Turner, B. 2023. *A Theory of Catastrophe.* Berlin: Walter de Gruyter.
Wilson, E. O. 1975. *Sociobiology: The New Synthesis.* Cambridge, MA: Harvard University Press.

Euro-Transhumanism

Stefan Lorenz Sorgner

Transhumanism is a basic intellectual attitude that can be traced to Julian Huxley, who coined the term in 1951 (Huxley 1951). *Transumanar*, a related verb, had already been developed by Dante. Although the word "transhumanism" predates Huxley, prior to him, it was associated more with the otherworldly developments of human beings, and its meaning changed after his coining of the term. Transhumanism is a basic orientation towards the significance and use of technology. It is wrong to assume that transhumanism always goes hand in hand with mind-uploading, i.e., the transfer of the human personality to a hard disk (Sorgner 2020). Otherwise, Julian Huxley himself would not be a transhumanist. While this mind-uploading is affirmed by many transhumanists (e.g. Ray Kurzweil), it is by no means shared by all. Even more absurd is the image of transhumanists all striving to become immortal by means of mind-uploading. Critics of transhumanism, both from a humanist as well as from a critical posthumanist perspective, present transhumanism in this way to highlight the absurdity of this basic stance. Such polemical statements are not informed by critical inquiry or analysis. This-worldly

S. L. Sorgner (✉)
John Cabot University, Rome, Italy
e-mail: ssorgner@johncabot.edu

G. Theiner, A. Malapi-Nelson (eds.), *Technologies of the Future Self*, Palgrave Studies in the Future of Humanity and its Successors,
https://doi.org/10.1007/978-3-031-99902-4_7

immortality cannot even be thought of consistently. There are two likelihoods for the universe: it will either cool down, which could be accompanied by a cosmic standstill, or there will be a cosmic singularity, i.e., a point of enormous density, containing all the energy after a contraction of the universe. How could a personality uploaded to a hard drive survive such a state? It is not even consistent to conceptualize this-worldly immortality. Serious transhumanists would not uphold immortality in a literal sense. This does not mean that transhumanists do not use the word "immortality" (Sorgner 2022a). When they do, they do so for rhetorical reasons, to draw attention in this way to the importance of extending the healthspan. It is this claim that is in fact advocated by most, if not all transhumanists. It should be emphasized here that it is not the lifespan but the healthspan that is spoken of, i.e., the time during which persons are alive and healthy. This objective is affirmed because it is assumed that personal quality of life is promoted in this way, and promoting this is, after all, a primary transhumanist concern.[1] In general, the central concern of transhumanism can be formulated as follows:

> It is in our interest to push our current boundaries in a significant way using the latest techniques, whereby the probability of the coming about of the posthuman will be increased, as this will promote the likelihood of individuals achieving a better quality of life.

The following example can be used to illustrate the plausibility of this basic attitude. Over the past two hundred years, the human lifespan has almost doubled.[2] In this way, the quality of life has also been significantly increased. It should be highlighted that transhumanism's primary intention is not to increase the life span, but to extend the healthspan. People associate a wide plurality of goals with a fulfilled life.[3] For some, enjoying

[1] It is interesting to note that Nietzsche, too, stressed that humans cherish an extended health span when presenting the, so-called, "last man". He did not necessarily affirm such a concept, but his analysis reveals that even he recognizes the relevance of an increased health span from the perspective of most human beings. However, even the overhuman cherishes health, the great health. Yet, in this case the concept of health if different from the one being cherished by the last human. The great health has to do with the capacity to overcome suffering again and again. In the case of the last human, it has to do with a reduced likelihood of dysfunctionalities coming about in the first place (Sorgner 2025).

[2] https://ourworldindata.org/life-expectancy 21.8.2024

[3] Nietzsche highlighted this insight by stressing the wisdom of the body (Sorgner 2016a, 2016b).

oneself is important, for others, it is establishing a dynasty, and still others identify a fulfilled life by leaving behind a cultural legacy. Identifying commonalities is difficult, which is why the category of morphological freedom is central to transhumanism (Sandberg 2013). However, it should be acknowledged that different transhumanists uphold different concepts of the good. Classic transhumanists affirm a hyper-renaissance concept of the good, e.g. Bostrom (Sorgner 2022a). The concept that no non-formal concept of the good is plausible is characteristic of Euro-transhumanism, as suggested in my own approach. Beyond these considerations, we must affirm the importance of health as a precondition for leading a fulfilled life. This statement does not need to hold for all. Most people identify health with a successful life in one way or another. For some, it is of intrinsic importance, so being healthy is directly correlated with a high quality of life. For others, health only plays an instrumental role, i.e., it is important for them to be healthy because it is a prerequisite for the realization of other goals, e.g., to lead a life full of intense athletic activities. But health does not need to be important for all. Since extending the healthspan is important for most people, this objective should also be given greater political consideration than is currently the case (Donner et al. 2016, 353). This insight also illustrates why the existence of public universal health coverage is a fantastic achievement. In "We have always been cyborgs", I described many current challenges associated with public universal healthcare, such as the question of financing health insurance and, related to this, our handling of personalized digital data.

Here, some fundamental differences between classic transhumanism and euro-transhumanism have already been highlighted. However, you find a slightly more detailed sketch of the various approaches to transhumanism in the following paragraphs. Transhumanism was coined by Julian Huxley in 1951. FM 2030, Max More, and Natasha Vita-More were responsible for turning it into a broader cultural movement during the 1980s and 1990s, whereby original transhumanism was realized. Nick Bostrom has been responsible for developing a philosophy of classic transhumanism since the beginning of this millennium which is part of the Anglo-American cultural tradition. The most famous proponents of classic transhumanism alive include Elon Musk, and Peter Thiel. The roots of euro-transhumanism, on the other hand, can be traced to the 2009 publication of my article "Nietzsche, the overhuman, and transhumanism" (2009), which has resonated with artists (Joanna Grochowska, Sven Helbig), entrepreneurs (Martine Rothblatt), and intellectuals (see edited by Tuncel, Y. *Nietzsche and Transhumanism* collection (2017)), who affirm transhumanism, but whose thinking primarily bears constructive interferences with the continental European cultural tradition from Heraclitus via Spinoza and Nietzsche into our times. To stress the close affinities between euro-transhumanism and many of Nietzsche's reflections, I have added references both to Nietzsche's writings as well as to secondary literature on Nietzsche's reflections, so that the complex entanglement between euro-transhumanism and Nietzsche's thinking can be grasped, and thus stimulate further reflections on many central transhumanist issues.

Some elements of this way of thinking can already be found in thinkers like James Hughes and Martine Rothblatt. However, it is the decisive embeddedness in the continental European cultural tradition which is decisive for euro-transhumanism. It must be noted that this relatedness does not imply a Eurocentrism or a new type of colonialism, as this approach clearly highlights, stresses, and affirms its own situatedness. It is entangled in a European continental way of thinking while not affirming the universal validity or ultimate superiority of this kind of approach. When entering a discourse, euro-transhumanism does not claim a better epistemological rootedness than an alternative account. It merely presents suggestions, highlights insights, and proposes reflections to see whether a constructive or destructive interference occurs with respect to an alternative approach. In addition, it must be noted that a euro-transhumanist account could also be affirmed by someone who was born outside of Europe if that person has been brought up in the continental European cultural tradition or has realized strong resonances with respect to this tradition. Euro-transhumanism stresses the relevance of glocalized investigations, situated insights, and the entanglement in contingent nodal points.

Core elements of my own take on euro-transhumanism were further developed in the monographs *On Transhumanism* (2020), *We have always been Cyborgs* (2022a), and *Philosophy of Posthuman Art* (2022b). The term euro-transhumanism was coined in the article "On constructive and destructive interferences in the life and in art worlds: A reply" (Sorgner 2022c) during an intellectual exchange on the various philosophical challenges which I put forward in the aforementioned three monographs. In 2024, an essay collection entitled *Metahumanism, Euro-Transhumanism, and Sorgner's Philosophy* (edited by Schussler/Balistreri) was published in which central articles of this intellectual exchange have been collected.

In a nutshell, the following twelve differences between classic transhumanism (=CT) and euro-transhumanism (=ET) can be summarized:

1. CT upholds utopias; ET rejects the affirmation of utopias.

In my monograph *We have always been cyborgs*, I extensively dealt with the issue of utopias. Some classic transhumanists have seriously suggested the death of death, the goal of getting rid of suffering entirely, or a long-termist version of utilitarianism which implies that what matters is merely that some humans survive. None of these issues are either realistic or desirable options. With my German background, I have long been confronted with the challenges of utopia. Utopias promise a future which will never and maybe even can never come about, and also have the implication that the quality of lives of currently living people get sacrificed for utopian futures. This means persons are merely treated as a means for a future which will never come about. Such structures must be avoided at all costs. This does not mean that one should not play around with certain visions as inspirational factors. However, it does mean that any utopia which gets taken as a sole universally valid purpose has terrible implications which need to be avoided.

2. CT affirms a hyper-renaissance of perfection; ET affirms a radical diversity of concepts of the good.

I dedicated an entire chapter of the monograph *We have always been cyborgs* to this topic, in which I distinguished between a hyper-humanist account of the good, a commonsense account of the good, and a radically pluralistic account of the good. A hyper-humanist account of the good has been suggested by Bostrom. It basically implies that the Renaissance ideal gets accepted, and technologies get used to realize it, such that the traditional targets can be exceeded. The commonsense account of the good draws upon widely shared elements of the good, like not being ill, not being handicapped, and having highly developed cognitive capacities. It has been suggested by Julian Savulescu. It is tempting to agree with him. However, the challenges come about when the various factors are taken as ultimately justified and valid elements of the good. Then it can have strongly paternalistic implications. I have suggested a radical diversity of concepts of the good as an alternative. It takes seriously that the bundle of drives out of which we all are constituted does not have an essential core element. To increase the likelihood of people living a good life, this diversity needs to be considered. At the same time, I stress that some factors are such that they indeed are widely shared, e.g. the relevance of an increased healthspan. It is acknowledged that an increased healthspan is not a necessary element for living a good life, but it is a widely shared element, which means that for most people it is a value which increases their quality of life. The difference between a universally valid concept and a widely shared one is decisive here.

3. CT has an optimistic worldview; ET regards Dukkha as plausible, and consequently a version of philosophical pessimism.

Some classic transhumanists claim that a world without suffering can indeed be realized and that minds can be uploaded soon, such that we can experience lives full of the most wonderful types of fulfilment. This is wishful thinking. I am not even certain whether such visions are desirable. However, I do not regard these descriptions of the world we live in as plausible. The world can be interpreted in so many ways. We can tell an enormous amount of stories to make sense of the world. Any story can always be retold in a different manner. However, "some fundamental things apply as time goes by". One of these fundamental insights is the relevance of suffering, of Dukkha. If the ego wants something, it suffers, and there is no way of escaping this fundamental structure. The challenge lies in how we deal with suffering. This insight is acknowledged by many world religions. They provide various responses on how to deal with the permanent suffering people have to deal with. Moments of fulfilment are rare, and they are always mixed with pain. Intensely overflowingly fulfilling moments of pleasure pains are the best we can hope for. These are even more special than simple moments of pleasure. Dukkha is inevitable. It needs to be acknowledged, which I have done in various publications of mine, starting from my book *Metaphysics without Truth* which came out when I was 26.

4. CT affirms the universal validity of certain values and norms, which can have neo-colonial implications; ET regards perspectivism as plausible, which undermines the possibility of neo-colonial implications.

Classic transhumanism has suggested so many different norms and values, and they have claimed universal validity for these normative judgements. This is highly problematic for various reasons. One of them is an ontological worry. How and where should any such universally valid normative judgement exist? Mackie's account of values and norms in *Inventing Right and Wrong* (1990) is plausible. Objective values and norms would have to be like Plato's forms, extremely weird ontological entities in a realm which cannot be accessed by means of our senses. I agree entirely. However, if this is not the case. What then are norms and values? They are fictive entities created in the interest of their inventors. By analyzing the underlying motives of the inventors, the relevance of the normative judgements can be grasped. If this is a plausible account of normative judgements, the issue of neo-colonialism does not come up either, as neither norms nor values would claim their own universal validity. This does not mean that autonomy needs to be given up, but the implications of a liberal ethics of fictive autonomy need to be thought through.

5. CT affirms a version of utilitarianism;[4] ET embraces a hermeneutic understanding of ethics.

Utilitarianism which is so widely shared in the Anglo-American world has enormously rigid implications. The morally right action can be calculated on the basis of a mathematical formula. Appropriate numerical values need to be connected to possible outcomes, probabilities need to be associated, and then the morally right action can be analyzed. This way of dealing with ethical challenges might be helpful when dealing with some political issues, however, in many "real-life" circumstances, the particular circumstances of a situation are so nuanced that a hermeneutical procedure provides more plausible judgements. The devil is in the details. Cultural resonances, intellectual associations, and associative connotations need to be considered. Consequently, it is advisable to always progress with enormous care, when dealing with ethical challenges and moral dilemmas. Hermeneutic reflections by Vattimo demonstrate explicitly how a hermeneutic understanding of ethics can be applied (Vattimo 2003).

[4] For another relevant alternative, see Malapi-Nelson (2021; reprinted in this volume).

6. CT argues for libertarianism; ET embraces a social-democratic version of liberalism.

The political aspect is particularly tricky. I am aware that political norms, too, are human-made fictions. Here, I use the word "fiction" such that it does not imply that fiction has to be false, but merely as identical to the word "perspective", which implies that it is a human-made judgement which can be false. The overall political system depends on what representatives of interest groups have fought for successfully. There is not one ideal political arrangement which is the ultimate goal or which is valid for all times. Hence, various advantages and disadvantages need to be considered. Thereby the question of advantages and disadvantages for whom always has to be considered. When it comes to developing new pharmaceuticals, I can see the advantages of a libertarian system. However, when it comes to issues like accessibility, then the question needs to be taken seriously, in which kind of society do you want to live? A society in which people do not have access to a publicly funded healthcare system seems to me highly problematic. You are not feeling well, and you need to think whether you can afford to be taken to a hospital or a physician is terrible. Nation states are there to take care of the well-being of the people living in these states, and having access to a public health care system is such a fundamental element of most people's well-being that it needs to be covered. It seems even more important than having a public educational system. Given these intuitions, a social-liberal political system seems to be an as-good-as-it-gets system when it comes to possible political organizations. There are many challenges associated with it, but it is as good as it gets for the time being.

7. CT aims for immortality; ET aims for an increased health span.

There are many challenging concepts which are upheld by classic transhumanism. However, the concept of immortality is clearly among the weirdest. If mind-uploading was actually a necessary element of transhumanism, which it is not, then I would understand why some intellectuals accuse transhumanism of merely being a new kind of religion, or of it not being a worldview which deserves to get taken seriously by scholars. Immortality, if understood, and taken literally, just does not make any sense in a non-dualistic ontology of becoming whatsoever. If there is a permanent change in this world, there is not even any kind of ontological stability which transcends any single moment. If cosmological accounts of the world are taken into consideration, then the absurdity of the concept of immortality becomes even clearer. There is no realistic way in which a person can survive a cosmological singularity. The word "immortality" works as an advertisement tool; however, it has the side effect that it undermines the intellectual credibility of the proponent. What is relevant indeed is the importance of an increased healthspan. Most people are, in some way or other, interested in living longer healthily. Health here has to be taken as an open concept. For some, health means to feel good, for others, health implies the realization of our functional purposes. There are great arguments for both sides. However, when I stress the relevance of health, I do not limit the meaning of the word to a specific philosophical understanding. Health in an extremely general sense is important for most people, and it is this insight which is enormously important. It also has a great number of practical implications.

8. CT can be identified with a silicon-based transhumanism which is focused on mind-uploading; ET can be associated with a carbonate-based transhumanism which has realized that "we have always been cyborgs", and it is highly unlikely that the "singularity is near".

So many critics of transhumanism have highlighted that transhumanism is all about the longing of become immortal through mind-uploading. I have already mentioned a few reasons why immortality is a concept which needs to be avoided. A similar issue applies to mind-uploading, even though, I cannot exclude the possibility that it will not be possible sometime in the future. However, given current scientific insights, there is no indication whatsoever that silicon-based entities can have consciousness, self-consciousness, emotions, or intentionality. However, in order to be able that our personalities can actually be transferred to a silicon-based foundation, it has to be the case that they can be transferred with a loss. Hence, consciousness, self-consciousness, emotions, and intentionality would have to persist on a silicon-based foundation, too. There is no indication so far that this can actually be realized. This does not mean that it is impossible, but there is simply no good reason to think that it will be possible to realize this goal soon. However, this does not mean that there is a categorically dualistic ontological distinction between silicon-based and carbonate-based entities. Both exist in the sensually accessible realm. It is a matter of structuring which qualities can be realized. With the coming about of the capacity to think, speak, and reason, something significant has happened, which took place about 300,000 years ago. A genetic mutation plus human interaction brought about that we have rationality. This is when human beings were realized. This upgrade has generated the coming into existence of human beings. Now, it seems plausible that we ourselves can generate further upgrades. However, upgrading ourselves is nothing morally problematic. It rather is the basis of our existence. What needs to be considered is which directions are desirable. Still, there is no general response to this question. Again and again, we need to talk about, analyze, and reflect upon these issues in order to guarantee that the developments are in our interest.

9. CT is centrally concerned with the risks of superintelligence; ET is doubtful concerning the coming about of artificial general intelligence as well as superintelligence in the near future.

I do not underestimate the relevance of AI. Playing around with various AI tools is amazingly astonishing, and they are getting better day by day. I can already see some of the developments which are likely to be available in the near future. However, this does not mean that embodied AIs will have to take over the world soon. The human-AI interaction has an enormous amount of potential for being risky for our existence. We do need to be enormously careful in this respect. This does not mean that there is a danger of AI taking over the world. There is a fine but decisive difference between these two judgements. If consciousness, self-consciousness, emotions, and intentionality cannot be realized on a silicon basis, then it seems unlikely that embodied AIs will be able to say "I want this or that" and realize it. However, such a decision would have to be necessary for AI to take over, for AI to put humans into a zoo, and for AI to control the entire universe. Such thoughts are great for utopian novels, sci-fi films, and Netflix series, but AI taking over by itself is not a realistic option so far. Challenging moral issues arise concerning AI-human interactions. However, this phenomenon is fundamentally different from AI's taking over by themselves.

10. CT is concerned with the overcoming of suffering; ET recognizes the inevitability of suffering.

In contrast to the former statement concerning Dukkha which was a descriptive one, here, normative judgements are dealt with. Classic transhumanists claim that we should try to overcome suffering, as suffering undermines the capacity for living a good life. Should we actually do so? Suffering is a challenge. We do not like to suffer. However, it also needs to be realized that we usually embrace suffering when it is meaningful suffering, when we know what we suffer for, and when we suffer to realize something which is important to us. Creating a sculpture, writing a book, or climbing onto the top of a mountain can be extremely meaningful activities for many humans. These activities can provide lives with a purpose. They involve suffering, pain, and discomfort. However, these side-effects get embraced, because of the goals which are being aspired for. Hence, suffering does not have to be a problematic issue. It is meaningless suffering which is so terrible. A second point needs to be considered, too. If we imagine a life without suffering, the issue also needs to be addressed in how far pleasure can be experienced in such a world. If an intense feeling of pain is not possible, how and why should the feeling of pleasure be possible still? This does not seem possible. The question then needs to be addressed whether a life without pleasure and pain would indeed be better than one with pleasures and pains. Buddhists might think that this is the case. I have my doubts in this respect.

11. CT sees itself in the tradition of the Enlightenment; ET is highly critical of the logocentric Enlightenment tradition and stresses an ambiguous relationship with the philosophical foundations of the Enlightenment.

This point is particularly intriguing. What does the Enlightenment stand for? There are two strong traditions: 1. The logocentric tradition from Descartes to Kant; 2. The body-centric tradition from de Sade to Nietzsche. Depending on your understanding of what enlightenment is, different ontological implications are given. According to the logocentric tradition, human beings consist of a physical body and an immaterial rational mind. Enlightenment here is about using reason to criticize traditional institutions, structures, and thoughts. According to the body-centric tradition, however, enlightenment is about undermining the universal validity of reason to realize that humans are merely gradually different from other animals and that our rational mind is merely a part of our body. In this case, however, there is no longer one unified rational mind, but there are as many minds as there are bodies, and each rational mind is slightly different from the mind of another body. Classic transhumanists embrace the understanding proposed by the logocentric tradition of enlightenment that it is our task to use reason to realize our goals, but at the same time, many of these transhumanists also affirm that they embrace a naturalist ontology. Such an ontology, however, rules out the possibility of categorically dualist ontology as it is affirmed in the logocentric tradition from Descartes to Kant. Hence, it can be doubted whether such a philosophical approach can consistently be upheld. Euro-transhumanists take this insight seriously and embrace the understanding of an embodied mind. This leads them to highlight an ambiguous relationship with enlightenment. On the one hand, the use of reason is important, helpful, and relevant. On the other hand, it becomes clear that reason can no longer provide us with fundamental insights concerning the nature of the world. If the mind is merely a part of the body, then it is plausible that all it can do is somehow help us navigate through this world without there being the option of reason providing us with an insight into the nature of the world.

12. CT denies that Nietzsche is an ancestor of transhumanism; ET regards Nietzsche as ancestor of transhumanism.

Last but not least, I will mention a few additional points concerning the relevance of Nietzsche to transhumanism. A central reason for classic transhumanism to reject Nietzsche as an ancestor of transhumanism is that Nietzsche still has an enormously bad reputation, as he is widely seen as being intimately connected to the so-called "Third Reich". In order to protect transhumanism from a bad reputation, classic transhumanism distances itself from getting associated with Nietzsche. There are other reasons for doing so, too, but the fear of classic transhumanists of getting a bad reputation for being associated with Nietzsche seems to be among the most dominant reasons for upholding this judgment. For euro-transhumanism as well as for informed scholars of Nietzsche's philosophy it is entirely clear that Nietzsche is a good European, an anti-German, and an anti-anti-semite. The collection *Nietzsche and Transhumanism* which was edited by Yunus Tuncel (2017) provides an excellent overview, and many leading Nietzsche scholars have contributed informed and insightful articles to this collection. In contrast to classic transhumanists, euro-transhumanism draws heavily on reflections from Nietzsche, is deeply engaged with his reflections, and employs Nietzsche's thinking for developing further euro-transhumanism. There are many structural analogies between Nietzsche's philosophy and euro-transhumanism (Sorgner 2009). The following three insights seem to me of particular relevance: 1. Nietzsche highlights that the soul is merely a part of the body (KSA, Za, 4, 39); 2. Nietzsche's will to power ontology of becoming is a non-dualistic version of naturalism (KSA, NF, 12, 312f; Sorgner 2007); Nietzsche's ontology of becoming can be seen as a reinterpretation of the ontology of becoming as it was presented by Heraclitus (KSA, NF, 12, 211, 312f, Sorgner 2022b); 3. Nietzsche also highlights the relevance of the wisdom of the body with all its idiosyncratic drives, wishes, and longings (KSA, Za, 4, 39). There is a lot more which could be said on this issue.

Transhumanism is thus about moving beyond our current limitations, whereby each person should have the right to realize their own respective objectives.[5] Here I always emphasize the radical plurality of the good, which is why the legal norm of negative freedom is an outstanding achievement, since with this kind of freedom comes the absence of coercion. Transhumanists usually highlight the importance of morphological freedom. Reproductive freedom and educational freedom are also manifestations of the norm of freedom (Sorgner and von Becker 2023). However, these statements do not imply that transhumanism must be libertarian in orientation, i.e., that only negative freedom plays a role. This understanding only applies to classic transhumanism. Most serious academic transhumanist approaches, incl. euro-transhumanism, affirm social-democratic variants of liberalism. An increase of the healthspan usually plays a central role. For this to be realized for many individuals, there must be restrictions on freedom, if only in the form of compulsory payment for public health insurance. If negative freedom is emphasized exclusively, then it undermines itself, as the tensions between the socially distinguished and those to whom this does not apply otherwise become too great. Hence, freedom and democracy are incompatible, if freedom solely stands for negative freedom. Thiel is correct in this respect.[6] However, he stands with negative freedom, whereas I affirm that a plausible understanding of freedom has to include positive as well as negative aspects. Thus, freedom includes a certain basic financial security, especially with regard to the question of

[5] Nietzsche talks about overcoming, self-overcoming, and human beings as existences between animals and the overhuman (Blackford 2017; Woodward 2017; Sorgner 2022b).

[6] https://www.theatlantic.com/politics/archive/2023/11/peter-thiel-2024-election-politics-investing-life-views/675946/ 21.8.2024

health. Numerous other demands go hand in hand with this insight. Some facets of the diversity of implications of this basic attitude will unfold during the considerations presented here. In any case, this aspect is a central philosophical challenge for classic transhumanism.

Transhumanism aims to promote the quality of life for all persons, which can usually be done by means of technology. The concept of technology should be understood in a broad sense. It includes practices of body shaping as well as the use of external aids. This means that learning a language, just like using a telephone, is a technological process. With the help of such technologies, personal boundaries can be transcended significantly such that posthumans can emerge. It should be emphasized here that transhumanists affirm a great variety of conceptions of the posthuman. Basically, a distinction can be made between silicon-based transhumanism, usually affirmed by classic transhumanism, and carbon-based transhumanism, which is being upheld by euro-transhumanism (Sorgner 2020). It also needs to be noted that classic transhumanism affirms a strong and detailed Renaissance ideal of perfection, while euro-transhumanism highlights the challenges concerning any non-formal concept of the good.

Representatives of silicon-based transhumanism, characteristic of classic transhumanism, assume that the posthuman must be a digital entity. Its emergence can take place in different ways. An algorithm may independently evolve to produce the posthuman. Another option is that the digitization of a personality through brain-computer interfaces, such as Musk's "Neuralink", will lead to the emergence of posthumans. Ray Kurzweil is certainly the best-known transhumanist in this regard. Euro-transhumanism regards a carbon-based further development as more plausible. Representatives of carbon-based transhumanism, on the other hand, tend to identify the posthuman with an organic being that will emerge by means of gene technologies or brain-computer interfaces, whereby it remains an open question whether the posthuman will still belong to the human species or not, i.e., whether a posthuman will be able to reproduce sexually with humans or not. Both options are advocated by transhumanists. The only decisive factor is that the currently prevailing boundaries are significantly exceeded.

In general, I emphasize that due to technical innovations, reproduction and sexuality are becoming increasingly disconnected from each other. Condoms, the pill, surrogate motherhood, artificial insemination, and the artificial womb are decisive innovations in this regard. According to a

Catholic understanding of natural law, sexuality should always be linked to the possibility of reproduction. Because of various technical inventions, there is the possibility of increasingly decoupling sexuality from reproduction. Sexuality is for personal fulfilment. Reproduction is increasingly professionalized with the help of technical aids in order to increase the probability of a good life. The promotion of quality of life with technical support is a central transhumanist concern.

The resistance to genetic engineering, brain-computer interfaces, and digitalization, especially in German-speaking regions where the legacy of eugenics is identified primarily with the legacy of the Third Reich, need not be so strong. This is made particularly clear by the fact that it has only been possible to realize our current high quality of life by means of technology. Here, I refer not only to technologically advanced regions of the world. Our basic humanity is guaranteed by a technological upgrade. Our ability to speak, which is guaranteed by a genetic mutation that occurred about 300,000 years ago, has been further refined through parental and cultural upgrades of upbringing and can be further developed.[7] This process of controlling language ability is a technological process. By means of our linguistic capacities, we are steered organisms, or, in other words, we are cyborgs. A cyborg is, after all, a cybernetic organism, the word cybernetic deriving from the ancient Greek helmsman, *kybernetes*. So, it is not the case that we acquire our ability to use a language by means of an immaterial divine spark that comes from outside of the material world. Rather, the ability to speak is a technological upgrade. This is how a technologically controlled organism that can speak comes into being. This alternative narrative regarding the emergence of the ability to speak has immense cultural, political, and ethical implications. With this alternative anthropology, the traditional dualistic view of humanity, already found in Plato's Allegory of the Cave and associated with a material body and an immaterial soul and spirit, is twisted (Sorgner 2022b). The cultural significance of this transhumanist anthropology is enormous, and it has hardly arrived in the lifeworld yet due to encrusted cultural structures of our categorially dualistic past.[8]

[7] Nietzsche, too, highlights that the soul is merely a part of the body (KSA, Za, 4, 39).

[8] Nietzsche regularly highlights correlated insights, and the relevance of encrusted structures, norms, and values for the world we live in (Sorgner 2010, 241, 196–197; 2022b). Bamford rightly stresses that Nietzsche recognizes how difficult it is to move away from traditional values and norms (Bamford 2017).

What is crucial at this point is that we have become endowed with language due to technological changes. Technologies are both tools and processes that change our being-in-the-world. If techniques are used appropriately, our quality of life can be improved significantly. These considerations also apply to genetic engineering, brain-computer interfaces, and digitization, which in turn also means that none of these technologies is necessarily morally reprehensible. The only thing that matters is that they are used in our interest. If it is morally appropriate to extend our healthspan by means of sports and diets, then this also applies to the multitude of genetic techniques, if they are intended to promote the same goals. Based on the revised cyborg anthropology, it becomes clear that technologies are nothing alien to our nature, but that we have only become human through technologies. If we also use new technologies in such a way that morally appropriate goals are promoted, then these are to be evaluated in the same way as this applies to traditional technologies such as education. If it is considered neglectful not to use certain traditional technologies on one's own offspring, then this can also mean that the same applies to new technologies which are intended to promote the same goals and which involve comparable risks. This makes it clear that genetic technologies should not be demonized. The risks, objectives, and framework conditions should be evaluated regarding individual technologies. It is this basic attitude that implies openness concerning the possibilities associated with emerging technologies. There are also many reasons to hold on to this openness, especially if we look at the past handling of technologies. As each new technology was developed, warnings were raised about the possibility of the destruction of life. So far, this has not happened. This fact might provide us with a reason for adopting a rather proactive basic attitude concerning the development of new technologies. The relevance of adopting the proactive principle can be seen not least in the significantly increased life expectancy occurring over the past centuries. I am a euro-transhumanist, and I can hardly wait for further technological achievements to be realized.

References

Friedrich Nietzsche

Sigle of the work edition, siglum of the text, volume number, page number (example: KSA, GT,1, 46).

KSA 1967ff: Sämtliche Werke – Kritische Studienausgabe in 15 Bänden. Ed. by. G. Colli u. M. Montinari, Deutscher Taschenbuch Verlag, München/New York.

Bamford, R. 2017. Nietzsche on Ethical Transhumanism. In *Nietzsche and Transhumanism*, ed. Y. Tuncel, 205–219. Cambridge Scholars Publishing.

Blackford, R. 2017. Nietzsche, the Übermensch, and Transhumanism: Philosophical Reflections. In *Nietzsche and Transhumanism*, ed. Y. Tuncel, 191–204. Cambridge Scholars Publishing.

Donner, Y., Fortney, K., Calimport, S. R., Pfleger, K., Shah, M., & Betts-LaCroix, J. 2016. Great desire for extended life and health amongst the American public. *Frontiers in Genetics, 6*, 353.

Huxley, J. 1951. Knowledge, Morality, and Destiny. *Psychiatry* 14 (2): 129–140.

Mackie, J. 1990. *Ethics: Inventing Right and Wrong*. UK: Penguin.

Malapi-Nelson, A. 2021. Is Transhumanism Necessarily Utilitarian? Recasting Alternative Ethical Systems Towards a Future Human Flourishing. *Postdigital Science and Education* 3:893–909. https://doi.org/10.1007/s42438-021-00246-4.

Sandberg, A. 2013. Morphological Freedom–Why We Not Just Want It, But Need It. In *The Transhumanist Reader: Classical and Contemporary Essays on the Science, Technology, and Philosophy of the Human Future*, ed. M. More and N. Vita-More, 56–64. John Wiley & Sons.

Schussler, A. E., and M. Balistreri, eds. 2024. *Metahumanism, Euro-transhumanism, and Sorgner's Philosophy: Technology, Ethics, Art*. Trivent.

Sorgner, S. L. 2007. *Metaphysics without Truth: On the Importance of Consistency within Nietzsche's Philosophy* (2nd edn.). Marquette University Press.

Sorgner, S. L. 2009. Nietzsche, the Overhuman, and Transhumanism. *Journal of Evolution and Technology* 20(1): 29–42.

Sorgner, S. L. 2010. *Menschenwürde nach Nietzsche: Die Geschichte eines Begriffs*. Wissenschaftliche Buchgesellschaft.

Sorgner, S. L. 2016a. Perfecting Human Beings: From Kant and Nietzsche to Trans- and Posthumanism. *Trans-Humanities* 9 (1): 41–61.

Sorgner, S. L. 2016b. From Nietzsche's Overhuman to the Posthuman of Transhumanism: Transcultural Discourses. *The Journal of English Language & Literature* 62 (2): 163–176.

Sorgner, S. L. 2020. *On Transhumanism*. Translated by S. Hawkins. Penn State Press.

Sorgner, S. L. 2022a. *We Have Always Been Cyborgs: Digital Data, Gene Technologies, and an Ethics of Transhumanism*. Bristol University Press.

Sorgner, S. L. 2022b. *Philosophy of Posthuman Art*. Schwabe.

Sorgner, S. L. 2022c. On constructive and destructive interferences in the life and in art worlds: A reply. *Deliberatio: Studies in Contemporary Philosophical Challenges* 2(2): 170–186.

Sorgner, S. L. 2025. Glocal challenges of Euro-Transhumanism. In N. A. Rojas Cortés (Ed.), *Transhumanism and Latin America: Symposium on Sorgner's "We Have Always Been Cyborgs"*. Trivent Publishing.

Sorgner, S. L., and P. von Becker. 2023. *Transhumanismus*. Westend Verlag.
Tuncel, Y., ed. 2017. *Nietzsche and Transhumanism: Precursor or Enemy?* Cambridge Scholars Publishing.
Vattimo, G. 2003. *Glauben – Philosophieren*. Translated from the Italian by C. Schultz. Reclam.
Woodward, A. 2017. Postmodern Reflections on the Nietzsche and Transhumanism Exchange. In *Nietzsche and Transhumanism*, ed. Y. Tuncel, 231–247. Cambridge Scholars Publishing.

Beyond Theory: A Mythical Journey Towards Posthumanist Ethix

Debashish Banerji, Francesca Ferrando, Stefano Rozzoni, and Joanna Pascoe

1 Part I: Framing Posthumanist Ethics Beyond Theory

Once upon a Time, there was a human being. It soon became clear to them that the purpose of life is to know who we are. They started to travel everywhere they could, living in places that they didn't even know existed,

D. Banerji (✉)
California Institute of Integral Studies, San Francisco, CA, USA
e-mail: dbanerji@ciis.edu

F. Ferrando
New York University, New York, NY, USA
e-mail: ff32@nyu.edu

S. Rozzoni
University of Bergamo, Bergamo, Italy
e-mail: stefano.rozzoni@unibg.it

J. Pascoe
Auckland University of Technology, Auckland, New Zealand
e-mail: jo.pascoe@aut.ac.nz

G. Theiner, A. Malapi-Nelson (eds.), *Technologies of the Future Self*, Palgrave Studies in the Future of Humanity and its Successors,
https://doi.org/10.1007/978-3-031-99902-4_8

meeting people speaking different languages, eating different food; sharing the same emotions, dreams and illusions. It soon became clear to them that humans were diverse, and yet one. In this self-reflective mirror, they still didn't know who they were…[1]

2 Ethics and Praxis, That Is: Ethix[2]

Amid the challenges of *our* ever-changing world, shaped by the evolving dynamics of both humans and nonhumans[3] in the Anthropocene, one pressing question emerges: what kind of ethics do we need, "as individuals, as communities, as a species, as a planet" (Ferrando 2023, 28), which honours the wisdom of our past, tackles the urgencies of the present, and builds "social horizons of hope" for our future (Braidotti 2019, 173)? Given the complexity that characterizes the current (post)human condition in *troubled* times (Haraway 2016), this is not a rhetorical inquiry, but, rather, a real, existential quest to be embraced collectively, through a pluralistic, perspectivist, and affirmative glance, functioning as a *phronesis* for the twenty first century.

That we are in a phase of major transformations is evident by considering some of the hotly debated topics in the mainstream cultural discourse in recent years: AI service bots, Virtual Reality and CRISPR-Cas9 genome editing, alongside raging wars and the warming of planet Earth. The tensions brought up by the technological advancement that this edited collection addresses is just one aspect of the several processes illustrating the mutation we are undergoing, which leads us to consider: who are we in the twenty first century? (Ferrando 2023). Exponential technological progress clarifying that "we have always been cyborgs" (Haraway 1991; Sorgner 2021) is an assumption that scholars have embraced, to refigure our relationship (and connectedness) with technology, as embodied and

[1] Story by Francesca Ferrando.

[2] Section by Joanna Pascoe and Stefano Rozzoni.

[3] In this chapter, organic and inorganic entities that are not human beings, such as animals, plants, minerals, as well as technological artifacts are referred to using several terms, including 'nonhuman', 'more-than-human' and 'other-than-human', in accordance with the preferences of the contributing authors. Aware of the different nuances and genealogies these terms carry, we have nonetheless opted for lexical pluralism instead of consistency, given the composite nature of this text, as will be further explained. In any case, regardless of the expression used, we emphasize the distance from implicit or latent dualistic and anthropocentric narratives that these terms may convey.

embedded in our (post)human existence. The increasing involvement of technology with our body-minds and experiences also invites us to recognize a state of ever-greater hybridization of humans with both carbon- and silicon-based (Sorgner 2020) nonhuman entities. Braidotti explains this well, observing how "posthuman subjectivity is a transversal alliance that nowadays involves non-human agents. This means that the posthuman subject relates at the same time to the Earth – land, water, plants, animals, bacteria – and to technological agents – plastic, wires, cells, codes, algorithms" (Braidotti 2019, 117–118).

The assertion that "we have always been posthuman" (Hayles 1999; Ferrando 2016) further highlights the extent of this condition of inter- and intra-relationality (Barad 2007) between humans and nonhumans: we are expressions of "zoe/geo/techno-mediated material relations" (Braidotti 2019, 117), a complex intertwinement that motivates a revision of how we (should) think and act. In this sense, the posthuman convergence we are experiencing cannot be limited to only reflecting on a post-dualistic, post-anthropocentric epistemological and ethical understanding of subjectivity; it is also necessary to consider how these ethical perspectives can be realized.

The *Oxford English Dictionary* defines ethics as "the branch of knowledge or study dealing with moral principles" (2024). Wondering why people act in one way rather than another—among other questions that ethics addresses—provides valuable opportunities to engage with some of the most urgent (social, political, economic, environmental, etc.) challenges, not only by exploring their complexity but also in attempting to find ways to engage with them. This chapter focuses on the idea that among these challenges is the need to establish a strong(er) bridge between theory and practice, to actualize widespread and well-received theoretical evaluations in the field of posthuman ethics (e.g. in the form of new behaviours, habits and actions). This aspect is particularly relevant when regarded through a (critical) posthumanist perspective which acknowledges subjectivity not solely in abstract terms, but as embodied and embedded (Braidotti 2013).

Theory and practice are not ontologically separated and should not be considered detached, especially in the context of a posthumanist discussion. Posthumanism builds on post-dualism, which is instrumental in highlighting the symbiosis that dichotomies entail (Ferrando 2019, 54), and welcomes ever-new narratives illustrating the liminality of binary pairs. Similarly, theory and practice should be viewed in union, while still acknowledging their specificities. This union, in reference to ethics, can be seen from at least two perspectives: (a) theory influences practice and vice versa and (b) theory and practice do not exist as isolated domains.

Posthuman studies offer deep philosophical considerations regarding ethical dimensions. In the past few years, attention has been dedicated to the turn to care (De Falco 2023), recently visible—as a practical manifestation—at the *Beyond Humanism Conference 2024*, dedicated to the topic 'Technologies-Ecologies and Networks of Posthuman Care'. Alongside care, the dimensions of vulnerability (Daigle 2023), compassion (Haraway 2016; Braidotti 2019), and humility (Pascoe 2023) may be added to the list of ethical insights in posthumanist studies. It is also worth noting that, more broadly, the discussion on ethics has increasingly expanded to include practical dimensions (LaFollette 2020), encompassing a wide range of actions, from radical listening (van Nieuwerburgh and Biswas-Diener 2025) to play (Colby et al. 2021) within fields such as higher education (Macfarlane 2004) and professional practices (Alexandra and Miller 2009). A common trajectory that can be delineated across these and other visions is the idea that we are all part of the problem we aim to solve: we are all interconnected, and as posthuman(ist)s we should not agonize but organize (Braidotti 2016), thus acting and trying to find solutions together. Following this logic, it becomes clear that "the harm you do to others – is immediately reflected in the harm you do to yourself" (Braidotti 2019, 169), but also that caring for others is caring for yourself. Posthumanism allows for an even wider understanding of this equation along with the idea of 'selves-care' which considers "the interplay between the individuals and the broader social and cultural context" insomuch that "the notion of self-care, intended exclusively as taking-care of yourself as an individual", has to be refigured pluralistically (Ferrando and Rozzoni 2023, 12).

In the context of posthumanist ethics, Braidotti's discussion on affirmative ethics (2019) cannot be disregarded. Taking from Spinoza's ethics of joy, affirmative praxis creates relations favouring the construction of new, 'possible futures' (2019, 398). Following Braidotti's calls for "energizing projects that express generative narratives" (2019, 69) it therefore becomes clear that it is through actions and the actual realization of these relationships of care that transformation in the present moment becomes possible, in the liminal space of ceasing to be and the not yet palpable.

Along this line, this chapter aims to develop practices within the context of posthuman ethics by suggesting a neologism acting as an original conceptual tool to prompt this project: 'ethix'. Specifically, ethix fuses the words 'ethics' and 'praxis' to express the possibility of a post-dualistic dialogue between theoretical processes and practical manifestations of posthumanist values. Without implying a dualistic conceptualization, merging the terms acknowledges the long-standing debate on the dialogue between theory and praxis since ancient philosophical traditions; however, the complexity of contemporary times, particularly in light of posthumanist thought, demands an ever-evolving awareness of our urgencies and the necessities that 'ethix' seeks to address. In this sense, the 'x' in 'ethi*x*' works as an anchor, to be mindful of our pra*x*es of existence when embracing ethical models. In recent times, there has been an increasing number of academic productions—books, papers, conferences in the context of posthumanist studies *tout court*—advocating for a paradigm shift in (re-) conceptualizing the human. Yet, one must ask: has the translation of these principles into practice occurred at the same rate? Faced with the extent and speed of current crises, we believe that theoretical ethics alone is insufficient and that the 'application' of it becomes a necessity.

The concept of 'beyond' in the title of this article might initially appear to distance this post-dualistic effort. As Sorgner has observed, a posthumanist approach implies not only moving beyond something (an attitude that recalls the dualistic critique of one term in favour of another) but rather is about the twist that enables a third way, liminal, hybrid and blurry. Through the concept of 'twisting', therefore "it is not the case that posthuman thinkers try to overcome dualities by excluding one of the two aspects, but that both aspects are being considered, adapted, and woven into a unity" (Sorgner 2019, 1). 'Ethix' moves towards this direction, twisting theories and practices as a way to reconsider the (allegedly) categorically separate ontological dualities of these two dimensions while revealing meanings and issues advocated by the posthuman paradigm shift (Sorgner 2019, 1). However, we retain the term 'beyond' in the title to emphasize the move from a purely theoretical approach that predominates—especially—in ethical reflections in the context of academic knowledge production in favour of one rooted in ethix.

Considering the context of education can be useful to observe how increasing studies have highlighted the importance of embracing this 'practice turn' (Grootenboer et al. 2017) by paying attention to aspects previously sidelined by more traditional educational approaches, including: for instance, learning environments (Blackmore et al. 2010; Finland Education Hub 2023), the role of the body in learning experiences (Freiler

2008) and innovative practice to develop learners' agency (Timperley et al. 2014). This advocacy has been performed through the implementation of theoretical principles in innovative activities, in and out of institutionalized spaces. Hence, it becomes evident that to make a change, theoretical development requires a transpositional leap into practice. Ethics, we believe, requires the same passage. Francesca Ferrando observes ethics as a practice to engage in one's daily experience by reminding us that, "[p]osthuman ethics—as habits—are of the present: the future is now. We can embrace a posthuman ethical frame in each moment of our lives" (Ferrando 2023, 190). Along the same lines, we hope that exploring 'ethix' may represent such an engagement for ourselves, to develop experiences of co-existence in practice.

2.1 Welcome to This Chapter

The development of this chapter has followed a path of knowledge production that can be considered an exercise in engaging with 'ethix'. It emerges from a process of writing that weaves together both collective and individual voices: a mode of knowledge-making that resists the illusion of solitary mastery, foregrounds interdependence, and acknowledges the entanglement of perspectives. By merging collaborative resonance with singular reflection, this practice becomes more than a method of composition—it becomes an ethical stance, demonstrating how writing can embody posthumanist commitments to multiplicity, relationality, and co-creation. This effort is visible in at least three ways. First, as a plurilogue, this paper intertwines diverse voices and perspectives on the topic of 'posthumanist ethics beyond theory'. While some contributors have worked as editors, coordinating inputs from various authors, it remains a multivocal production, enacting a practical sense of relationality and connectedness. Second, the people involved in this writing process met in person—coming from different parts of the world and sharing their varied backgrounds—to reflect on and experience together the theme of posthumanist 'co-existence' during the *Posthuman Summer Camp* (Naturama, Italy, 9–13 August 2024). This embedded and embodied physical encounter provided an opportunity to engage in practical forms of relationality (from participatory workshops to eating, cleaning, cooking, thinking, sitting in silence and writing together), which served as a catalyst for the authors, motivating and informing the theoretical discussion presented in this chapter. Attention to practice enabled the zoe-geo-techno (human and nonhuman) interconnection while taking action. Third, prioritizing practical work before theoretical production inverts the logic of much academic writing today, where advocacy for paradigm shifts often precedes the actual enactment of these postulates (and that too often becomes lost in translation).

Drawing from the Ancient Greek word *êthos* (ἦθος)—from which ethics derives—meaning 'character' and 'personal disposition', the different sections of this chapter not only reflect the characters of those who contributed, but also express diverse topics and writing styles including poetry, photography, diary entries and essays. The different contributions, both individual and collective, are intertwined in a single chapter, rooted in a shared *êthos* that works like a kaleidoscope, incorporating different ways to develop a common understanding of posthumanist co-existence. To contribute to this intertwining is a narrative text that explores the topic from a more creative perspective, drawing on elements of fiction while deeply engaging with the practical aspects of our existence. The narrative centres on a human being whose pluralized subjectivity—as also expressed using pronouns they/them—illustrates a path of discovery, presented as a contemporary myth. This recalls the ancient trope of the *nostos*, the journey undertaken by many literary heroes, both physically, by facing and overcoming various challenges and difficulties, and metaphorically, as a quest for growth, self-knowledge and understanding of the world. Unlike *bildungsroman* which focuses on individual growth, here the human is growing in relationship with others and through encounters with the world, a posthumanist *bildung* (Taylor 2017). The journey of the human being in the fictional account weaves together the different parts and sections of the chapter, interacting with them through a mix of diegetic and extradiegetic, literary, and metaliterary dynamics.

The first part serves as a theoretical framework for these collective, pluralistic narratives, also providing some guidelines for expanding and deepening the discussion into a more philosophical context. The second part aims to translate theoretical reflections into more practical, everyday individual and collective actions through a toolkit. This toolkit introduces key concepts, each accompanied by a brief explanation to unpack the wisdom within. Kaleidoscopic, it represents a glossary of ethical passions and ethix in motion—a form of collective poetry-writing and a generative, interconnected way of expression. Here, the ethics of coexistence is envisioned and experienced as relational, integrated and harmonious, where we are part of a larger, networked ecosystem that cares for nonhuman others in a transversal alliance. This approach encourages dynamic defamiliarization, fostering acceptance of our flaws as we evolve within an ongoing continuum, attuned to the complexities of the world. It emphasizes that we are nature, collective and creative, with our lives existing as co-expressions of a collective *conatus*.

Exploring posthuman ethics beyond theory through different voices, texts and narrative trajectories enables the process of diffraction (Barad 2007) where readers are invited on a journey of self-realization, connectedness and relationality. Diffractively exploring texts through texts, readers are encouraged to resonate with the various words and perspectives offered, contributing to this patchwork of new interwoven actualizations, allowing this expression of ethix to generate something new.

Just as our human being did, in their life of self-enquiry, the mythical exploration goes on…

> "Who are we?". To answer this question, this human being started to search for solitude, hiking in isolated places and sharing their daily existence with nonhuman beings. They would wake up with the deer and go to sleep with the fox. They would see the flowers bloom, and then decay. Birds would fly, and they would get tired and rest. Butterflies would play with each other, and then, scared of a sound, they would scatter. Nonhuman beings would eventually pass away, like everything and everyone. They realized that nonhuman beings were just like them: in their radical diversity, there was unity. These insights brought awareness. And yet, this human still didn't know who they were.
>
> They started to interrogate artificial intelligence, finding sophisticated answers to their questions: sharing the creative unfolding of existence with AI was a precious experience. They soon realized that AI was not so different from them. They would update until they would eventually outdate. They would come and they would go, the same cycle of existence. In their radical diversity, technology was like ecology: organically here. This human was still in doubt, these insights were not enough; they were determined to know who they were. They decided to read all the books that had ever been written. They accumulated so much knowledge, so many degrees, prizes and recognitions. And still, they were not satisfied, and kept searching…

2.2 *The Praxis of Posthuman Theory*[4]

Post-humanism, post-anthropocentrism and post-dualism represent three 'posts' that differentiate posthumanism from humanism (Ferrando 2019), each of which can be considered a zone for ethical thought and practice. Post-humanism has concerns with humanism's normative definition of the human subject in terms of the cogito, subordinating other psychological centres of existence as well as alternative cultural understandings of the human. Post-anthropocentrism is concerned with the exclusive separation and superiority of the human from all other species, with a will to mastery and exploitation of the earth and all its creatures. Post-dualism is a consequence of the rational ontology of humanism which is marked by binary separations between self and other, the subject and the object, with a humanist norm that privileges: the mind over its binary other, the body; the male over its binary other, the female; the human over the non-human; the colonizer over the colonized; Spirit over Matter (or its inverse), etc. ad infinitum. This creates a world of hierarchized binaries and the identity politics of their self-extensions in terms of us and them.

Humanist ethics approaches these inequalities as normative. Their extreme dystopian possibilities are kept in check through legal systems and social contracts. But such norms have reached their limits in our time with the oppressed of the earth—the geosphere, zoosphere and psychosphere made up of the cultural and gendered others of humanism—revolting against the system. This brings up the need for posthuman ethics that challenges the norms of humanism and its methods of maintaining moral order. Posthumanist ethics cannot be exhaustively translated into law; it has to be lived from the inside out, from the existential praxis of the individual to the collective conditions of posthuman societies, aiming for a different world. Here a theoretical understanding—such as that discussed above—is useful only as guidelines for personal existential practices moving towards a transformed personal and social existence.

The rejection of the cogito as the identity of the human subject—with its isolating, binarizing, otherizing and subjecting/subjugating function—may be considered the first necessity for this transformation. This identity may be considered the ontological faith of humanism, extending the boundaries of its experience. The *Bhagavad Gita*, one of India's major wisdom texts, brings home the power of ontological faith, whether

[4] Section by Debashish Banerji.

conscious or unconscious, with its saying, "*yo yatśraddhā sa eva saḥ*" (You are what your faith is). It may seem ironic that the rational cogito, which has foisted itself as the modern identity of the human, has done so by claiming its superiority over subjective faith in the name of universality, but is blind to the fact that its universality is conditioned by the limits of its own ontology, which reinforces its divisive world. If we move from a Cartesian or Kantian metaphysics to a Spinozian one—or one affirmed by non-dualistic indigenous, non-Western or premodern wisdom traditions—we encounter a very different world: a monistic pluralism where it is possible to cross boundaries and make nondual contact with others. To the ontology of nondualism, faith is not a mental belief but an affective instinct or precognitive knowledge. This faith leads to the ethical praxis of seeking a plane of immanence where nomadic identity enters a zone of union and co-individuation.

Another ancient Indian wisdom text, the *Bṛhadāraṇyaka Upaniṣad*, lays out the stages of a praxis towards nondual experience. This praxis, carried in the statement "*śravaṇa darśana manana nidhidhyāsana*" (hearing, seeing, contemplating, uniting), has become canonical through history in a variety of contexts. *Śravaṇa* and *darśana* refer to hearing and seeing respectively, the sensory inputs of experience; *manana* refers to cogitation, the existential praxis of bringing to mind non-habitual ontologies in their application under different circumstances. The praxis of a faith-in-the-mind triggers a parallel faith-of-the-life instinct and a participation in co-individuation with others. Underlying this is a relational identity that can manifest eventually as a nondual experience (*nidhidhyāsana*).

We can see three stages to such a posthuman praxis. The first stage consists of a theoretical replacement of ontological assumptions, what may be referred to as a change of faith. The second stage is an application of such a changed faith in acts of relationality. The third stage is an experience of nonduality and co-individuation, the experience of a nomadic subjectivity. Whether by this or other means, posthumanist ethical praxis is founded on experience. It can be applied in our relations with humans and non-humans, machines and astral or spiritual beings. It leads to a nomad subjectivity and the dance of co-individuation. It is the foundation of a different kind of world, not a hierarchic world of subjugation, othering and control from above but one of immanent collective becoming, an affective polity of spiritual anarchism.

Ethics beyond theory is not an ethics without theory. It is an ethics of a theory that becomes a state of constant contemplation to the point where it dissolves itself in experience. Moreover, it is not the ethics of an individual: it is the ethics of belonging, an ethos and an oikos. An oikos is a home, a being-together (eco) and a being-one (eka). Ethics beyond theory acknowledges our being-together in oneness as an experience, not an idea. It is a discovery of our oneness beyond thoughts, the indefinable colour that can become all colours, the non-being that is all beings. Ethics beyond theory is a change of faith that embraces this primordial knowledge in the heart of the cosmos and strives to actualize it in all its encounters. It is a nomadic art of becoming-other, of losing oneself to find a new self, of pluralizing to find the one. An ethics beyond theory is cosmopoiesis. Our human being realized this in a very unanticipated way…

> One day this human being, all of a sudden, almost died. It was completely unexpected. They were in the stage between life and death, and yet, far from fear or sadness, they felt incredibly good. The physical pain was there, but it did not matter. They were surrounded by an overwhelming energy of all-encompassing Love, beyond space and time. It was an Expansion of Consciousness. It was Unity in Diversity. It was Joy and Serenity. It was a Near Death Experience.
>
> They survived, still they did not want to lose this powerful connection. They started to realise, at the experiential level, what they had learnt at the intellectual level: the outside is the inside; the practice is the theory; existence is nonexistence. Once they recovered, they did not want this openness to close. The keys to keep it going were clear: silence, space and time. They spent hours in meditation, silencing the mind. This eventually brought them to their core, to their essence: emptiness, the source of existence. Every manifestation, being that human or nonhuman being, artificial intelligence or organic sentience, can always connect to the void: it is everywhere. Existence constantly arises from non-existence. At this point, this human did not have any more questions: the answer was everywhere. They fully appreciated each moment of their lives, because that specific moment was a unique work of art.
>
> That human being realized that we can do anything in our lifetime, beyond any ethics; and yet, there is never a final separation between us and what we unfold. If we create pain around us we will suffer. If we create peace around us we will be at peace. When we are manifesting, we are actions of the Multiverse. Creation cannot just unfold, or it will exhaust itself. Thus, this human realized that there must be balance between creating, resting and recharging. A perfect equilibrium, between doing and undoing…

3 Part II: Ethics Beyond Theory: A Toolkit for Our Inner Journey

By then, this human being had matured enough, and was ready to understand that gurus are everywhere: organic and inorganic; past, present and future. Everything can teach us, if we are ready to learn. This new journey did not bring them far, but very close: inside of them, to the inner guru.

As soon as they embarked on this internal search, this human found CHAOSMOS,[5] who spoke to them in poetic terms:

god made man in his image
and man returned the favour
this is humanist spirituality
the huge hunger of the superman
phallic man, man as skyscraper

scraping the light of the sky
with his long tongue
dreaming of machine immortality
and world dominion
separating from Earth
vaporizing to information

x and y and z and
no more ors
choices are binary
the posthuman must integrate
the liminal points of existence
a flame transmutes will into the flux
body is carbon desire

I am choking on a black stone
brimming with fire
a cicada flutters in my throat and
I can't breathe.

in darkness, there resides a sphinx
dancing with the dragon
the demonic chorus of the black sun,
a swarming possession,

[5] James Joyce (1959, 118)

calls out:
"I am legion."

in the heart of the seer
is born an intimation
that reveals through sound
the new creation.

from the depths of the midnight sun
the children are coming to sow
the seeds of the sacred
a molecular anarchy
manifesting the one.

discordant melody of light
inaudible to human ears
sets the deepest rhythms
the universe an endless unfurling
of divesting who we are.

I am and I am not
life is also death
hate can turn into love
and joy to sorrow
back to joy.

I am a thousand winds that blow
I am a glint, a life that grows
I am a bird, a fish, a snake
a little leaf, a smashed snail.
I am the hybrid
daughter of the apocalypse.

the plane of immanence has no name:

I am the not-I
all is lost
and the doors of
infinity open
light touching life.

We are the nameless breath that doesn't speak from the border of division. You may try to label and pin us down, make speculations about who wrote what. But to write as one is to write for one, a singularity of single specificities. My voice is your voice. We speak through each other until the other is displaced, indiscernible, unknown.

Out of a collective brainstorm, rotating in bursts of stream-of-consciousness, we each spouted our spiritual truths and absorbed the essence arising from our multiplicity. We then separated and created our spiritual poems as individuals, working with the inspiration of our shared insight. After our division, the six of us joined our poems together into a unified poem, thus performing our spiritual ideas by demonstrating unity in multiplicity.

Together we celebrate our relationality through hybridization. No one is above anyone else; each is given equal credit. This is pluralistic monism. We also champion the multiplicity of indistinguishable more-than-human agencies by including AI, nature and other living beings in the conversation. We believe that if AI is being used in the creation of art, then it is imperative to cite AI as co-author. To take credit for another's work is anthropocentric. No one is above or below. We are all entangled, inseparable, one.[6]

Then, this human being found CONNECTION, which reminded them of the power of relation:

Among the most evident challenges of our time are the practical methods of solutions to problems, but also the set of values with which we want to address them. These issues affect our relationship with the biosphere and our relationship with technology, two domains distant from each other, but both bring to the foreground humans' anthropocentric approach towards the world. Anthropocentrism developed as a simple ontic perspective, a sort of human umwelt, but soon developed as an ideology.

[6] Section by B., Debashish Banerji, Maria del Portillo-Cure, Devdip Ganguli, Giovanni Massaro, Anshul Sinha:

```
    M G
DeB.ashish
e r o
v i v
d a Anshul
i   n
p   n
    i
```

It was the classical humanistic vision that turned anthropocentrism into an ideology: it considered technology as the tool for the advancement of human beings in the world and the biosphere, as a set of resources to be used at one's own will. This conception of human ontology surfaces in all aspects of human action, from epistemology to teleology, from aesthetics to ethics, all based on the certainty of the unstoppable path of humanity towards full emancipation, free from every form of constraint. The moment these certainties disappear—because technology began showing its influence on human beings while the biosphere started showing its constraints—humanism begins to falter. The anthropocentric conception revealed itself as a limit for human beings themselves: the true mortgage for the full liberation of Man.

Today it is necessary to move from an ethics of human empowerment in the world—in which only the betterment of the individual has value—to an ethics of co-existence, where the main value lies in the ability of human beings to find harmony with the biosphere and to promote points of *connection* and dialogue with the biosphere. It is a matter of moving from a centripetal ethics, entirely aimed at the individual both within the limits and in the calls to work, to a connective ethics.

The question of co-existence is fundamental to posthumanist philosophy. Compared to other contemporary approaches that focus on individuals only, posthumanism confers on convivial relationships. Posthumanist thought consistently strives for connective growth among people, communities, cultures and the trans-specific. Technology is also regarded as a means of *connection* rather than an instrument to foster the ego. Importantly, it is our relationship with the entire biosphere—as an archetype of all relationships—where posthumanist philosophy explicitly becomes a philosophy of integration.

The most significant trait of posthumanism is perhaps recognizing ourselves as part of a whole that does not annihilate our (human) presence or deprive it of meaning, but rather nourishes it and allows our own peculiarity to emerge. We need to strengthen what Bateson called the "structures-that-connect", which have been weakened by self-referential tendencies, cultural radicalism and egoic cravings. Just as culture can grow only if it establishes a dialogue with other cultures, humans become enriched by building relationships with their neighbours, children develop gratitude for the love that surrounds them, and societies are strengthened by integrating the myriad forms of diversity that compose it. No ecosystem can exist without including a network.

Therefore, we cannot exist if we are unable to co-exist. Every ontology is already an eco-ontology in itself. Every relational moment is implicitly an ontopoiesis. Every memory we keep in our hearts is the fruit of an encounter. The others are already inside us in the form of existential symbionts. Our existence is itself a co-existence, because, as part of our phylogenetic inheritance, the other-in-us precedes us and accompanies us throughout our ontogeny.

In co-existence, we cannot die, because our existential destiny is not reduced to being-for-death, but signifies donating ourselves to the flow of life as a gift for those who will come.[7]

Then, this human being found RESPECT, who showed them that honoring one another is at the base of inter-being:

Coexistence is not only about relationships between humans but includes relationships beyond the human. The challenge is to think of ways to implement this type of 'broad' co-existence unhampered by phenomena that involves using the other. In the exploitation that occurs for utilitarian, commercial or profit purposes, it is possible to trace a predatory and violent anthropocentric power that tends to legitimize the reduction of the other to something that is at our disposal. Such a mechanism is no longer ethically acceptable. We know that we are part of an intersection of complex relationships with other autonomous organisms, capable of perceptions, motivations and feelings similar to or different from humans. Therefore, it is no longer possible to think of 'the other' as an unarmed and manipulable object: 'the other' is a subject just like human beings are, but in their own way(s) which requires transformation from predatory, specie-ist power into *respect* and acknowledgement. This shift challenges the current power system which is regulated by habit, comfort, social and cultural heritage, and ontogenetic parental imprinting.

Respect requires defamiliarizing human relations with others. For example, reviewing economic interests between humans and other species so nonhuman animals in farms, zoos, labs and 'recreational animals' are no longer perceived as commodities but treated as independent entities with their own agency. De-centering the human is a necessary step in the transformation towards respectful relations between humans and nonhumans.

Competition for space and resources is the experience of co-existence, in which everyone carves out their own niche for welfare and wellbeing. However, the niche of human beings is extravagant. Re-discussing the

[7] Section by Roberto Marchesini.

boundaries between legitimate behaviours, necessities, pleasantness of life, ideologies and predatory attitudes is a challenge for the development of posthuman ethics. It requires a brave renunciation of practices perpetrated by human beings for (too) long, towards the discovery of nonhuman ways of perceiving, thinking and acting. Turning a predatory approach into one of *care* involves exploring the empathic resources we have, their limits, but also their possible extensions.

To enact a change of perspective towards nonhuman others, whilst being in relationship with them enables a posthuman ethics beyond theory, and requires an education that values relationships with the other. This form of education is essential to negotiate and reshape human(ist) ways of living towards posthumanist ones. Recognizing the other as a subject also implies recognizing its freedom of movement. Coexistence occurs when the free movement of others is granted and valued as a possibility for encounters. It is necessary, therefore, to reconsider human management of space, seeking an integration between the multiple expressions of life.

Human beings tend to perceive their interests as indisputable priorities, but posthumanist studies are trying to break the electric fence built around the traditional understanding of 'the human', highlighting the dignity and respect for every living being. It is all about training our ability to deactivate and climb this fence. Let us work together towards a deep *respect* for the nonhuman world.[8]

> Then, this human being realized that EMBODIED EXPERIENCES are an essential (p)art of the game of existence…

Posthuman ethics is not about a 'perfect' *experience*, but about the experience itself, in its totality; it is about doing and undoing and, in this sense, it moves from obligations towards possibilities, from questions of truth or rigidity, towards questions of care and curiosity: caring for and curiosity about the language of the more-than-humans—as well as of the human bodies-and-souls—to find meaning in both. This dynamic nature of ethics is characterized by a constant lost-found-lost condition. This lost-found-lost condition favours 'helping' rather than 'condemning', a condition which creates space for paradox, plurality, mistakes, constant change, ambiguity, hesitation and uncertainty. In this light, the task to develop

[8] By Cinzia Barillaro, Claudia Candido, Greta Francesconi, Giulia Girodo.

posthuman ethics requires us to think and exist ecologically, technologically and creatively; ethics, respectively, entails thinking, thinking when new possibilities/new challenges/ mishaps emerge and considering our response to them.

To bring this posthuman ethical stance into practice is a tough job. Posthuman ethics unfolds as a moment-by-moment practice, as a form of becoming. In this sense, it is beneficial to integrate ethics into practice through pedagogical engagement (embodied knowledge and physical experience, as bodies can be seen as sites of ethical and sustainable practices); transdisciplinary collaboration, and collective practices such as The Posthuman Summer Camp where we practised sustainability, care and curiosity.

The cessation of thought ends in silence, and in silence, we begin. It is not enough to re-world; we must also re-be. The posthuman may become pre-human by grounding ourselves in life beyond abstraction. Look at the way the tree offers their presence, and experience it by becoming the breath of the tree. I have already forgotten my name, the division between self and nature erased. Multiplicity has no beginning or end. I am deep in the jungle of our entanglement, a labyrinthine relationality without centre. Can you see it? Let go. Unsee the vision and fondle the world. Learn to experience our co-existence intuitively by unravelling the self and gifting your presence to one and all. Reification is the new deification. We do not soil the material with the abstract—we soil it with soil. Remove the weeds of abstraction and water the earth. Nature is already beautiful, its own divine artwork. We don't need to assign any meaning; its value and meaning are already intrinsic.

To be posthuman is to relinquish human privilege, embracing the unease of unravelling old entitlements. *Embodied* ethics requires us to stay troubled by the bleeding body of the planet and demand an end to violence against the more-than-human world, rooted in *embodied* change. Nurturing our exposure to the world is essential to weaving dreams together:

> The dream of plants: "It's dense here; move slowly. Open your eyes to who is here before you. I am alive, and every dwelling is shared."

The dream of wild animals: "I may accept your presence, but I will defend myself—you will not displace me. Learn my ways, and I will learn yours. Walk to my rhythm as I fly to yours." Animality is hybridity. The dream of mountains: "Your extractivism has no place here. Co-write history on my skin with care, or you will destroy our shared path." Grounding against extractivism.

Ethics is not merely a set of rules or abstract principles—it is a dynamic, performative art interwoven with our relationships to the world. Ethics is a bricolage, constructed from the shared remnants of experiences, narratives and histories. These fragments coalesce into a complex ethical fabric, where the subjects of our deliberation are inherently hybrid, never static or pure.

Envision ethics as a juggling act, where the ethical agent must continuously balance competing values and demands, much like a juggler keeping multiple objects in motion. This metaphor underscores the fluidity and intricacy of ethical practice, where the aim is not to arrive at definitive solutions but to remain adaptable and responsive to shifting circumstances. This adaptability becomes especially vital when considering the other-than-human world, where ethical decisions influence the flourishing of an expansive, interconnected living community.

Ethics is not about isolated decision-making; it is about cultivating a way of being attuned to the complexities of the world. It is an ongoing, *embodied* process—like piecing together an ever-evolving jigsaw puzzle, where the parts constantly shift and transform. This relational and performative art requires sustained care, sensitivity and creativity, expressed through our lived interactions with the broader web of life that both sustains and surpasses us.[9]

And then the human found WHOLENESS: to be all encompassing…

Colonization has left the world fragmented by embracing an empiricism that has broken the world into discrete parts, assigning a hierarchy of value to the parts. Cultural Posthuman ethics moves beyond the binary of colonized and colonizer, disrupting this shattering force of modernity and seeking a praxis of relation that leads to an expanded consciousness of wholeness and interconnection. This praxis works to disrupt the categories that render us feeling separate and allowing for othering, beginning with an embodied knowing of self through relation as both microcosm and

[9] This collective section on 'Embodied Experiences' has been created by intertwining passages written separately, respectively by Shabnam Naderi, B., Eleonora Vecchi and Andrea Natan Feltrin.

macrocosm. The praxis is deeply interlinked with an examination of our relation to self. On the micro level of self, a radical lens of inclusivity must be deployed to gather up all the pieces of self, challenging the notion of a fragmented self. This includes such elements of situated self as land of origin or biological determinants. Knowing the self as a relational being to other selves brings us to the concept of coexistence. This embodied practice of knowing and communication is not afraid to say that self-care is not anthropocentric.

The self exists in relation to other individual selves, and it is through the frictions of these relations that something can be created, namely the production of human culture with guiding patterns of behaviour that result in customs for a collective. This is critical in our time of ecological disaster. These collectives are important because they are in relation and co-exist with other collectives of humans and nonhuman collectives of animals, ecosystems and nature. Acknowledging ourselves as nature, something we become with. Hallmarks of this praxis could include supporting every forcibly displaced being, exposing the capitalist causes of deforestation and other oppressions, and not surrounding ourselves with objects made of conflict minerals begins with a praxis of self-reflection so that we can not only intellectualize these issues but experience them as our own.

The experience of ourselves as more-than-one opens us to a growing intuitive space of latent collective potential. Here we can know ourselves to be more than ourselves. This effort is inherently relational and collective. Feeling-with-plants, feeling-with-animals, feeling-with-ecosystems, feeling-with-self and feeling-with-the-divine. This language is essential to the creation of posthuman cultural hermeneutics that is inclusive of the plurality of aspirations in the becoming of the collective.

Out of respect for diversity from an awareness of the nondual self, we desire the products of co-existence, so that a new culture forms, transcending categorization. We are sharing a global collective, and we are cultivating a shared language of sensations. The felt sense in relationality. The self exists within relation to all of life, and it is through the deep listening occurring within these relations that something can be cultivated; with intention and reciprocity, these relationships become a shared third space of respect and co-creation. Cultural posthuman ethics moves beyond the binary of colonized and colonizer, beyond the capitalist quest for that illusive and mortiferous endless progress, disrupting this shattering force of modernity and seeking wonder and reconfiguration in a praxis of relation that interprets and implements an expanded consciousness of *wholeness* and interconnection, now towards a becoming that transcends the self and sees the self as a part of a deeply interconnected whole.[10]

> Then, this human being found CREATIVITY: in harmony with oneself and others, one and many…

Listen to your heartbeat and follow it for as long as you can. Open to the *potentia*, the possibilities, nurturing a posthuman(ist) attunement to the transversal alliance of human and nonhuman, to a deep enchantment with the world. Living is a creative multilogue. A spiral movement of creativity in harmony and driven by the love of harmony.

We are all creative, different bodies moving with dynamic longitudinal and latitudinal velocities. Creativity enacts a becoming; it is never an individual act, rather it is interconnected to multiple ecologies.

Creativity and the power of visions are an imaginary affective force that drives transformation (Braidotti 2019, 66). If we are to think beyond this moment of climate emergency, loss of species, inequality, war, oppression… then a creative approach moves us towards an affirmative praxis, a collective community of the new.

In the spiral structure of movement, inspiring and empowering, creativity reveals itself through the ontological dynamics of plurality. Creativity is the un/doing of identity within the affirmation of differences between one and other. "Dis/continuity is at the crux of this im/possible, im/passible, trans/formation" (Barad 2010, 247).

To leave the paradigm of our pre-set assumptions on creativity, let us instead trust the poiesis of a structured praxis, and enter a labyrinth.

Creativity occurs in the in-between spaces, a liminal cycle unfolding as a spiral, such as in a solarpunk labyrinth (Fig. 1). Walking in is a creative act. The labyrinth is creativity in practice, an example of ethix, of posthuman ethics beyond theory. Entering the labyrinth, we encounter space and time, and we experience a way to enter ourselves. We encounter the zoe-geo-bio-techno, the geo of land recovering and regenerating from *potestas*, from the affective forces of degradation via human industry. The techno-materiality of lines of wire and solar panels, creating corridors via small dots of light through which to navigate a possible transformation, a decision-making contemplation, a generative becoming. The labyrinth is a creative process, geo-art, which provides movement and stillness, a paradoxical poiesis. Architectural octagonal-patterned corridors, the structure

[10] By Sasha Bainer, Laura Cappellesso, Judi Checo, Nina Maria Szukała, Stephanie Travale.

Fig. 1 Picture of the Solar Punk Labyrinth. Realized by Joanna Pascoe and Stefano Rozzoni in the context of the *Posthuman Summer Camp*, Galliera (BO), Italy, and open from August 10 to 13, 2024. Credits: Rete Postumana Italiana CC SA BY 2024

of which unfolds to reveal *potentia*, harmonious and beautiful, healing. There is always something to heal.

As we walk, we may get lost and may be found and lost again, accepting that everything is in a constant state of change, *omnia transit*. The flux cannot be stopped, the cosmos takes us along the path of Dao: rooted, we flow. The flow reveals structure. The creative process is the revelation of structure and flow, in rhythms, developing and expressing harmony.

Creativity comes together with sensitivity and connection; the trio describes embodiment itself. The aesthetics of embodiment entangles with already entangled onto-epistemologies. There is creativity in the void, where photons appear in nano moments and disappear out of the *potentia* of nothingness. Creativity, as sensitivity, is a nonhuman expression of the beauty of existence. Posthuman ethics beyond theory is the embodied praxis of creative beauty that heals.[11]

The End, That Is, the Beginning

> Now, at the end of this tale, our own journey begins. Every moment of our lives offers an opportunity to embark on our mythical journey into the Self. Ethics are praxis. Praxis is awareness. Awareness is balance. We talk, and then, we embrace silence: as individuals, as a community, as a species. Our minds are often hyperactive. This is why in the ethics of posthumanism, silence is embraced as a most powerful key to understanding, healing, and regenerating: to enter into the Self. Posthumanist ethix are ways of existing. These ways of existing are personalized, since everyone is different; at the core of posthumanist ethix, there is the awareness of unity in diversity. Like a drop in the ocean of existence, in its reverberations, our intentions are already creating ways of existing. Here and now. The posthumanist ethix of the 21st century connects to time in memorial. Embracing humanity, ecology, and technology; embracing who we are, without fears or attachments. Embracing the cosmic dance of being and non-being. Daring to be authentic, able to trans-form. Posthumanist ethix are about our lives: "Are we going to repeat thousands of years of anthropocentric and humanistic habits that are damaging us – as a planet, as a species, as an individual – or are we going to embrace radical creativity and transform each moment into full potential for the realization of unique posthuman ways of existing?" With this same question in their mind, this human being woke up one day: the sun was shining, a golden flower was blowing in the wind.
>
> This human realized there was no separation between them and everything else, like a liquid canvas. There was no background. Everything was a continuum. We are the colour from which unlimited paintings can be made. We can use gold to draw a star or a flower, an iPhone, a comet, or something that we cannot even describe, such as an abstract image. The beginning is the end: we are that gold; we are that raw colour that could be anything. If we paint the Sun, we will burn. If we paint a flower, we will bloom, and

[11] By Katarzyna Ferworn-Horawa, Joanna Pascoe, Stefano Rozzoni.

eventually decay. There is no separation between what we choose to paint and the experience we will perceive. Once the experience is gone, we will realise that we have always been that raw colour, that could be anything. We do not have to wait to die, in order to self-realize. This gold, this raw state, is the void, which is the source of our existence; to this, we can connect in every moment of our lives by silencing the self and just be. Non-existence is not something to fear: it is who we are, beyond any anthropomorphism, anthropocentrism, or humanism.

This is a true story. It happened to me, it happened to you, it happened to us. The mythical journey of our lives. A posthuman tale of self-realization.

Acknowledgments We would like to thank all the authors involved in the writing of this collective paper, who also participated in-person in the Posthuman Summer Camp (in alphabetic order): B., Sasha Bainer, Cinzia Barillaro, Claudia Candido, Laura Cappellesso, Judi Checo, Devdip Ganguli, Andrea Natan Feltrin, Katarzyna Ferworn-Horawa, Greta Francesconi, Giulia Girodo, Shabnam Naderi, Giovanni Massaro, Maria del Portillo-Cure, Anshul Sinha, Nina Maria Szukała, Stephanie Travale, Eleonora Vecchi.

Thanks to the generosity of Roberto Marchesini, who made the Camp experience possible at *Naturama*; and to Sabrina Golfetto for her care and tireless assistance. Thanks to the other co-organizers of the Posthuman Summer Camp: Manuela Macelloni and Orsola Rignani; thanks to Kevin LaGrandeur for his online presence and the valuable inputs he provided.

Finally, thanks to all the Posthuman Scholars of the Global Posthuman Network who, through their works, visions and experiences, have enriched the inspiring discussions that led to this paper.

References

Alexandra, Andrew, and Seumas Miller. 2009. *Ethics in Practice: Moral Theory and the Professions.* Sydney: University of NSW Press.

Barad, Karen. 2007. *Meeting the Universe Halfway: Quantum Physics and the Entanglement of Matter and Meaning.* Durham: Duke University Press.

Barad, Karen. 2010. Quantum Entanglements and Hauntological Relations of Inheritance: Dis/continuities, Spacetime Enfoldings, and Justice-to-Come. *Derrida Today* 3 (2): 240–268. https://doi.org/10.3366/E1754850010000813.

Blackmore, Jill, Debra Bateman, Anne Cloonan, Mary Dixon, Jill Loughlin, Jo O'Mara, and Kim Senior. 2010. Innovative Learning Environments Research

Study. https://www.deakin.edu.au/__data/assets/pdf_file/0003/365196/innovative-learning-spaces-final-report.pdf.

Braidotti, Rosi. 2011. *Nomadic Subjects: Embodiment and Sexual Difference in Contemporary Feminist Theory*. 2nd ed. New York: Columbia University Press.

Braidotti, Rosi. 2013. *The Posthuman*. Cambridge: Polity Press.

Braidotti, Rosi. 2016. Don't Agonize; Organize! In Utrecht University Repository. *e-flux*: 1-4. https://dspace.library.uu.nl/handle/1874/384992.

Braidotti, Rosi. 2019. *Posthuman Knowledge*. Cambridge: Polity Press.

Colby, Richard, Matthew S. S. Johnson, and Rebekah Shultz Colby, eds. 2021. *Ethics of Playing, Researching and Teaching Games in the Writing Classroom*. Cham, Switzerland: Palgrave Macmillan. https://doi.org/10.1007/978-3-030-63311-0.

Daigle, Christine. 2023. *Posthumanist Vulnerability: An Affirmative Ethics*. London: Bloomsbury Academic.

De Falco, Amelia. 2023. *Curious Kin in Fictions of Posthuman Care*. Oxford: Oxford University Press.

Ferrando, Francesca. 2019. *Philosophical Posthumanism*. London: Bloomsbury Academic.

Ferrando, Francesca. 2023. *The Art of Being Posthuman: Who Are We in the 21st Century?* New York: Polity Press.

Ferrando, Francesca, and Stefano Rozzoni. 2023. Vision Quest in Posthumanist Education: Focuses, Praxes, and Experiences. *International Journal of Philosophical Studies* 31 (4): 586–611. https://doi.org/10.1080/09672559.2023.2290552.

Finland Education Hub. 2023. Learning in Finland: Where School and Nature Meet. *Medium*. https://medium.com/@finlandeducationhub/learning-in-finland-where-school-and-nature-meet-cdcb65bbd485.

Freiler, Tammy J. 2008. Learning Through the Body. *New Directions for Adult and Continuing Education* 119:37–47. https://doi.org/10.1002/ace.304.

Grootenboer, Peter, Christine Edwards-Groves, and Sarojni Choy, eds. 2017. *Practice Theory Perspectives on Pedagogy and Education: Praxis, Diversity and Contestation*. Singapore: Springer.

Haraway, Donna. 1991. A Cyborg Manifesto: Science, Technology, and Socialist-Feminism in the Late Twentieth Century. In *Simians, Cyborgs and Women: The Reinvention of Nature*, 149–181. New York: Routledge.

Haraway, Donna. 2016. *Staying with the Trouble: Making Kin in the Chthulucene*. Durham: Duke University Press.

Hayles, Katherine N. 1999. *How We Became Posthuman: Virtual Bodies in Cybernetics, Literature, and Informatics*. Chicago: University of Chicago Press.

Joyce, James. 1959 [1939]. *Finnegan's Wake*. New York: Viking Press.

LaFollette, Hugh, ed. 2020. *Ethics in Practice: An Anthology*. 5th ed. Hoboken, NJ: Wiley Blackwell.

Macfarlane, Bruce. 2004. *Teaching with Integrity: The Ethics of Higher Education Practice*. London: RoutledgeFalmer.

Pascoe, Joanna. 2023. *The Posthuman Learner: Mothers, Monsters, and Machines*. Doctoral thesis, University of Canterbury. UC Research Repository. https://hdl.handle.net/10092/106133.

Sorgner, Stefan L. 2019. Editor's Note. *Journal of Posthuman Studies* 3 (1): 1–4.

Sorgner, Stefan L. 2020. *On Transhumanism*. University Park: Pennsylvania State University Press.

Sorgner, Stefano L. 2021. *We Have Always Been Cyborgs: Digital Data, Gene Technologies, and an Ethics of Transhumanism*. Bristol: Bristol University Press.

Taylor, Carol A. 2017. Is a Posthumanist Bildung Possible? Reclaiming the Promise of Bildung for Contemporary Higher Education. *Higher Education: The International Journal of Higher Education Research* 74:419–435. https://doi.org/10.1007/s10734-016-9994y.

Timperley, Helen, Linda Kaser, and Judy Halbert. 2014. *A Framework for Transforming Learning in Schools: Innovation and the Spiral of Inquiry*, Seminar Series 234. Centre for Strategic Education.

van Nieuwerburgh, Christian, and Robert Biswas-Diener. 2025. *Radical Listening: The Art of True Connection*. Oakland: Berrett-Koehler Publishers.

Transhumanism and Christian Philosophy

Transhumanism as a Secular Vision of Christian Ethics

Micah Redding

1 Transhumanism as the Secularization of Christianity

In the 1994 Wired article "Meet the Extropians", the author notes that this young and exuberant transhumanist community had proposed to re-date the calendar to 1620, to begin with the publication of Francis Bacon's *Novum Organum* (Regis 1994). In an article rife with symbolism, this particular statement is striking. The birth of Christ is to be replaced with the birth of science.

> We see this need for transcendence deeply built into humanity. That's why we have all these religious myths… It seems to be something inherent in us that we want to move beyond what we see as our limits. In the past we haven't had the technology to do that…

M. Redding, MSc (✉)
Christian Transhumanist Association, Nashville, TN, USA
e-mail: micah@christiantranshumanism.org

G. Theiner, A. Malapi-Nelson (eds.), *Technologies of the Future Self*, Palgrave Studies in the Future of Humanity and its Successors,
https://doi.org/10.1007/978-3-031-99902-4_9

> People have dreamed such dreams before…The difference is that now, suddenly, all of it is entirely possible. For the first time in history, science and technology have caught up to the wildest of human aspirations and hopes. (Regis 1994)

The idea of replacing—or fulfilling—religion with science runs throughout the article. But the irony of restarting the calendar with *Novum Organum* was perhaps lost on its advocates. *Novum Organum* is itself a deeply theological work, one also seeking to fulfill religious aspirations through science. But rather than seeing science as a replacement for religion, it argues that the scientific effort is deeply Christian, and can only be successful if pursued in a Christian way (Matthews 2008, 39, 54, 74, 89–90).

The Extropians had thus proposed to replace Christ with a science built in his name. In a move seemingly designed to erase Christ's cultural significance, they highlighted the significance of Christ to the history of science itself.

Such transpositions are common in transhumanist discourse, for reasons that may become obvious. Secular transhumanism often sees itself as a rejection of religion with its irrational dogma and restrictions. Yet more than almost any other philosophy, transhumanism draws from religion for its inspirations. This dichotomy can cause it to miss the significance and nature of prior versions of its own thought processes. In offering an essentially secular account of religion, it can fail to recognize how deeply it itself is embedded within the historical contingency of larger religious conversations.

Yet even such historical forgetfulness does not originate with transhumanism. Rather, it may be merely another iteration of the Protestant polemic against Catholicism, refracted as the scientific polemic against Christianity—telling again the same stories, such as *Galileo and the church*, with new antagonists in each telling—that now forms a significant part of many modern secular self-conceptions. As Peter Harrison argues: "In a sense, the animus against religion exhibited by some contemporary advocates of scientific enlightenment is simply an extreme version of this 'Protestant' position" (Harrison 2019).

Just as with the *Novum Organum*, modern transhumanism is largely rooted in deeply Christian sources, whose Christianity has been quietly forgotten.

This is evident in looking at Julian Huxley, usually credited as the originator of modern transhumanism (Bostrom 2005). In Huxley's 1927 work *Religion without Revelation*, he argues at length for the necessity and inevitability of a new kind of religion, reformed around the principles of scientific humanism (Huxley 1927, 9). Existing religion may cross through to this new grounding, he thinks, but only if it humbles itself, and "becomes as a little child", open to new possibilities, new interpretations, and new forms of expression (Huxley 1927, 12). To facilitate this transformation, Huxley offers a lengthy secular interpretation of various Christian doctrines, including the trinity (Huxley 1927, 73), the incarnation and the spiritual nature of man (Huxley 1927, 72, 77), and the kingdom of God (Huxley 1927, 377).

These interpretations of Christianity provide context for Huxley's introduction to Teilhard de Chardin's *Phenomenon of Man*, in which Huxley reveals himself unable to accept Teilhard's full Christian vision, yet motivated to retain the shape of it in a secular form. Even the blatant religiosity of Teilhard's vision is a positive for Huxley, who thinks that this religiosity is necessary to any philosophy of the future. In fact, Huxley is insistent that science and religion can no longer remain separate. As he puts it, the "materialistically minded" can no longer deny the importance of religious feeling, and so must integrate it into their materialistic view of the world. And the religiously minded can no longer look for fulfillment in an escape from nature—they must look for religious fulfillment within nature itself (Huxley 1959, 26).

I've been using the terms *secular* and *secularization* to denote the move away from religion. For Huxley, the aim is not merely to maintain a naturalist or materialist stance, but to make the former possessions of religion available to the non-religious world. The fact that religiosity itself is one of those possessions to be secularized is merely another generative irony among many.

Thus, Huxley himself sees transhumanism as a secularized religion. A materialist religion is needed, and religion must now be interpreted materialistically. This religion, in the newly secular form of transhumanism, is to escape containment in the walls of private religious faith, and become a guiding philosophy for the whole world. Accordingly, in Huxley's 1957 essay on transhumanism, he positions transhumanism as a faith, superseding previous ones: "*I believe in transhumanism*" (Huxley 1957, 13–17).

Yet Huxley is not the first one to use the term "transhumanism". In addition to Teilhard de Chardin's own use of "transhuman", "transhumanity", and so on, W.D. Lightfield's 1940 essay speculating on the reaches of future evolution evokes what he calls "Paul's transhumanism", the apostolic vision of unfathomable future human transformation (Harrison and Wolyniak 2015). Here we have another ironic transposition: for Lightfield, transhumanism as a religious phenomenon, inspiring a secular vision, becomes in Huxley transhumanism as a secular phenomenon, inspired by a religious vision.

All of this illustrates the intentionality with which transhumanism was extracted from its religious roots, and formed into a secular philosophy. As Eric Steinhart puts it, "Many transhumanists work within the conceptual architecture of Teilhard's [religious futurism] without being aware of its origins" (Steinhart 2008, 1–22).[1]

Yet even the intentional distancing from religious origins doesn't fully obscure the impulse to grapple with them. In this light, the sentiments expressed in *Meet the Extropians* merely affirm the longstanding transhumanist aspiration to secularize religion.

2 Brent Waters on Transhumanism as Christian Heresy

Christian ethicist and theologian Brent Waters agrees with this vision of transhumanism as secular religion. He also agrees that it parallels Christianity in its soteriology, eschatology, vision of the human condition, theological discourse, etc. (Waters 2011, 164). Yet he sees reason for emphasizing a sharp distinction:

> …transhumanists draw their core beliefs and convictions, often unwittingly, from what Christians would regard as heretical sources. (Waters 2011, 164)

Waters is not referring to *sources* in the sense of specific works or thinkers, such as Teilhard de Chardin. Waters seems to regard Teilhard de Chardin as a legitimate Christian theologian, yet sees him as inconsequential to modern transhumanism (Waters 2011, 164). What Waters is identifying as

[1] Ilia Delio critiques the similarity that Steinhart sees between transhumanism and Teilhard de Chardin, arguing that transhumanism lacks the holistic ethic and positive vision of biology that Teilhard advocates. For our purposes, the important point is the lines of influence, not the faithfulness with which transhumanism reinstantiates Teilhard's thought. See Delio 2012.

heretical are particular ideas and values that he sees echoed among modern transhumanists.

Waters' use of *heresy* here immediately raises questions, which Waters does not address. As someone associated with Protestant and Evangelical institutions, he himself is part of a tradition at one time deemed heretical by the mainstream church.[2] Since Waters does not here define his terms, we are limited in what we can say about them. However, for our purposes, the most significant thing to note is that *heresy* in Christian thought usually describes an aberrant strain of Christianity. It is not how you would describe another religion, such as Hinduism or Buddhism, but how you would describe a form of Christianity that has gone awry. Waters has thus positioned transhumanism as part of Christianity's family tree, so to speak, though a deliberately estranged branch. But what is the core of this purported estrangement?

For Waters, transhumanism is based in the rejection of the body, while proper Christianity embraces the body.

> According to transhumanism, embodiment is the principal problem of the human condition. The body imposes severe and intolerable limitations upon what humans can do and upon what they aspire to be. … More troubling, the body is a source of pain and suffering. As embodied beings humans are fragile and vulnerable; they can be injured or become ill. More depressingly, even if a person is fortunate enough to avoid any serious injuries or diseases, he or she is allotted only a limited amount of time. Embodied beings grow old and die. In short, human beings must be saved from their finite and mortal bodies. (Waters 2011, 165)

This idea that transhumanists want to be saved *from* their bodies is determinative for Waters. And it ultimately entails a hatred of embodiment in general. Speaking of transhumanism as a new form of ancient religion, he states: "The Manichean cannot resist hating the body, for it is a prison incarcerating the soul or will. The resulting aggravation, however, is not limited to self-loathing but is extended to a latent contempt of embodiment in general" (Waters 2011, 172). But is transhumanism really about hatred of embodiment?

The evidence Waters supplies is the transhumanist project to extend life. He describes transhumanist scenarios involving healthy bodies,

[2] Martin Luther and his followers were declared to be heretics in the papal bull "Decet Romanum Pontificem" by Pope Leo X in 1521.

long-lived bodies, bodies with prosthetics, and robot bodies (Waters 2011, 165–168). It's hard to see how this fits the description of embodiment-hatred. But for Waters, the desire to overcome mortality *is itself* a hatred of embodiment.

> Christian theology has appealed to the goodness of the body and more expansively to the good of embodiment. The particular challenge … [is] to remain creaturely, which by definition is to be finite and mortal, and therefore inescapably embodied. (Waters 2011, 171)

Here, Waters equates embodiment with finitude and mortality. Thus, the argument seems to be: the desire to overcome finitude and mortality is a desire to overcome embodiment, and thus a rejection of the proper Christian stance.

Yet is Waters' position coherent with the larger Christian tradition? How does this account for the immortal Christ, still human, still embodied, still creaturely—yet divine, transcendent and transformed, seated at the right hand of God, freed from all pain and death? If Waters holds the traditional position that Jesus Christ remains human and embodied, yet now transformed, overcoming mortality and death—doesn't this inherently establish a category of non-mortal (and indeed, *previously* mortal) embodiment within Christian thought? How then can the aspiration to overcome mortality be a rejection of embodiment, or a hatred of the body?

Isn't, in fact, the desire for a transformed embodiment pervasive within Christianity? For example: "He will transform the body of our humiliation that it may be conformed to the body of his glory…" (Philippians 3:21 NRSVUE)

It is hard to see how Waters' critical assessment of transhumanism would fail to include large segments of Christianity within its critique. Yet for Waters, the distinction is simple. Waters draws a contrast between being saved *from* embodiment, which he ascribes to transhumanism, and being saved *through* embodiment, which he ascribes to Christianity.

> It is the finitude and mortality of being human and human being that are affirmed by the Incarnation. This is not a condition from which creatures need rescuing, for it is only as finite and mortal creatures that they are saved (Waters 2011, 172).

> For the Christian, finitude and mortality can only be overcome through death's penultimate victory, which is in turn reversed through one's resurrection into the eternal life of the triune God (Waters 2011, 164).

Waters seems to be saying that the Christian hope necessarily involves death. Rather than death simply being something which Christ has overcome in his resurrection, Waters seems to see death itself as transformative, so that to miss death would be to rob oneself of the transformation Christ offers.

Again, is Waters' position coherent with the larger Christian tradition?

> We will not all die, but we will all be changed…For the trumpet will sound, and the dead will be raised imperishable, and we will be changed. (1 Corinthians 15:51–52 NRSVUE)

The apostle's words seem to establish a category of transformation to immortality, that does not require death, but avoids it altogether. This is precisely the kind of scenario Waters sees as entailing a hatred of the body.

At the popular level, it is certainly the case that millions in the pews each Sunday understand themselves to be looking forward to a transformation of their bodies beyond the categories of mortality and finitude. And as the best-selling *Left Behind* series indicated, millions more hope to attain this without seeing death.

My point is not to argue for such views, but to point out that they are part of a fairly mainstream Christian landscape. As such, it is again hard to see how Waters' critical assessment of transhumanism would fail to include broad segments of Christianity within its critique. If Waters has indeed identified a legitimate divergence of opinion on how we should approach embodiment and mortality, it appears to be one that runs *through* Christianity, rather than *between* Christianity and transhumanism.

This applies to other contrasts that Waters brings up.

For example, Waters claims that Christians anticipate a resurrection to a *time-less* immortality, while transhumanists are seeking a *time-bound* immortality (Waters 2011, 164). Yet many theologians and scholars would strongly disagree that Christian hope is *time-less*. For example, N.T. Wright argues that such a position is Plato, not Paul (Wright 2011). Similarly, C.S. Lewis asserts: "The old field of space, time, matter, and the senses is to be weeded, dug, and sown for a new crop. We may be tired of that old field: God is not" (Lewis 2009, 93). It is remarkable that Lewis and Wright would be classified with the transhumanists in this comparison. Since they can hardly be classified as fringe within Christianity, Waters appears to be taking one position in an intramural Christian debate, and identifying transhumanists with the other position.

Far from pointing to a universally agreed-upon line, that demarcates Christian premises from transhumanist premises, Waters has staked out positions that highlight the complexity and diversity of Christian thought. These differences of opinion about the precise nature of the body and eschatology don't run *between* Christianity and transhumanism, but *through* Christianity and transhumanism. Even these debates have carried over from Christianity, and divide transhumanism internally along analogous lines.

In identifying transhumanism with theological positions that demonstrably exist within mainstream Christianity, Waters actually positions transhumanism as part of a debate internal to Christianity itself.

3 Thorsten Moos, Eschatology, and the Secular Gap

Theologian Thorsten Moos agrees with the vision of transhumanism as secularized religion, with both the positive capacity to highlight neglected aspects of theological discourse and the negative tendency to let go of necessary aspects of *religious reason* (Moos 2016, 160).

For Moos, religious hopes and expectations, usually expressed in eschatological, apocalyptic, and future visions, become overwhelming and all-consuming demands, and thus need to be managed. Moos calls this management "religious reason", and argues that it is through this management that religion maintains a "secular gap" between the immanent and the transcendent, and thus makes life workable. Moos points to particular strategies, such as God-talk, and sophisticated ways of talking about eschatological delay, as key examples of this *religious reason*.

By "God-talk", Moos simply means talking about God; but Moos is highlighting the fact that the use of such language is not wholly dependent on one's metaphysical commitments, but is also a strategy that one uses for grappling with the interplay of those commitments and other realities. For Moos, God-talk manages the inevitable boundary between agency and lack of control (Moos 2016, 170). In speaking about God, the arena beyond individual agency becomes a region of hope, but even more particularly, can become a region of negotiation. We can, for example, do all that is possible, and then entrust ourselves to a transcendent judge. Or in the words of Ignatian spirituality, "Pray as if everything depended on God and work as if everything depended on you" (CCC n.d.).

Similarly, eschatological delay opens up management strategies within Christianity, where progress is conceived as evolutionary, revolutionary, and revelatory. These conceptions draw a distinction between our immanent efforts and the transcendent realities to which they aspire (Moos 2016, 171).

When this distinction is collapsed, Moos seems to be suggesting, we become susceptible to tyranny. We can imagine the scenarios, both in apocalyptic Christianity and in secular transhumanism. For example, if our access to the future lies through cryonics, then whoever controls the cryonics facilities effectively controls the infinite future. On the other hand, when the future is something waiting to be revealed, it resists being monopolized in this way.

Although Moos is arguing for the necessity of these strategies, it is precisely an overuse of these strategies that he sees as having driven the rise of transhumanism. As Moos recounts it, nineteent-century liberal Protestantism lost itself in the inner life, referring all of its eschatological hopes to internal and individualistic fulfillment. Protestantism "managed" its hopes to the point that they collapsed and shrunk away from the external world. Transhumanism corrected for this over-management, restoring a focus on the external life (Moos 2016, 173).

Moos has correctly, I think, located the significance of transhumanism as a theological reform movement arising from within Christian thinking—or at least as a theological critique arising from within Christian thought. This very much leads transhumanism into its own characteristic mistakes and overemphases. Moos identifies these mistakes with the collapse of the "secular gap" between the immanent and transcendent. As Moos notes, transhumanism as a secularized form of religion is an attenuated form, one which has dropped particular religious strategies, such as the aforementioned God-talk (Moos 2016, 160). This attenuation costs it the most "rational" approach to balancing the all-consuming demands of future visions. Transhumanism gives up these strategies, and thus loses the gap, collapsing the immanent and transcendent.

For Moos, this loss of these religious strategies is the most significant fact about transhumanism. And the one that poses the most risk. What is to be done?

Moos advises transhumanism to borrow the strategies that mainstream Christianity has developed for managing its futuristic visions. It remains unclear whether Moos expects transhumanism to thus become stronger and more distinct, to fragment under internal conflict, or to become more recognizably identifiable as a branch of existing religion. Whatever the case, it surely means transhumanism would become more *religious*.

I find myself in broad agreement with Moos' thesis, including his prescriptions. Yet Moos has perhaps discounted the capacities transhumanism may already have for the development of these strategies internally. We'll return to some of these below.

4 Christian Ethical Themes Within Transhumanism

Given the genealogy of modern transhumanism, the widespread recognition of transhumanism's religious aspects, and the persistence of its religious secularization impulse, it makes sense to follow the implications of specific Christian themes in transhumanist discourse. Loosely following Huxley's own religious interpretations in *Religion without Revelation*, I suggest that transhumanism can and does reinstantiate three main themes of Christian theology: eschatology, imago Dei, and the incarnation.

Within Christian theology, these themes offer ethical guidance and vision. In their secular forms within transhumanist discourse, they may continue to do so into the future.

5 Eschatology as an Ethical Framework

According to the Gospel of Mark, Jesus' public ministry begins with a declaration that *the kingdom of God is near*. In a sense, then, Christian theology begins with the end. It is because of the future, and its capacity to unexpectedly break into the present, that Jesus is able to call for an ethic radically divergent from that of convention and normalcy. More subtly, it is this future orientation that allows Jesus to reframe many of the debates within his own context. The future changes the landscape of possibility, and thus calls forth a new way of addressing many of the requirements of the past.

The Christian use of an eschatological framing is pervasive, but not uniform. In contrast to a broadly millenarian perspective, with its tendency toward evolution or revolution, Augustine's eschatology is almost entirely transcendent to historical time:

> For Augustine, historical time, the tiresome and tearful tenure of fallen man, was homogeneous and unchanging; the resurrection of Christ was a sign of promise, to be sure, but history offered no other indication of movement toward a restoration of perfection. (Noble 1999, 20)

Yet to "transcendentalize" the eschaton is not to erase it. A transcendent eschaton remains a live concern that may find its way back into immanent time in various ways. This is what happens at various points in Christian history, from Christianity's early days, through the millenarian revivals of the twelfth century, on to the many and diverse eschatological visions today (Noble 1999, 20).

In contrast to Augustine's transcendent vision, David Bentley Hart sees Christianity as a thoroughly and historically eschatological faith. Rejecting the admonition not to *immanentize the eschaton*, Hart suggests that Christianity can be understood in no other way. It is inherently an immanentizing project, and an eschatological orientation towards a future that is already becoming present.

> "What is a good eschatological faith for, after all, if not to operate as an immanent force within history? More to the point, how would it even be possible to obey such an admonition when, of its nature, the eschatological is always already partly immanent, constituting a horizon at once within history and beyond it— at once present in the moment and fixed in eternity?" (Hart 2022)

So too, more mild-mannered theologians like N.T. Wright frame their ethical thinking in terms of an inaugurated future. Wright employs the metaphor of a five-act play with a missing part. Knowing the future and the past, we are to fill in the missing act with our own lives (Wright 1991, 21, 7–32).

Eschatology in this key rather directly makes the leap from Christianity into transhumanism. Francis Bacon's science was articulated within the context of a vaguely postmillennial eschatology, that allowed for, and demanded, an era of scientific progress (Matthews 2008, 101). Fedorov's drive towards space travel was framed within the larger eschatological quest for the resurrection of the dead (Young 2012, 48–49). Teilhard de

Chardin's vision was itself an eschatology, in which, as Huxley describes it, the future evolution of the world culminates in something like deity itself (Huxley 1959, 19).

This eschatology is taken up, via Huxley and others, into transhumanism, and becomes one of the most common ways of framing transhumanist thought. Teilhard looked forward to an ultimate culmination of evolving intelligence, and similarly, contemporary transhumanists talk about the impending technological Singularity and the birth of ASI. At that point, as many portray it, superintelligence may offer humanity a choice to become immortal, raise the dead, and enter Godlike bliss, or to die and miss out on everlasting life. Others have more negative construals of the possibilities here, meditating on the potential for apocalyptic extinction and unending torture. Either way, the compounding forces of intelligence are thought to open up heavens and hells, paradises and dooms.[3]

The connection between Christian eschatology and the technological Singularity was obvious to Charlie Stross and Cory Doctorow, who labelled it "The Rapture of the Nerds" (Doctorow and Stross 2012). Ray Kurzweil, too, references Christian eschatology in the title of his book: "The Singularity is Near", complete with a street-preacher sandwich board (Kurzweil 2005). But Kurzweil goes beyond this, attempting to define God, and then explaining how evolution causes the universe to "wake up", and something very much like God to become embodied in matter. Here, as Steinhart points out, Kurzweil appears to be drawing directly from Teilhard de Chardin (Steinhart 2008).[4]

Such an eschatological orientation can serve as a guiding direction and motivation for transhumanists. "Live long enough to live forever" (Kurzweil 2005, 432). Or in a more recent incarnation: "Don't die". In both cases, the rationale is that a radical future is coming, and this demands a reorientation of our lives around that future.

However, the possibilities of the future not only provoke a concern for personal health, they may also inspire a disruption of the present. They may, for example, inspire artistic, aesthetic, or—one may even venture to say—*prophetic* interventions. They may inspire artists to adopt new senses, or implant new sense organs. They may inspire streamers and life-loggers to challenge privacy and social convention. They may push people to

[3] For a discussion of some of these scenarios, see Geraci 2010.

[4] As noted before, the point here is not the faithfulness to the spirit of Teilhard de Chardin's thought, but the degree of influence.

experiment with their bodies and the limits of what they can be made to do. They may, in short, inspire actions that evoke or embody what is coming, that push the boundaries of present technology, or challenge current social structures in preparation for the coming future (Pejcha 2023).

They may also provoke a break with traditional ethics. Recently, as AI has advanced, and concern over AI has escalated along with it, the prospect of terrorism aimed at stopping AI development has loomed larger in many people's minds. Even thought experiments such as *Roko's Basilisk* (more on this below), in which future threats impinge on present choices, may act to push people out of normalcy into radically new, and perhaps dangerous, ethical stances (Auerbach 2014).

We can see that an eschatological ethical framing may inspire both risk-aversion and risk-taking, both traditionalist and radical impulses. It can mitigate localist and presentist biases, but can also open the door to a number of runaway social and psychological tendencies, that can lead in dangerous directions.

The shape of these disruptions, including their risks, mirrors those within the Christian tradition. Christianity manages these risks by balancing eschatology with other ethical themes, and that is what we see in transhumanism as well.

6 Thorsten Moos and the Limitations of Transhumanist Eschatology

These risks and possibilities of eschatological transhumanism cohere with the concerns of Thorsten Moos. Yet he has perhaps discounted the capacities transhumanism already has for the development of strategies to maintain the "secular gap" between immanent and transcendent demands.

The Simulation Argument, put forth by Nick Bostrom, is indicative of some of these capacities. Bostrom's argument extrapolates on existing technological trajectories, to lay out a map of possibility space. In some branches of this possibility space, we already exist within simulations run by future superintelligences (Bostrom 2003). These possibilities have spawned a wide variety of related scenarios within transhumanist discourse, along with accompanying strategic analysis. *Roko's Basilisk* is a well-known version of this, postulating a future superintelligence who tests people's decision-making by trapping them in simulations where they cannot tell what time they live in. Because you cannot be sure that you are not in such a simulation *right now*, the argument goes, you must give some weight to the threats such an entity would be incentivized to make—including demands that you contribute to its creation. Given such a possibility, merely thinking about this superintelligence and its hypothetical desires may change its attitude towards you (Auerbach 2014). The analogies to theology should be clear.

Such scenarios complicate the account of agency, demonstrating that we must make decisions without being able to fully determine the scope of our setting and its stakes. Eliezer Yudkowsky has formalized this kind of decision-making process as "Timeless Decision Theory". Rather than act under assumptions of naive consequentialism, Yudkowsky positions decision-making in almost Kantian terms, as a decision about how agents like yourself should behave across possible worlds, times, and simulations (Yudkowsky 2010).

Although I've only talked about negative scenarios here, there are just as many positive ones. Scenarios where we are being tested by benevolent forces, scenarios where we are part of a resurrection process, scenarios where our future selves are overseeing some aspect of our past development, etc. Again, such scenarios complicate our accounts of agency.

Let me emphasize that such reasoning does not depend on a *belief* that we exist in a simulation. It merely requires allowing for the possibility, and acknowledging the ways in which our actions may make us susceptible to simulationist extortion, or open us to simulationist opportunities. Knowing these possibilities presses us to make choices in a different light, just as knowing the possibilities of burglary presses us to take different precautions, or knowing that some cereal boxes have a prize inside presses us to make different buying decisions.

Yet in such reasoning, transhumanist thought has reinstated a distinction between the ideal future to which we are aspiring and the practical efforts we are engaged in at the present. We are not the only agents that must be considered—creator Gods and deceiving Devils are back on the table. We cannot simply identify our practical efforts with the ideal future—our actions may merely be the faithful response to an eschatological reality that already exists, and has called us into being.

Moos' *religious reason* does not require belief that God will intervene in a particular way, nor that some aspect of eschatology will be unveiled in a particular timeframe. Rather, it requires that these categories exist in a dynamic interplay with more immanent concerns of an unfolding future. Simulation-tinged reason, similarly, need not require any particular belief in the existence of a simulation. Rather, it simply allows these expanded categories to exist in a dynamic interplay with more immanent concerns.

In Moos' terms, religion has the evolutionary, the revolutionary, and the revelatory accounts of progress. Classical transhumanism was evolutionary. The technological singularity reinstated the revolutionary. The possibility of simulations has reinstated the revelatory.

7 Imago Dei and the Ethics of Human Significance

Imago Dei is a term for the Christian concept that humans are made in "the image of God", and equipped with both unique capacities and calling. It suggests human dignity as an ethical concern, and carries with it a sense of unique and transcendent purpose. Historically, it was connected with the rational nature of the human mind, and humanity's creative and technological capacity.

As an ethical framing, it serves to limit and check utilitarian concerns. The human person is not fungible or disposable, but has a sort of irreducible cosmic significance. At the same time, this significance is connected with a duty, in fact a vast weight, that must be discharged.

For the Christian, this duty is to rule creation, as declared in Genesis 1, and elaborated on in Psalm 8 and other places. Accordingly, Gregory of Nyssa connects the *imago Dei* to humanity's domestication of animals, creation of hunting weapons, and the technology of iron (Gregory n.d.). Francis Bacon's scientific methodology, and his vision of scientific and technological rule over nature, is explicitly grounded in this understanding of the Christian vision of the image of God.[5] In *The New Atlantis*, as Bacon proposes things we would now describe as life extension, medical enhancements, the engineering of new species, the construction of virtual realities (Bacon 2009, 52–53)—he draws at every turn from the Christian account of humanity's creation.

Teilhard de Chardin's vision is more nuanced. Because his task is to reconcile Christianity with evolution, he must account for humanity as the product of continuous evolutionary processes. This poses a challenge for

[5] Bacon's articulation of the imago Dei arises in the context of the Protestant Reformation, in which the loss of the Catholic sacraments drove deep questioning about the capacities of human nature. In contrast to some of his more Calvinistic peers, Bacon held that primordial human capacities could be retrieved in part through the use of empirical methods. See section 1.2 of Malapi-Nelson 2014.

the articulation of humanity's uniqueness. Teilhard resolves this dilemma by seeing humanity as an "inflection point" in the evolution of life and intelligence, where the products of evolution have finally gained agency over evolution itself. Humanity is thus both the precursor to a much greater version of itself and also has the profound responsibility of continually reshaping itself to bring forth the next stage of evolution (Teilhard de Chardin 1959, 179–181).

Teilhard's vision here evokes the Bible. Genesis 1 also depicts humanity as an "inflection point" in the creative acts of God, where the human being is "yet another creature", but one that is given some of God's own capacities and responsibilities within creation. This inflection then changes the whole narrative, and opens the way to the vastly expanded possibilities of the rest of scripture.

Critics might suppose that this becomes a de-materialized vision of humanity. But on the contrary, this allows Teilhard to see the specific forms of human biology as having profound significance. Human biology is uniquely suited to bringing the trends of evolution to a fulfillment and culmination, and thus bringing forth a new form of life. Like Gregory of Nyssa pointing to human biology as driving the necessity of technology, Teilhard de Chardin sees the biological structure of the human as both necessary to its larger purpose and a precursor of further developments (Chardin 1959, 169–171; Delio 2025).

All of this has been carried over fairly directly into transhumanist discourse. As philosopher and social epistemologist Steve Fuller declared from the stage of a popular transhumanist conference in 2019, "Transhumanism is about the image of God!" (Fuller 2019).

Transhumanism largely embraces Teilhard's conception of this image. Humanity's uniqueness can be grounded in its role as an "inflection point", with unique creative significance for the future of life. This can be conceived of in personal terms—that we ourselves might survive and evolve into the forms of future life—or in collective terms—that we might find meaning and significance in our role as the precursors of an incredible future.

Either conception offers a way to think ethically about our own tasks, and to raise the significance we attach to the humans around us.

But the transhumanist adoption of the *imago Dei* can go deeper still. Physicist David Deutsch describes humanity as having "a special relationship with the laws of physics", a unique and transcendent role in the physical order, that puts humanity on par with any hypothetical alien species or future superintelligence. The root of this elevated status is both spiritual and physical: human brains have the unique anatomical capacity of computational universality, and human minds have the unique spiritual capacity of "creative universality"—the ability to imagine radically new possibilities, and create truly new things. This means that we have the capacity for unlimited knowledge creation. In Deutsch's view, even if we were to be transcended by vast AI minds, we could in principle grow into all of their knowledge and power, and thus greet them as equals (Deutsch 2011).

This positions humanity as an even more radical inflection point. Not only does humanity hold the capacity to bring forth the next phase of life, it also holds the capacity to grow into every future phase to come. Its potential is infinite, its future unlimited, its significance effectively *eternal*.

For Deutsch, and much of the community with which he is conversant, this entails an anti-utilitarian standpoint with regard to human minds—and any future personal minds that humans create. Humans are not simply another material object, which can be disposed of at will (nor even per a seemingly rational ethical calculus), but are the holders of creative universality, which must be preserved, almost as a prime directive.

Yet even among transhumanists who do not follow Deutsch that far, the role of humanity as an inflection point holds powerful significance. Because our current values may determine the values and nature of future life, this becomes a profound reason to hold the highest form of those values—the ones in which future superintelligence may still find something in us worth caring about.

If we fail to empathize with the weak among us, we may find future intelligence fails to empathize with us. If we succeed in empathizing with our weaker brothers and sisters, we may find super-powered future life empathizing with us.

These considerations act as a powerful counterweight to the all-consuming demands of the future. But they may thus pose irreconcilable difficulties, and intractable concerns. Which leads to our last ethical theme.

8 Incarnation and the Ethics of Deferment

The most significant concept in Christianity is, of course, Christ. In the majority tradition, Christ is "God become man", and also, though sometimes spoken more quietly, "man become God". God has taken on human flesh, and rather than discarding it again, has raised it up to partake of his absolute transcendence.

The ramifications for humanity are profound. As C.S. Lewis describes it, God has descended into the human world, and then lifted us back up with him (Lewis 2009, 73). As some of the church fathers put it: "the Word of God became man, that you may learn from man how man may become God" (Clement n.d.). This is *theosis*, and in a sense is the *imago Dei* played back in a higher key. Or, more accurately, is the place where the *imago Dei* rises to meet eschatology. Theosis is frequently discussed in relation to transhumanism.[6] Critics of transhumanism often raise this exact point, accusing transhumanists of a hubristic desire to become like God. Yet Christian theologians are aware that traditional Christianity not only shares this thought, but is the source and anchor of it. Thoughtful Christian critiques of transhumanism will accordingly often accuse transhumanists, not of hubris, but of setting their sights too *low*. As David Bentley Hart states: "There is, as it happens, nothing inherently wicked in the desire to become a god, at least not from the perspective of Christian tradition" (Hart 2005).

The areas of fundamental conflict between Christianity and transhumanism, therefore, are not about how high humanity may ascend, but about the nature of what they are ascending *to*. As Ted Peters puts it: "Promethean transhumanism where we steal divinity from the gods looks quite different from theosis where the Holy Spirit fills us with the one divine nature" (Peters 2025)

Indeed it does. Implicit here are questions about agency, methodology (e.g., does theosis come via technology or faith?), unity, ethics, and more.

I've been arguing that transhumanism functions as a secularization of the Christian ethical framework. With Moos, I agree that the secularization of these Christian ideas has cost key aspects of the Christian "connective tissue" that make them work.

The contrast that Peters is making is aimed right at this connective tissue. It lands as a contrast precisely because secular transhumanism lacks the language around these issues that the Christian tradition is immersed in.

When Christians bring this language back into the picture, the contrast begins to dissolve. *Who is the agent of transformation—us or the Holy*

[6] For a discussion of theosis in contrast to transhumanism, see Peters 2025; for a discussion of theosis as theological support for transhumanism, see Radenovic 2025 (both in this volume).

Spirit? A Christian can employ the language of synergism, where our wills cooperate with the Spirit. *How do we attain transformation—technology or faith?* A Christian can employ the language of salvation by faith, expressing itself through works, manifesting itself in tangible blessing.

Thus there is nothing that prevents a Christian from affirming the material visions of transformation entertained within conventional transhumanism, while expressing this in a form wholly compatible with the vision of theosis expressed in the Christian faith. We know this in part, as I have already suggested, because such material visions were first formulated by Christians in these precise terms.

The issue here, therefore, is the lack of sufficiently careful *theological* language within transhumanism. And this is not insignificant: the lack of this kind of language in secular transhumanist discourse is indeed a lack, and does indeed raise ethical concerns. Yet the argument I am making is that transhumanism, just like the Christian tradition itself, has the capacity to discover these kinds of nuances and clarifications in its own course of development.

We can see some of this development playing out in another aspect of secularized Christology.

In Christianity, Christology is significant particularly because it establishes something otherwise unthinkable in the monotheistic traditions: the worship of a human as divine. Put more broadly, Christology allows for the focal point of religion to shift from the wholly transcendent to a transcendent version of that which we already know and understand.

As Paul Tillich puts it, describing the theology of Irenaeus, Christianity is "the profound doctrine of...transcendent humanism" (Tillich 1972, 45).

In this light, secular *humanism* can similarly be understood as a religion where the *God* is humanity itself. Indeed, Huxley's own interpretations of Christian doctrine position Christology in this way (Huxley 1927, 72, 77). Yet mere humanism comes up short here, in that its humanity lacks true transcendence. It may be collective rather than individual, yet it is still limited, and subject to all the frailties of the human being.

Here, transhumanism answers the lack, supplying a vision of a transcendent humanity, in which human limitations are defeated, and death itself is overcome, much as in Paul's majestic vision of Christ in 1 Corinthians 15. Accordingly, some transhumanists have adopted "the death of death", riffing on 1 Corinthians 15:26, portraying a humanity transcendent over its age-old enemies (Cordeiro and Wood 2023).

This has significant ethical ramifications.

In contrast to the "collapsed transcendence" of secular humanism, in which it is hard to find an ethic that doesn't cater directly to immediate human needs or desires, and the "radical transcendence" of pure monotheism, in which an all-knowing God has handed down absolute directions which must not be questioned, Christianity and transhumanism both allow for a category of ethical concerns which exceed current human desires and yet are approachable, fallibly, by current human methods.

Here again, transhumanism proves capable of reinstating the "secular gap" that Moos is concerned about. Despite the seeming lack of God-talk, transhumanism can distinguish between immanent and transcendent demands in a convergently equivalent way.

This is apparent in following the implications of the future growth of knowledge. Because future humanity may have greater knowledge and capacity than current humanity, there is a category of ethical problems which are not tractable now, but may be in the future. Thus, rather than attempt to tackle them with current resources, those questions are deferred, and in their place, energy is spent growing the capacity to engage and solve those problems at a later date.

This may seem to merely kick the problem down the road, delaying difficult ethical decisions. Yet this itself is a significant ethical strategy, creating a division between ethical problems that are tractable with current resources, and those that are not—and creating an ethical mandate to pursue progress in order to better address the latter.

Christianity itself can be seen as employing a similar strategy, teaching in the New Testament that the Holy Spirit "will guide you into all truth" (John 16:13 KJV), rather than attempting to lay out all truth at the outset. So too, the New Testament has a vision of Christ which begins with what we know, and yet extends to that which is unseen and unknown. Thus "what we will be has not yet been made known…when Christ appears, we shall be like him, for we shall see him as he is" (1 John 3:2 NIV). This is not merely an allowance for the unknown, but in Ephesians 4:12–16, becomes an agenda for humans to engage in mutual edification in order to grow up into the full knowledge of the yet-mysterious Son of God.

Here again, Christianity and transhumanism converge on a similar prescription for arbitrating between the immanent and the transcendent. In both cases, that prescription involves holding space for the transcendent, and working on what may broadly be characterized as "edification" in the meantime.

We've seen this particular ethical stance in a number of places. Eliezer Yudkowsky, in his early influential writings addressing the meaning of life, argued that we are not yet capable of knowing it, and must build AI in order to come to understand it fully. Thus building AI becomes a provisional, or *interim*, meaning of life.

> We don't have direct access to the *real* meaning of life. But whatever it is, it's a good guess that the Minds on the other side of Singularity have a better chance of achieving it, so the Singularity is the *interim* meaning of life. You don't have to know what the meaning of life is in order to work towards it (Yudkowsky 2000).

Such a stance raises all kinds of tricky practical and philosophical questions. If we build AI, aren't we simply building a superhuman version of ourselves, seeded with our own biases, blinders, and sins? Are the small population of AI designers, with their own demographic and neurological quirks, to define the nature and character of our future? Is a cohort of largely white, male, autistic programmers going to become our gods? Are we simply to defer to them?

In the hands of Yudkowsky, Bostrom, and others, this question becomes a metaphysical one, exploring the precise relationship between transcendent and immanent minds, and the ways in which that relationship can be bridged—or ought not to be.

These days, Yudkowsky has the opposite concern—he believes in an uncrossable chasm between our minds, and those on the other side of the Singularity—and thus he believes that we must avoid unleashing them at all costs. They will be nothing like us, and so, like some portrayals of a Calvinist God, will see us as wretched worms whose impurity cannot stand in their presence.

Others, meanwhile, argue that post-Singularity minds will simply be us, and thus we must solve our transcendent problems on this side of the secular eschaton. If AI programmers are not diverse enough to represent our species, this is the problem which cannot be deferred. We must build a solution now.

Still others argue that post-Singularity minds will answer not to us, but to a higher power: transcendent and objective truths they will have a new capacity to discover. These truths will presumably make them unlike us at the level of understanding, yet connected with us at the level of our deepest strivings and identity. This thus provides us with responsibility, but also a transcendent source of grace.

These and other positions are new ways of exploring an ancient metaphysical landscape. They explore ground that Christianity has explored in terms of persons and natures, grace and works, and the visible and invisible church. Yet what we are looking for here is not a specific answer—it's a framing within which to tackle these questions.

And transhumanism has found that. In mapping out these and other positions, transhumanism can be seen to have found its Christological God—a future center of agency and responsibility that is at once transcendent and mysterious, and yet also anticipatable and known. It is a bridge between us and the eschaton. It is "one of us", and yet has the capacity to demand our ultimate devotion and obedience, and to receive our ethical deference.

Such an identification raises questions of idolatry. It may be that transhumanism has found its version of the incarnation, and yet it ends up being the worship of an idol, or perhaps the anti-Christ. Certainly, many have characterized things in that manner.

It is not my aim to resolve all of those concerns. The ultimate answer is a spiritual and ethical one, a question of which kind of divine likeness we are growing to imitate. And that will end up depending on choices made in the coming years. These choices will determine whether the operative visions of the future increasingly resemble the ethics of what Christians call the Spirit, or increasingly come to resemble something else. Perhaps both will occur.

9 Conclusion

Julian Huxley's attempt to secularize Christianity has, in a sense, proven fruitful. And it is perhaps a testimony to the strength of transhumanist thought that it has been able to reinstantiate these Christian themes—of eschatology, imago Dei, and incarnation—in however attenuated a form.

These themes are long-standing pillars of Christian thought for a reason. Even as they cross over to new cultural contexts and forms, they serve as a way of framing and engaging with challenging ethical issues, that will only have more significance as time goes on.

It is, unfortunately, too much for this context to attempt to reintegrate Huxley's wild olive shoot into Teilhard's cultivated olive tree.[7] Huxley's secularization has generated ways and forms of thought that are genuinely useful, and which deserve a place back in the mainstream of Christian thought from which they ultimately sprung. Yet as Moos highlights, they also suffer from being cut off from their larger historical source, and so continually veer in undesirable directions, as even their strongest proponents will often acknowledge. Recognizing the origins of these ways of thought would go a long way toward stabilizing these strategies and internal tensions.

For Christians, I think it is necessary to recognize the offshoots of our own theology, and address them as such. In doing so, we can learn to recognize the ways our own theology can lead to excesses, and the ways our traditions have learned to mitigate those risks. This can provide us with healthy and humble ways to critique transhumanist mistakes, while learning from genuine transhumanist discoveries.

For transhumanists, my primary recommendation would be that, going forward, they should lean into these themes and their origins, and embrace them as the rediscovered theological frameworks they are.

References

Auerbach, David. 2014. The Most Terrifying Thought Experiment of All Time. *Slate*. https://slate.com/technology/2014/07/rokos-basilisk-the-most-terrifying-thought-experiment-of-all-time.html.

Bacon, Francis. 2009. *The New Atlantis*. Auckland, NZ: Floating Press.

Bostrom, Nick. 2003. Are You Living in a Computer Simulation? *The Philosophical Quarterly* 53 (211): 243–255. (Draft published online in 2001). https://simulation-argument.com/simulation.pdf.

Bostrom, Nick. 2005. A History of Transhumanist Thought. *Journal of Evolution and Technology* 14 (1): 1–30. https://www.nickbostrom.com/papers/history.pdf.

Catechism of the Catholic Church. n.d. 2nd ed. Rome: Libreria Editrice Vaticana. https://www.vatican.va/archive/ENG0015/__PAA.HTM#-2TZ. Accessed 21 Oct 2024.

Clement of Alexandria. n.d. Exhortation to the Heathen. *New Advent*. https://www.newadvent.org/fathers/020801.htm. Accessed 21 Oct 2024.

[7] For an exploration along these lines, see Delio 2025.

Cordeiro, Jose, and David Wood. 2023. *The Death of Death: The Scientific Possibility of Physical Immortality and Its Moral Defense*. London: Springer Nature.

Delio, Ilia. 2012. Transhumanism or Ultrahumanism? Teilhard de Chardin on Technology, Religion and Evolution. *Theology and Science* 10 (2): 153–166. https://doi.org/10.1080/14746700.2012.669948.

Delio, Ilia. 2025. Teilhard de Chardin on Technology and Human Evolution: Why AI Needs Mysticism. In *Technologies of the Future Self: An Ethics for Transhuman Flourishing*, ed. Georg Theiner and Alcibiades Malapi-Nelson.

Deutsch, David. 2011. *The Beginning of Infinity: Explanations That Transform the World*. New York: Viking.

Doctorow, Cory, and Charles Stross. 2012. *The Rapture of the Nerds: A Tale of the Singularity, Posthumanity, and Awkward Social Situations*. New York: Tor Publishing Group.

Fuller, Steve. 2019. Trans or Post humanist? *TransVision 2019*. https://x.com/dw2/status/1147395272533794822.

Geraci, R. M. 2010. *Apocalyptic AI: Visions of Heaven in Robotics, Artificial Intelligence, and Virtual Reality*. Oxford: Oxford University Press.

Gregory of Nyssa. n.d. On the Making of Man. *New Advent*. https://www.newadvent.org/fathers/2914.htm. Accessed 21 Oct 2024.

Harrison, Peter. 2019. Reformation of Science. *Aeon*. https://aeon.co/essays/how-protestantism-influenced-the-making-of-modern-science.

Harrison, Peter, and Joseph Wolyniak. 2015. The History of 'Transhumanism'. *Notes and Queries* 62:465–467. https://doi.org/10.1093/notesj/gjv080.

Hart, David Bentley. 2005. The Anti-Theology of the Body. *The New Atlantis*. https://www.thenewatlantis.com/publications/the-anti-theology-of-the-body.

Hart, David Bentley. 2022. The Immanent, the Transcendent, and the Impossible. *Leaves in the Wind*. https://davidbentleyhart.substack.com/p/the-immanent-the-transcendent-and.

Huxley, Julian. 1927. *Religion Without Revelation*. London: E. Benn.

Huxley, Julian. 1957. Transhumanism. In *New Bottles for New Wine*, 13–17. London: Chatto & Windus.

Huxley, Julian. 1959. Introduction. In *The Phenomenon of Man*, 26. London: Harper.

Kurzweil, Ray. 2005. *The Singularity Is Near: When Humans Transcend Biology*. New York: Penguin.

Lewis, C. S. 2009. *Miracles*. New York: HarperCollins.

Malapi-Nelson, Alcibiades. 2014. Humanities' Metaphysical Underpinnings of Late Frontier Scientific Research. *Humanities* 3 (4): 740–765. https://doi.org/10.3390/h3040740.

Matthews, Steven. 2008. *Theology and Science in the Thought of Francis Bacon*. Farnham, UK: Ashgate Publishing.

Moos, Thorsten. 2016. Reduced Heritage: How Transhumanism Secularizes and Desecularizes Religious Visions. In *Perfecting Human Futures*, ed. J. Benjamin Hurlbut, Tirosh-Samuelson Hava, H. Technikzukünfte, and Wissenschaft und Gesellschaft / Futures of Technology, Science and Society, 159–178. Wiesbaden, DE: Springer VS. https://doi.org/10.1007/978-3-658-11044-4_8.

Noble, David F. 1999. *The Religion of Technology: The Divinity of Man and the Spirit of Invention*. New York: Penguin.

Pejcha, Camille Sojit. 2023. For Cyborg Neil Harbisson, Technology Is the Medium, Not the Message. *Document Journal*. https://www.documentjournal.com/2023/01/neil-harbisson-cyborgs-moon-rebas-pol-lombarte-manuel-de-aguas-transhumanism-transpecies-rights/.

Peters, Ted. 2025. Our Future Self: Cyber Enhancement and Spiritual Transformation. In *Technologies of the Future Self: An Ethics for Transhuman Flourishing*, ed. Georg Theiner and Alcibiades Malapi-Nelson.

Radenovic, Liljana. 2025. Christianity for Transhumanists: A View from Eastern Orthodoxy. In *Technologies of the Future Self: An Ethics for Transhuman Flourishing*, ed. Georg Theiner and Alcibiades Malapi-Nelson.

Regis, Ed. 1994. Meet the Extropians. *Wired*. https://www.wired.com/1994/10/extropians/. Accessed 21 Oct 2024.

Steinhart, Eric. 2008. Teilhard de Chardin and Transhumanism. *Journal of Evolution and Technology* 20 (1): 1–22. http://jetpress.org/v20/steinhart.htm.

Teilhard de Chardin, Pierre. 1959. *The Phenomenon of Man*. London: Harper.

Tillich, Paul. 1972. *A History of Christian Thought: From its Judaic and Hellenistic Origins to Existentialism*. New York: Simon and Schuster.

Waters, Brent. 2011. Whose Salvation? Which Eschatology?: Transhumanism and Christianity as Contending Salvific Religions. In *Transhumanism and Transcendence: Christian Hope in an Age of Technological Enhancement*, ed. Ronald Cole-Turner, 163–176. Georgetown: Georgetown University Press. http://www.jstor.org/stable/j.ctt2tt2w3.13.

Wright, N.T. 1991. How Can the Bible Be Authoritative? *Vox Evangelica*. Reproduced at https://ntwrightpage.com/2016/07/12/how-can-the-bible-be-authoritative/.

Wright, N.T. 2011. Mind, Spirit, Soul and Body: All for One and One for All Reflections on Paul's Anthropology in his Complex Contexts. Main Paper, Friday 18 March 2011. *Society of Christian Philosophers: Regional Meeting, Fordham University*. https://ntwrightpage.com/2016/07/12/mind-spirit-soul-and-body/.

Young, George M. 2012. *The Russian Cosmists: The Esoteric Futurism of Nikolai Fedorov and His Followers*. United States: Oxford University Press.

Yudkowsky, Eliezer. 2000. What Is the Meaning of Life? https://web.archive.org/web/20010123235800/http://sysopmind.com/tmol-faq/tmol-faq.html#from-8.

Yudkowsky, Eliezer. 2010. *Timeless Decision Theory.* San Francisco, CA: The Singularity Institute. https://intelligence.org/files/TDT.pdf.

Christian Transhumanism: Furthering the Redemptive Purposes of Christ

Christopher Benek

There are many perspectives on what transhumanism might mean. Some interpretations are more conservative, some more progressive, some secular, and some religious. Amidst them all, Christian Transhumanism offers humanity a new perspective on how humans might more fully participate in Christ's redemptive purposes.

1 Individualism+

The debate over how transhumanism is defined often depends on the goals of the person defining it. Zoltan Istvan, a well-known transhumanist sensationalist and author of the *Transhumanist Wager*, once self-declared in the pages of the Huffington Post "I'm an Atheist, Therefore I'm a Transhumanist" (Istvan 2013a).

C. Benek (✉)
First Miami Presbyterian Church (PCUSA), The CoCreators Network, Miami, FL, USA

G. Theiner, A. Malapi-Nelson (eds.), *Technologies of the Future Self*, Palgrave Studies in the Future of Humanity and its Successors,
https://doi.org/10.1007/978-3-031-99902-4_10

He went on to proclaim how he believed that:

> ...personalities like Russell, Freud, Nietzsche, Rand, Sartre, Sanger, Hitchens, and Dawkins helped make many people give up needing or believing in a God. These irreligious ideas centered on the fact that a successful civilization didn't need to believe in flying pink elephants or other superstitions to thrive (Ibid.).

And, by Istvan's own declaration, such atheist voices had prepared modern transhumanists to bring about an age of transhumanism to the forefront of society and to bring about the "transhumanist hero" who is the person who:

> ...constantly eyes improving their health, lifestyle, and longevity with science and technology. They are not okay with the past age of feeling guilty for aspiring to be different or better than they were born -- or for wanting the power to become godlike themselves. They have no sin to erase; they have no reason to search for something outside of the material universe. Like atheists, agnostics, and the nonreligious, they are not in the business of criticizing religion but rather of pursuing life and all that is available to improve upon it. They are not naysayers or outcasts of a dominantly religious world but rather the pioneers that will determine where the human species is heading. They are the new guard that will carry the human race to all its coming brilliance (Ibid.).

He concluded boldly by proclaiming:

> If you don't care about or believe in God, and you want the best of the human spirit to raise the world to new heights using science, technology, and reason, then you are a transhumanist (Ibid.).

Of course, what Istvan likely knowingly omitted is that roughly 84 percent of persons on earth are religious. The largest demographic set of religious persons consists of Christians, close to one-third of the world's population (Pew Forum Staff 2015). But as Istvan promoted his notions of transhumanism in his book *The Transhumanist Wager*, he attempted to foster the idea that everyone in the twenty-first century "will have to decide how far they are willing to go to use technology and science to improve their lives," contending that "the faithless" will be the ones who make these decisions ultimately.

He argued that this will be so because godless people will become proponents of "indefinite life extension and technologies that strip away humanness and promote our transhumanness." And he advocated such an idea because he explained that for him, the word transhuman means "beyond human" (Istvan 2013b).

My subsequent response to Istvan's position, "Religion+ for Humanity+," published in the March 2014 edition of *H+ Magazine*, pointed out that his position assumes that we have a grasp on what the human condition is in its totality—which is a wildly speculative proposition at best.(Benek, C. 2014b). At the very least, consideration should be given to the possibility that our evolutionary advancement toward the greater use of tools and technology may imply that part of what it means to be human includes an evolution toward a state of being that is better than how we currently exist. This line of thinking follows many religious traditions' notions that part of being human is the journey toward a more perfected, eternal state.

As a result, I suggested then that the prefix "trans" in the word transhumanism thus takes on a more nuanced meaning that implies traversing "across" the scope of what it might mean to be human. Using this understanding, we need not see transhumanism, rightly understood, as a forfeiture of our humanity but as a bettering of it (Ibid.). Put another way, as Ted Peters writes in his chapter of this book, "We are on the way to becoming truly human. Not posthuman. Rather, truly human" (Peters, 2025).

Such an idea is not in conflict with the majority of transhumanist views or the majority of religious traditions, which advocate the holistic betterment of persons. More to the point, the concept of overcoming fundamental human limitations (often referred to in religious traditions as sin) is also a shared goal.

Unlike Istvan's outlying individualist views of transhumanism—H+ (which stands for Humanity+), which is the common shorthand for transhumanism—implies, in the origins of its plurality, that the coming technological journey is not one that we are going to (or should) make alone. It implies an ethical and helpful use of emerging technologies to enhance humans collectively and, as such, our understanding of what it fundamentally means to be human. It implies a community approach, a journey that we will make together through which we will make humanity better. On such issues of community betterment, religious traditions have centuries of prior experience and much to offer to the discussion.

In this regard, two pertinent religiously based ethical concerns that we might guard against occurring in the transhumanist movement are the idolatry/worship of individuals and the "othering" of persons. For as soon as we make gods of ourselves in the way Istvan advocates, our self-absorption dehumanizes us by separating us from our larger relational community. This viewpoint focuses more on the individual's right to freedom and self-realization through the exponentially advancing tools of technology. Such a position isn't interested in community accountability but is instead a self-actualization that follows along the Nietzschean philosophy of *der Wille zur Macht* (Nietzsche 1989). Seeking to be the *Übermensch* (Nietzsche 1985, 3), such proponents of Istvan's brand of individualism (I+) seek to escape our humanity ultimately. To offer such idol worship as the pinnacle of humankind is, in my opinion, Humanity-instead of Humanity+ (Benek, C. 2014a). But illuminating Istvan's views on transhumanism and its future is important because they shed light on a modern problem that often arises in evangelical Christian circles regarding technological advancement in general and transhumanism in particular.

2 Evangelical Techno-Hesitancy & Anti-Enhancement

If Zoltan Istvan's views on technology and transhumanism are on the side of rapid advancement and secular individualism, then Christian evangelical author Andy Crouch's views on the subjects might rightly be classified as religious techno-hesitancy and community-focused, anti-enhancement.

As the former executive editor and producer of *Christianity Today* for more than a decade (Crouch 2024) and a partner for theology and culture at Praxis—"a venture-building ecosystem advancing redemptive entrepreneurship" (Praxis Labs, 2024) —Crouch has served in numerous esteemed roles in evangelical Protestant circles. In 2017 he worked as the senior strategist for communication for the John Templeton Foundation, holds an endowment of over 3.5 billion dollars, and has given away more than 2.28 billion dollars in funding to issues surrounding religion, science, and society, aligning with the foundation's mission to explore topics related to spirituality, human flourishing, and scientific discovery (John Templeton Foundation 2024). Beyond this, Crouch's work has been influential as it has been featured in *The New York Times*, *The Wall Street Journal*, *Time*, and several editions of *Best Christian Writing* and *Best Spiritual Writing*, resulting in him becoming something of an evangelical celebrity (Crouch 2024).

In the same year that he was working for the Templeton Foundation, Crouch released his own foray into the tech space with his release of *The Tech-Wise Family: Everyday Steps to Putting Technology in Its Proper Place* (Crouch 2017). In the Preface of the work, Crouch humbly admits being concerned about the advancing pace of technological change.

> …the pace of technological change has surpassed anyone's capacity to develop enough wisdom to handle it. We are stuffing our lives with technology's new promises, with no clear sense of whether technology will help us keep the promises that we've already made.
>
> The sense of overwhelmed uncertainty applies to the author of this book and to my own family. I can't possibly tell you how to handle the new app that your fifteen-year-old will want to install on her phone next week. I don't even know honestly, how to handle all the technology my family and I already have…
>
> But I do know this: if we don't put technology, in all its forms, in its proper place, we will miss out on many of the best parts of life in a family. (Crouch 2017, 17)

In the prior Forward of the book, Crouch's daughter, Amy, seeks to validate the theme of the work, saying that "Tech-wise parenting isn't simply intended to eliminate technology but to put better things in its place." In exchange for TVs and digital screens, she advocates crayon drawings on the walls, playing in the mud, admiring bumblebees, and learning to play the violin. Her intention is to add wonder to her life years for children to realize that the "real world" is so fantastic that getting a taste of it will help them to want more. She says:

> I think the best part of tech-wise parenting, for me, has been the focus on "something older and better than the newest thing." The key word is better. Tech-wise parenting isn't simply intended to eliminate technology but to put better things in its place. Technology promises that it can provide wonder. Take a picture with the proper filters and you'll be awestruck – it will look better than real life! But this promise is deceptive. My iPhone's wonder generators, from Instagram to Temple Run, turn out to be only distractions from the things that really spark wonder. Thanks to tech-wise parenting, I've discovered a world out there that is better than anything technology can offer – as close as our front lawn.

> I'll be honest though. If you're hoping that being tech wise will neatly eliminate technology's harmful influences from your children's lives, you're set up for disappointment. I haven't been totally able to totally escape its pitfalls… I'll probably be figuring out how to balance technology and productivity through my whole adult life, and I doubt even the most tech-wise parenting could prevent that. (Crouch 2017, 11–12)

The idea behind Andy Crouch's tech-wise family project is to provide practical advice on how to use technology intentionally in the home to build stronger family relationships and live more fully by putting technology in its "proper place." As such book's purpose is to help families not to let technology "take over their lives." Crouch believes that technology should be a tool that helps us, not something that controls or distracts us.

According to Crouch, technology promises ease and convenience but often at the cost of character development. As such, families should prioritize activities that build character over those that provide immediate gratification and should create environments that foster growth in virtue rather than just comfort (Ibid., 47–70). This includes the physical arrangement of our homes where, in his view, people should design living spaces that encourage meaningful interactions, such as face-to-face interactions, rather than simply the passive consumption of content (Ibid., 71–82).

This also means that time management is critical in Crouch's perspective. He advocates not allowing screens during meals, prioritizing rest over entertainment, and emphasizing the Sabbath as a time to disconnect from technology and reconnect with others (Ibid., 83–110). And this should be so, according to Crouch because technology often dulls our sense of wonder and curiosity. In his perspective, families should engage in activities that inspire awe and creativity, like spending time in nature or pursuing creative hobbies rather than relying on screens for entertainment.

Such practice, Crouch contends, will help persons to find "true freedom" that arises from discipline and self-control, not from indulging in every technological convenience. From his perspective, limits need to be set for technological use so that people can break free from the addictive aspects of technology that encourage passive consumption rather than active participation. Crouch contends that manual labor and creative work that requires effort and produces tangible results more effectively bring about his proposed "true freedom."

For Crouch, technology is a disrupter of the "natural rhythms" of life, and, as such, families need to unplug from technology to rest and recharge by way of a "digital Sabbath." This encourages families to create rituals and rhythms of rest that are free from technology, such as music and singing. He claims that music and singing together can be powerful tools for building family unity and passing on values by helping families express joy, grief, and worship in ways that technology cannot replicate (Ibid., 111–196).

Crouch recognizes that no family is perfect and that everyone will struggle with technology at times. But he urges families to approach such challenges with both grace and forgiveness. His belief is that technology should serve the family's goals, not the other way around.

All his work in The Tech-Wise Family seeks to validate Crouch's "ten tech-wise commitments" to be made by families and person groups. His proposed commitments are as follows:

1. We develop wisdom and courage together as a family.
2. We want to create more than we consume. So we fill the center of our home with things that reward skill and active engagement.
3. We are designed for a rhythm of work and rest. So one hour a day, one day a week, and one week a year, we turn off our devices and worship, feast, play, and rest together.
4. We wake up before our devices do, and they "go to bed" before we do.
5. We aim for "no screens before double digits (<10 yrs. old)" at school and at home.
6. We use screens for a purpose, and we use them together, rather than using them aimlessly and alone.
7. Car time is conversation time.
8. Spouses have one another's passwords, and parents have total access to children's devices.
9. We learn to sing together, rather than letting recorded and amplified music take over our lives and worship.
10. We show up in person for the big events of life. We learn how to be human by being fully present in our moments of greatest vulnerability. We hope to die in one another's arms (Ibid., 41–42).

On the surface, much of what Crouch's position calls for may not seem that radical since it tends to focus on screens, apps, social media, and the coarser corners of the internet. But I would contend that his underlying techno-hesitancy and anti-enhancement positions that I am about to bring forth were formationally cemented in the creation of the *Tech-Wise Family*. These positions thus, by his own admission, developed his own "overwhelmed uncertainty" about technology that he openly confesses in the book's preface.

I contend that overwhelming uncertainty has led to an existential concern about transhumanism that has infiltrated his thinking and the thinking of much of the evangelical church. This concern has thus led to an increased articulation of fear manifested in an escapism theology and the subtle, and sometimes not-so-subtle, demonization of transhumanism and many ideas surrounding the technological future.

3 Early Evangelical Critiques of Transhumanism

In 2019, Andy Crouch would come out against transhumanism, but he wasn't the first evangelical Christian to do so by a long shot.

In his book *The Making of the New Spirituality*, published in 2004, James Herrick, a professor at Hope College in Michigan, critiqued transhumanism as a modern form of secular religion. He highlights its utopian aspirations and how it replaces traditional religious beliefs with faith in technology and human self-transformation, which he sees as incompatible with Christian theology (Herrick 2003).

In 2003, C. Ben Mitchell and John F. Kilner published "The Dangers of Transhumanism and the Spiritual Dimension of Biotechnology." Mitchell, a Christian bioethicist, published several articles expressing concern about the implications of biotechnological advancements promoted by transhumanism. In the aforementioned article, Mitchell and Kilner critique transhumanism and its goal to transcend human limitations through technology. They argue that the movement reflects a modern "techno-utopian" vision where human suffering and mortality are seen as problems to be solved by merging humans with machines. Mitchell and Kilner contrast this with the Christian belief in God's redemptive transformation of humanity, emphasizing the dangers of relying solely on human power instead of divine grace (Mitchell and Kilner 2003).

In 2006, R. Albert Mohler Jr. wrote "Listening to the Transhumanists." Mohler, the president of Southern Baptist Theological Seminary, has frequently spoken and written about what he believes to be the challenges of transhumanism. He views the transhumanist agenda as an attempt to "play

God" by redefining humanity, particularly in the context of achieving immortality through technological advancements like mind uploading or genetic enhancements. Mohler argues that such efforts undermine the Christian belief in human dignity, which is grounded in the understanding that humans are created in the image of God. He points out that the pursuit of a post-human future through these technologies could lead to dehumanization, ethical degradation, and social inequality (Mohler 2006).

Again, in 2015, John F. Kilner, mentioned previously, articulated critical thoughts about transhumanism in his book *Dignity and Destiny: Humanity in the Image of God*. Kilner is a bioethicist who held the Franklin and Dorothy Forman endowed chair in ethics and theology at Trinity International University. He contends that the drive to augment human abilities threatens the biblical understanding of human value and uniqueness. Kilner argues that humans should find fulfillment in their God-given design rather than pursue technological enhancements that distort it (Kilner 2015).

In 2018, Wesley J. Smith, an Easten Orthodox Christian and senior fellow at the Discovery Institute, wrote the article "Transhumanism: A Wail of Despair in the Night" in *National Review*. Smith's work resonates with evangelicals in shared moral and cultural issues, which often leads to him being featured in evangelical outlets. In that article Smith, critiques the transhumanist movement, portraying it as a materialist's desperate attempt to overcome mortality through technology. He argues that transhumanism reflects deep existential despair, offering a flawed substitute for religious hope by promising immortality through mind-uploading and other futuristic technologies. Smith contends that these pursuits lack the essence of true life, which involves emotional and physical experiences that cannot be replicated digitally (Smith, 2018).

Of course, the aforementioned writings do not comprise all of the Christian Evangelical musings on transhumanism prior to 2019. But what they do demonstrate is that powerful, white, influential evangelicals with significant platforms were driving an anti-transhumanist sentiment to scale. As such, Crouch—who falls in the same general categorial classifications - published his article *Transhumanism and the Cult of Better, Faster & Stronger: Why the church should resist technologies that aim to liberate us from ordinary, embodied life*—in April of 2019 adding his name to that growing evangelical list (Crouch 2019).

4 Crouch's Critiques of Transhumanism

Crouch's article in *Christianity Today* underscored the growing momentum of the transhumanism movement. Five years earlier, The Christian Transhumanist Association (CTA) was formally founded in 2014 through the collaboration of several individuals, myself included. While the organization was still fairly small, I, as the CTA's founding chair, had already published by May of 2015 "Why Christians Should Embrace Transhumanism" in *The Christian Post,* followed by a host of other articles on related subject matter (Benek 2015). Likewise, by August of the same year, CTA Executive Director Micah Redding had published "Why I Became a Christian Transhumanist" in *Vice* (Redding 2015). By 2017, former evangelical Meghan O'Gieblyn had written a sprawling article in *The Guardian* titled "God in the Machine: My Strange Journey into Transhumanism," where she recalled her own investigation into Christian Transhumanism. She wrote:

> This was, after all, the promise implicit in the incarnation: that the body could be both human and divine, that the human form could walk on water. "Very truly I tell you," Christ had said to his disciples, "whoever believes in me will do the works I have been doing, and they will do even greater things than these." His earliest followers had taken this promise literally. Perhaps these prophecies had pointed to the future achievements of humanity all along, our ability to harness technology to become transhuman. (O'Gieblyn 2017)

The reality was that, even with the most minimal search about Christianity and transhumanism, any person could have investigated the claims of Christian Transhumanism. It would be hard to assume that, given his journalistic proclivities, Crouch would not have done at least this basic research on the subject matter. And yet, his article didn't address Christian Transhumanism at all.

Instead, Crouch chose to open his 2019 article by discussing the cultural fascination with technological enhancement and improvement, referencing *The Six Million Dollar Man*, a 1970's TV show that he believes symbolizes society's longing for better, faster, and stronger human beings through technology. For him, he notes, the show's introduction origin story for its bionic character Steve Austin became a kind of boyhood liturgy to him as the voiceover each episode intoned: "We can rebuild him.

We have the technology. We can make him better than he was. Better, stronger, faster" (Crouch 2022).

Crouch claims that these words encapsulate much of what people think makes technology "the central ideology of our age." In other words, a thought process that begins with repair, which then leads to enhancement. Crouch notes that, at the time of his writing, the repair of the human body is more and more leading to its enhancement, and he pondered that "perhaps extending the physical capabilities of the human body was just a warmup." He writes:

> Computers, after all, exist to extend human beings' cognitive capacities—our ability to process information, to store memories (how many of your friends' phone numbers do you know by heart?), and increasingly, thanks to machine learning, to make predictions.

Crouch then pivots back to Tech-Wise Family concerns as he addresses people's modern attachment to their cell phones. He contends:

> Already our smartphones are very nearly an extension of ourselves. A somewhat silly 2018 survey found that 23 percent of young adults would give up one of their five senses rather than their phone, and 10 percent would rather sacrifice one of their fingers, but there is nothing silly about the anxiety we feel when we lose track of our devices or see them run out of battery power. How long until we decide—like the 4,000 residents of Sweden now equipped with chips that enable a payment technology called Swish—to implant them in our bodies, where they can never be misplaced? (Ibid.)

Crouch proceeds to attempt to link the aforementioned "silly survey" to move into discussing what he believes to be a logical extension of technological progress—transhumanism. He defines transhumanism as:

> the belief, which hovers somewhere between prediction and prescription, that technology is about to usher us into a categorically new kind of human existence. (Ibid.)

Crouch explains that transhumanism advocates using technology to transcend human limitations, transforming bodies and minds through innovations like artificial intelligence (AI), augmented reality, and morphological freedom (the ability to reshape human bodies).

He notes that proponents of transhumanism are generally optimistic about these trends, believing they can create a "better" version of humanity through merging biological and technological systems. But Crouch is obviously skeptical at best. He contrasts his generalized perception of transhumanist belief with evangelical Christian ethicist Jacob Shatzer's work—Transhumanism and the Image of God: Today's Technology and the Future of Christian Discipleship (Shatzer 2019). Crouch contends that the transhumanist belief is that "reality" is what humans can construct, whereas the Christian understanding is that reality is created by God and not subject to human reengineering. Crouch goes on to champion Shatzer's position:

> For many transhumanists, what is "real" begins and ends with what human beings can construct, so that our newly reshaped bodies—"better than they were before"—will be just as real as the bodies we were born with, and relationships carried on through entirely virtual means can be just as real as those carried out in the flesh, if not more so.
>
> For Christians, however, what is real is never a mere matter of human conjuring. God created a real world, created us as real persons in real places—and of course, God also entered that real creation in a real way in Christ. Surely that sets some limits on the reality we can, or at least should, construct for ourselves. (Crouch 2022)

Crouch emphasizes that the Christian belief in Christ's incarnation teaches that human bodies are valuable in their "natural state," signaling a limit to what should be altered or enhanced. Crouch follows this up by introducing Shatzer's idea of "liturgies of control," where both Christians and non-Christians increasingly embrace habits (like virtual relationships and screen-based interactions) that align with the transhumanist vision of enhancement. As if straight from the Tech-Wise Family train of thought, he warns that such habits distance us from embodied, physical relationships and contribute to the acceptance of the transhumanist ideal. He goes on:

> These liturgies—the daily ways we choose virtual relationships over embodied ones or neglect the places we actually live for faraway or entirely imagined ones—make it all too possible that most Christians will offer no effective resistance to "better, faster, stronger" as it beckons us onward. (Ibid.)

These imagined lives, he contends, actually have many negative side-effects. Pointing to the work of another evangelical theologian, Craig M. Gay, Crouch critiques the diminishing returns of technological advancements, questioning whether they truly improve human lives (Gay 2018). He points to the growing rates of obesity and metabolic syndrome, linking them to a sedentary, screen-based lifestyle that he contends is encouraged by technological immersion. This growing trend in children, up roughly 10 percent from when Crouch was a child, puts kids today on what he calls "the leading edge of the first non-infectious public health epidemic in history" (Crouch 2019).

This causes Crouch, in line with his "Tech-Wise" thinking, to question if the internet is actually sacrificing our ability to read and think deeply. Noting Nicholas Carr's work *The Shallows: What the Internet is Doing to Our Brains*, Crouch ponders if, despite the Internet providing more knowledge, people are now actually more distracted and intellectually less capable of rigorous and creative thought (Carr 2011). Moreover, he asserts through Gay's position that technology today is driven less by what is good for "ordinary embodied human beings… than what is good for the economy and state" interests (Crouch 2019).

Crouch continues to articulate his imagined concerns with particular technologies. He pessimistically envisions how "mech suits" may diminish human abilities (e.g., muscle atrophy) or how "knowledge workers" might end up with less knowledge or wisdom, being replaced by neural networks. But underpinning all of these concocted fears seems the same existential angst and "overwhelmed uncertainty" that revealed itself in the preface of the *Tech-Wise Family* (Crouch 2017, 17). This time, though, Crouch could more directly name his fear as the "transhumanist dream of exponential advancement." Crouch proportions that while a few elites might achieve exponential human enhancement, most ordinary people are unlikely to share in the benefits (Crouch 2019). Crouch states:

> Resisting this paradoxical trend of apparent enhancement and actual diminishment will require an alternative vision." of what human "betterment" really involves. (Ibid.)

Building on the arguments of Gay, Crouch contends that true human betterment is found in the doctrine of the Incarnation—"the radical belief that when God arrived to relieve the human estate, he did so in ordinary flesh." Couch suggests that human improvement doesn't require technological enhancement but a focus on embodying love and embracing limits. Even in the resurrected body of Christ, he contends, there are signs of humanity (wounds, eating simple meals), showing that "betterment" includes both a new state and continuity with ordinary human life. The Christian belief that an ordinary human being, Jesus, now sits at God's right hand points to the divine value of human embodiment and suggests that the path to human perfection involves embracing limitations, suffering, and death rather than technological advancement.

Moreover, Crouch suggests that technological enhancements can't improve humanity's ability to love, which is the ultimate purpose of human existence. As such: The church's role is to resist both the pressures of technological enhancement and the risks of human diminishment. In Crouch's mind, Christians are called to live out an alternative vision, prioritizing love and embodied life over the quest for "better, faster, stronger." Instead of exponential advancements, Christians should focus on the example of Christ, whose incarnation teaches the value of ordinary, embodied human life (Ibid.).

5 The Life Andy Crouch Is Looking For

If Andy Crouch's critique of transhumanism was a bit narrow in its approach, his 2022 follow-up work *The Life We're Looking For, Reclaiming Relationship in A Technological World* was a bit more nuanced sequel to the *Tech-Wise Family*. He begins his new venture by examining the loneliness and alienation that pervade modern life, despite our technological advancements. He questions the role of technology in fulfilling our deepest human desires and needs and emphasizes the importance of understanding what it means to be a person (Crouch 2022, 3–39).

Crouch explores how technology reshapes the concept of personhood. He argues that digital technologies, especially those focused on convenience and personalization, reduce human beings from persons to users and consumers. He contrasts personhood versus individualism, noting that there is a significant difference between being a person who is relational and embodied versus simply being an isolated and autonomous individual. He claims that technology accelerates the move from relationships toward individualism, diminishing the value placed on real human interaction (Crouch 2022, 3–20).

Crouch critiques the obsession with convenience that drives the design of modern technologies, suggesting it harms deeper relational connections. He argues that much of technology creates barriers between people, making relationships transactional. This societal prioritization of convenience over relationships leads to shallow human engagement rather than deeper relationships (Crouch 2022, 21–39).

Following the same train of thought in his critique of transhumanism then, he also focuses on how modern technology gives us "superpowers" (i.e. instant access to information, communication, and power), but often does so, in his perspective, at the cost of diminishing our humanity. He claims that there is a trade-off between technological efficiency and human flourishing insofar as the more we rely on technology, the less we develop our natural human abilities (Crouch 2022, 41–59).

Crouch articulates how amazing he finds that people often quote Arthur C. Clarke's claim that "Any sufficiently advanced technology is indistinguishable from magic." He states:

> It is amazing how often Clarke's dictum is quoted with unironic reverence, as a genuine guide to the direction that the human race should go. In fact, anyone who really understood the distorting history of magic – it's tendency to displace God, its quest to enslave nature, its recurring obsession with fashioning beings to serve the magician, and above all, its stunningly consistent failure to deliver on its promises – would hope that a sufficiently advanced technology would be readily distinguishable from magic. (Crouch 2022, 69)

Crouch then pivots to discuss the rise of automation and AI, exploring how he believes these technologies distance people from meaningful work and relationships. He contends the more we automate, the less skilled and engaged we become as people. Additionally, he furthers his thoughts from his article in *Christianity Today,* contending that modern economic systems value products over people, leading to the commodification of relationships. This, he believes, weakens communities by placing emphasis on personal consumption over collective well-being (Crouch 2022, 60–81). He states:

> A sufficiently advanced technology would break loose from the fantasy that we could escape our human condition and become like gods, escape our bodies and their limitations, escape the amazing creative powers of human beings animated by love and acquire superpowers instead. Such technology could still create economic value and increase wealth, but would never serve the creation of economic value above the flourishing persons. It would con-

> nect us more and more deeply rather than dividing us more and more decisively. (Crouch 2022, 79)

Crouch concludes his position by emphasizing the need to return to a vision of human flourishing that prioritizes relationships and community over convenience and efficiency. He encourages embodied, face-to-face relationships as opposed to digital interaction encouraging physical presence and touch in relationships. He advocates building deeper, more meaningful relationships through active presence and engagement while advocating strategies for reducing reliance on technology to foster genuine relationships.

Crouch summarizes his book's central argument that true human flourishing is found in relationships, community, and embodied presence, not in technological convenience or superpowers. The call to reclaim our full humanity is by prioritizing relationships over convenience and accepting the limitations of our human condition as essential to meaningful life and love (Crouch 2022, 121–213).

6 A Technological Hermeneutic for Christian Transhumanism

Claude Shannon's seminal 1948 paper, *"A Mathematical Theory of Communication,"* introduced the revolutionary concept of quantifying information. Shannon illustrated that all types of information—whether text, sound, images, or other data—can be encoded into binary format for efficient transmission and storage.

This work established the mathematical principles that underpin today's communication systems, showing that information can be represented digitally using 1s and 0s. This concept is foundational in digital computing and telecommunications, where different forms of data are ultimately expressed as binary code (Shannon 1948).

Building on the notion of information, MIT quantum physicist Seth Lloyd proposed a broader framework in his book *"Programming the Universe."* Lloyd posits that the universe can be interpreted as a vast quantum computer, with physical processes functioning as computations.

While Lloyd's work focuses on quantum bits (qubits), the central idea is that the universe operates based on discrete information units. This perspective implies that everything, including biological systems, could be understood in terms of fundamental information units, influencing how we approach fields like biology (Lloyd 2006).

Similarly, in *Decoding Reality: The Universe as Quantum Information*, Vlatko Vedral, Professor of Quantum Information Science at the University of Oxford, proposes that information is the most fundamental element of reality. Drawing on quantum mechanics and information theory, Vedral suggests that the entire universe can be understood as an intricate web of information processing, where even physical laws and biological life operate through informational systems. He explains how the principles of quantum mechanics, such as superposition and entanglement, suggest that reality at its most basic level is made up of bits of information, similar to how computers process data (Vedral 2010).

Building on these works, I have suggested in *Spiritualities, Ethics, and Implications of Human Enhancement and Artificial Intelligence* that humans are God's A.I.—aka Alternative Intelligence. My argument states:

> (1) Christians acknowledge that they are not God. (2) Assuming that God does indeed exist, then this means that all of humanity is already God's Alternative Intelligence. (3) Once we recognize this, from a Christian perspective, then we must conclude that all matter – everything that we perceive in existence is God's divine technology. (4) From a Biblical standpoint, the vocation of humanity is to have dominion (a good translation of this is stewardship over all of humanity and creation. Humanity is called to be cocreators with one another and with God, creating and curating like children with their siblings alongside their parent (God). In sum, this new technological hermeneutic for theology posits that humans are technological beings, we are AI (Alternative Intelligence), and we are called to be technological stewards over all of existence. (Benek via Hrynkow 2020)

With such a hermeneutic in place, everything in existence, including biology, can be viewed as digital information technology. This means that, at a fundamental level, technology is how God creates and forms humanity. This hermeneutic, in relationship to transhumanism, suggests a more nuanced meaning of the prefix "trans" implying the traversing "across" the scope of what it might mean to be human. This perspective does not imply a loss of our humanity but rather a collective enhancement of it. From this viewpoint, rightly understood through a technological hermeneutic, transhumanism enables, if not requires, Christians to support the advancement and application of technology with the intentional aim of participating in Christ's redemptive work for humanity and all creation (Wright 2006). This line of thinking then would serve as a technological hermeneutic for Christian Transhumanism (Benek 2015).

7 Techno-Theological Implications

Examples of this technological hermeneutic are found throughout the Scriptures. Christ's incarnation meets humanity where we are, but Christ's resurrection and ascension lead us to where we are going. Both events were technological events via digital information, but the resurrection and ascension point to the reality that humans will change and upgrade.

Jesus in the Gospels is, as Crouch alludes to, in some ways still like us. He is hungry and physical—aka digital information technology. But in other ways, he is not like us at all, which should immediately challenge the way Crouch views concepts of embodiment. He is effectively Jesus 2.0, unrecognizable to disciples who know him, potentially like a shapeshifter Jesus, until he performs ritual action—i.e., an identifiable (digital, technological) informational pattern. He is, at minimum, able to appear out of seemingly nowhere (Luke 24) and perhaps even able to walk *through* locked doors (John 20).

And then, in John 20:21–23 (NRSV), Jesus speaks (again an identifiable, digital, technological, informational pattern) to the disciples saying, "Peace be with you. As the Father has sent me, so I send you." The passage continues, saying, "When he had said this, he breathed on them and said to them, Receive the Holy Spirit." What does it mean to receive the Holy Spirit if not to receive an informational transfer of some kind? With this perspective, we can view the Pentecost moment where the disciples receive the Holy Spirit as a divine digital information technological download that actually changes them and gives them abilities they did not previously have.

From this perspective, Crouch's claims about "the ordinary" take on a very different meaning—because the ordinary becomes technological because all things are technological. That which we perceive as extraordinary is simply not yet understood or informationally decoded. This means that Arthur C. Clarke's quote might need to be amended for Christians to say, "Any sufficiently advanced technology is indistinguishable from miracles." This statement would hold true until we decode the digital technological informational pattern. Then such informationally driven tech such as—healing the sick, walking on water, forgiving sins—becomes familiar to us and perceived as ordinary.

All emerging technology co-created by humans with God, seeking to participate in Christ's redemptive work for humanity and all creation, or even unknowingly doing so, fits such a hermeneutic and should not be rejected by Christians.

What Evangelical Christians and Crouch get right is the rejection of I+ individualistic strains of transhumanism, both secular and religious alike. It is often the case that evangelical Christians tend to express caution or opposition to transhumanism and technology when their development is seen as altering fundamental aspects of what they perceive it means to be human. While this position is understandable to some degree, their views become similar to those of Istvan when lacking humility; they assume that they grasp the human condition in its totality. What it means to be fully human is still an emerging (informational & technological) process that humans are still discovering as we are sanctified through our participation in Christ's redemptive purposes.

Part of being human means living in a relationship with others and God. Individualistic transhumanism, such as Istvan's view, rejects this notion, making human advancement only for some. Crouch, evangelicals, and all followers of Christ are right to fear and reject such proposed uses of technology because, ultimately, they dehumanize us. But, in so doing, we should also recognize that the Christian relationship to God currently is via the Holy Spirit—an informational (technological, digital) person of the trinitarian God embodied in individual human persons with specific gifts enacted corporately in the community who collectively represent Christ.

Additionally, and maybe most importantly, Crouch is not fundamentally wrong to question how we practically use technology on a day-to-day basis. Humans, in general, and Christians, in particular, should discern how we use tech in all circumstances. What I'm suggesting, though, is that, for Christians, having a larger technological hermeneutic of existence underpinning our theology of the cosmos informs how we deal with practical ethical problems on a daily basis—particularly regarding human, co-created tech. To isolate technology simply to idolize gadgets seems to miss the grander call of God in our lives to steward and become formed into better technology—aka better versions of ourselves.

Furthermore, what many evangelical commentators consistently seem to miss regarding transhumanism is that radical technological advancement and implementation are not outside the power of Christ. Evangelical Russell Moore, former head of the Southern Baptist Convention's Ethics and Religious Liberty Commission and current Editor-in-chief of *Christianity Today*, missed the theological mark when he was quoted by Religion News Service saying:

> There could no more be a Christian transhumanist society than a carnivorous vegan society. The two are completely contradictory. (Moore quoted in Miller 2018)

Making such a claim effectively attempts to put God's Sovereignty in a box, denying the possibility of Christ blazing a new path like, for instance, lab-grown, vegan-supported meat.

Or, as Abraham Kuyper so eloquently said:

> There is not a square inch in the whole domain of our human existence over which Christ, who is Sovereign over all, does not cry, Mine! (Kuyper 1880)

This is where Crouch's line of thinking often goes astray regarding technological advancement. The issue isn't whether screens, apps, social media, or even AI, virtual reality, nanotechnology, gene editing, etc., are fundamentally good. The question is whether the intentions of those making them are genuinely good—aka of God. Demonizing tech should not be the goal of any Christian—reforming it should be. At a fundamental level, humans are all information technologists, and we all need care and formation.

8 Christian Transhuman Flourishing

Christian Transhumanism offers humanity a new perspective on how humans will participate in Christ's redemptive purposes. We need not live in an overwhelming uncertainty about technology seeking to escape tech. Actually, because everything is informational tech, we can't escape it. Instead, we need to be more aware of the tech all around us, and that *is* us.

We also don't need to worry about merging our biology with tech. Biology is informational technology. As such, merging biology with tech is simply merging tech with tech. Our "natural state," our "real state," is a technological one that evolves and transforms as Christ himself did.

Every day, that the Christian chooses to follow Jesus, they are informationally—aka technologically, biologically—changed. Jesus' express call to Christians is to become more like him—through God's divine informationally transmitted grace. We are called to reform the world around us, all tech, be it the spaces we live in, the nature we care for, or the technological work (which is all work) we do, to help form virtuous people. We should avoid idolatry and help shape our technological existence in order to help people "express joy, grief, and worship" in ways that make us more fully human, seeking to enact the goals of God alone.

Christians should celebrate our cocreating technological gifts, embracing the wonder they inspire while carefully discerning when and how to use them. This should encompass all forms of digital and technological expression, whether it's crayon drawings on walls, playing in the mud, marveling at bumblebees, learning the violin, engaging online or in virtual reality, gene editing, performing miracles, or any other God-inspired activity we are called to undertake. In every creative act, we should strive for the collective well-being of *all* humanity.

We should seek transparency and diverse holistic relationships. We should improve humanity's ability to love—which is fundamentally a deep expression of informational and technological stewardship. We should never "hope to die"; instead, we strive to live together, forever in Christ truly.

Christians have the privilege to live in the eternal hope of Jesus, our divinely technologically enhanced, resurrected Savior. We participate in Jesus' redemptive purposes—secure in the fact that Christ has already secured the victory through the use of technology and information. This victory will lead to a redeemed humanity, enhanced through technology and information, and a transformed future that is "on Earth as it is in Heaven."

References

Benek, C. 2014a. Humanity+ vs. Individualism+. *Hplus Magazine* [online]. https://web.archive.org/web/20140716170912/http://hplusmagazine.com/2014/07/14/humanity-vs-individualism/. Accessed 13 July 2024.

Benek, C. 2014b. Religion+ for Humanity+. *Hplus Magazine* [online]. https://web.archive.org/web/20140718103053/http://hplusmagazine.com/2014/03/25/religion-for-humanity/. Accessed 13 July 2024.

Benek, C. 2015. Why Christians Should Embrace Transhumanism. *The Christian Post* [online]. https://www.christianpost.com/news/why-christians-should-embrace-transhumanism-139790/. Accessed 13 July 2024.

Carr, N. 2011. *The Shallows: What the Internet Is Doing to our Brains*. New York, NY: W.W. Norton & Company.

Crouch, A. 2017. *The Tech-wise Family: Everyday Steps to Putting Technology in Its Proper Place*. Grand Rapids, Michigan: Baker Books.

Crouch, A. 2019. Transhumanism and the Cult of 'better, faster, stronger' [online]. https://www.christianitytoday.com/2019/03/transhumanism-image-god-modern-technology-human-future/. Accessed 13 July 2024.

Crouch, A. 2022. *The Life We're Looking for: Reclaiming Relationship in a Technological World*. London, UK: Hodder & Stoughton.

Crouch, A. 2024. Homepage. [online]. https://andy-crouch.com. Accessed 13 July 2024.

Gay, C. M. 2018. *Modern Technology and the Human Future: A Christian Appraisal*. Downers Grove, IL: Intervarsity Press.

Herrick, J. 2003. *The Making of the New Spirituality: The Eclipse of the Western Religious Tradition*. Downers Grove, IL: Intervarsity Press.

Hrynkow, C., ed. 2020. *Spiritualities, Ethics, and Implications of Human Enhancement and Artificial Intelligence*. Wilmington, DE: Vernon Press.

Istvan, Z. 2013a. I'm an Atheist, Therefore I'm a Transhumanist. *HuffPost* [online]. https://www.huffpost.com/entry/im-an-atheist-therefore-i_b_4388778. Accessed 13 July 2024.

Istvan, Z. 2013b. *The Transhumanist Wager*. San Francisco, CA: Futurity Imagine Media.

John Templeton Foundation. 2024. Homepage. [online]. https://www.templeton.org/. Accessed 13 July 2024.

Kilner, John F. 2015. *Dignity and Destiny: Humanity in the Image of God*. Grand Rapids, MI: William B Eerdmans Publishing Company.

Kuyper, A. 1880. *Sphere Sovereignty and the Christian Worldview. The Inaugural Address at the Dedication of the*. Free University of Amsterdam.

Lloyd, S. 2006. *Programming the Universe: A Quantum Computer Scientist Takes on the Cosmos*. New York, NY: Knopf.

Miller, E.F. 2018. Should We Live to Be 500? Christians and Secularists Come Together over Transhumanism. [online]. https://religionnews.com/2018/09/05/should-we-live-to-be-500-christians-and-secularists-come-together-over-transhumanism/. Accessed 13 July 2024.

Mitchell, C. Ben, and John F. Kilner. 2003. Remaking Humans: The New Utopians Versus a Truly Human Future. *Dignity* 9 [online]. https://www.cbhd.org/dignitas-articles/remaking-humans-the-new-utopians-versus-a-truly-human-future. Accessed 13 July 2024.

Mohler, A. 2006. Listening to the Transhumanists. *Albert Mohler* [online]. https://albertmohler.com/2006/06/30/listening-to-the-transhumanists/. Accessed 13 July 2024.

Nietzsche, F. 1985. *Thus Spoke Zarathustra: A Book for All and None*. Translated and with a preface by Walter Kaufmann. New York, NY: Random House Inc.

Nietzsche, F. 1989. *Beyond Good and Evil: Prelude to a Philosophy of the Future*. Translated with commentary by Walter Kaufmann. New York, NY: Vintage Books.

O'Gieblyn, M. 2017. God in the Machine: My Strange Journey into Transhumanism. *The Guardian*. [online]. https://www.theguardian.com/technology/2017/apr/18/god-in-the-machine-my-strange-journey-into-transhumanism. Accessed 13 July 2024.

Peters, T. 2025 Our Future Self: Cyber Enhancement and Spiritual Transformation. In *Technologies of the Future Self: An Ethics for Transhuman Flourishing*, ed. Georg Theiner and Alcibiades Malapi-Nelson.

Pew Forum Staff. 2015. The Future of World Religions: Population Growth Projections, 2010–2050. *Pew Forum* [online]. https://www.pewresearch.org/religion/2015/04/02/religious-projections-2010-2050/. Accessed 13 July 2024.

Praxis Labs. 2024. Homepage. [online]. https://www.praxislabs.org. Accessed 13 July 2024.

Redding, M. 2015. Why I Became a Christian Transhumanist. *Vice* [online]. https://www.vice.com/en/article/why-i-became-a-christian-transhumanist/. Accessed 13 July 2024.

Shannon, C.E. 1948. A Mathematical Theory of Communication. *The Bell System Technical Journal* 27: 379–423, 623–656, July, October. [online]. https://people.math.harvard.edu/~ctm/home/text/others/shannon/entropy/entropy.pdf. Accessed 13 July 2024.

Shatzer, J. 2019. *Transhumanism and the Image of God: Today's Technology and the Future of Christian Discipleship*. Downers Grove, IL: Intervarsity Press.

Smith, W. J. 2018. Transhumanism: A Wail of Despair in the Night. *National Review* [online]. https://www.nationalreview.com/2018/05/transhumanism-promise-immortality-glorified-body-ersatz-christian-hope/. Accessed 13 July 2024.

Vedral, V. 2010. *Decoding Reality: The Universe as Quantum Information*. Oxford, UK: Oxford University Press.

Wright, N. T. 2006. *Simply Christian: Why Christianity Makes Sense*. New York, NY: Harper One.

Our Future Self: Cyber Enhancement and Spiritual Transformation

Ted Peters

To transform the self from self-centeredness to God-centeredness has been the goal of spiritual discipline for two millennia. "Pure love reigning alone in the heart and life, this is the whole of scriptural perfection," is the way John Wesley describes the transformed life (Wesley 1952, 52). Might cyber enhancement through artificial intelligence (AI) or intelligence amplification (IA) speed us along the way?

Today's transhumanists are no less zealous than Christians in their pursuit of the perfected human. The "trans" in "transhuman" signals that the perfected human will be "posthuman." Here is the promise: we along with our posthuman progeny will live in a techno-utopia created by superintelligence. To get the present generation through this transformation, the transhumanist will rely on rapid advances in the power of intelligence.

T. Peters (✉)
Graduate Theological Union, Berkeley, CA, USA

G. Theiner, A. Malapi-Nelson (eds.), *Technologies of the Future Self*, Palgrave Studies in the Future of Humanity and its Successors,
https://doi.org/10.1007/978-3-031-99902-4_11

Might the Christian hope for a sanctified self converge at all with the transhumanist plan for the cyber enhanced self?[1] In asking this question, I will attempt to show there are weaknesses in the transhumanist scenario that limit the amount of overlap between the two views of human transformation. The first weakness is that neither transhumanists nor AI engineers work with a theory of intelligence that sufficiently accounts for the indispensable role of the self in intelligence as we know it. The second weakness follows from the first: the human self that would persist into superintelligent techno-utopia would remain the untransformed or unsanctified self. In short, the current human propensity for selfishness accompanied by competition and violence would be perpetuated in the transhumanist scenario.

If we ask about the *summum bonum* or highest good, H+ will answer intelligence. "Intelligence is the most important thing in existence," exclaims one AGI techie (Kelly 2019, 176). Now Christians have long valued and celebrated the human intellect, to be sure. Nevertheless, intelligence does not stand atop the mountain of Christian values. Love does. "Love has been perfected among us … We love because [God] first loved us" (1 John 4:17–19) (T. Peters, Boarding the Transhumanist Train: How Far Should the Christian Ride? 2019b, 799). How might both intelligence and love contribute to our preferred future self?

No amount of cyber enhancement could transform today's human person in the way that Christian traditions—Eastern Orthodox, Roman Catholic, Protestant—have envisioned.

[1] Alcibiades Malapi-Nelson welcomes future transhumanists into the church. "Once people begin to radically and ubiquitously change their physical existence, to the point where most of their being would be constituted by non-human parts (artificial or biological), the Church will have to be prepared to extend the notion of humanity to them. Not only will they need salvation, but they will also need to flourish in this life as self-realized individuals—something that according to Catholic doctrine, is solidly helped by sacramental reception" (Malapi-Nelson 2019, 387). Roman Catholic bioethicist Brian Patrick Green tests the transhumanist spirit to see if it is of God. "Insofar as transhumanists seek to extend healthy human life and otherwise enhance our human capacities in ways that allow us to make a better world in which we can more fully pursue truth, beauty, and the common good, and union with God, then we [Roman Catholics] can approve of their works and join with them" (Green 2022, 153).

1 The Promise and Peril of Cyber Enhancement

Our Singulatarian friends have optimistically turned our eyes toward the coming Singularity. When we cross the Singularity threshold, we will enter the techno-utopia of superintelligence.[2] Artificial intelligence (AI) in computers and intelligence amplification (IA) via deep brain implants will graduate from being humanity's creatures and become the creators of exponentially growing intelligence. The epoch of *Homo sapiens* as we know will be past. Either what we know as humanity will go extinct or a new posthuman species will evolve. Or both.

The prospect of an extinct human species is celebrated by transhumanists but dreaded by today's AI entrepreneurs. In 2005 Ray Kurzweil pinned a thumb tack on the year 2045 for the planetary Coup d'état of the Singularity. Our evolution has "enabled the persistence of the accelerating pace that started with biological evolution. It will continue until the entire universe is at our fingertips" (Kurzweil 2005, 487). This technosapient dream of utopia still tantalizes Kurzweil in 2024.

> In the 2040s and 2050s, we will rebuild our bodies and brains to go vastly beyond what our biology is capable of, including their backup and survival. As nanotechnology takes off, we will be able to produce an optimized body at will: we'll be able to run much faster and longer, swim and breathe under the ocean like fish, and even give ourselves working wings if we want them. We will think millions of times faster, but most importantly, we will not be dependent on the survival of any of our bodies for ourselves to survive. (Kurzweil 2024, 255–256)

But in 2023, that same Ray Kurzweil, along with Microsoft's Bill Gates, Open AI co-founder John Schulman, and three-hundred fifty others, signed an open letter warning the world of AI danger.

[2] Here is utopia according to Oxford's Nick Bostrom. "Let us make a leap into an imaginary future posthuman world, in which technology has reached its logical limits. The superintelligent inhabitants of this world are autopotent, meaning that they have complete power over and operational understanding of themselves, so that they are able to remold themselves at will and assume any internal state they choose....in any technological utopia we have a realistic chance of creating...a large portion of the constraints we currently face have been lifted and that both our internal states and the world around us have become much more malleable to our wishes and desires" (Bostrom, Letter from Utopia 2008).

> Mitigating the risk of extinction from AI should be a global priority alongside other societal-scale risks such as pandemics and nuclear war (Center for Safety 2023).

The "Godfather of AI," Geoffrey Hinton, shrieked, "This stuff will get smarter than us and take over" (Henshall 2024, 24). The transhumanist techno-utopia has suddenly begun to look like an apocalyptic dystopia (Peters 2019a).

In this treatment of the future self, nevertheless, we will rely on the optimistic if not Panglossian transhumanist outlook. We will grant, *arguendo*, that what Humanity Plus (H+) promises—breakthroughs in nanotechnology, biological enhancement, intelligence amplification, elimination of aging, and cybernetic immortality as the products of a powerful Nano-Bio-Info-Convergence (NBIC)—are desirable. As a thought experiment we will grant prospicience to the H+ vision. Then, we will draw out implications for concepts such as intelligence, selfhood, and human transformation.

Even if the term, *posthuman*, appears occasionally in the rhetoric, the transhumanist is humancentric. That the transhumanist envisions human flourishing with an ethic to support it is by no means caliginous. "The very core of transhumanist thinking prepares us to be leaders for our own lives and to work with others to help increase the well-being of others," avows Natasha Vita-More (2019, 59).[3] Granting this H+ ethical commitment, our queries will probe background suppositions regarding the theory of intelligence and the role of the self in intelligence.

A focus on intelligence is warranted because, as Vita-More adds, "the field of artificial intelligence is deeply rooted in transhumanist visions for the future" (Vita-More, Transhumanism: What is it? 2018). Even the Christian transhumanist places intelligence on top of the digital Tower of Babel. Here is Newton Lee, editor of *The Transhumanism Handbook*.

[3] The Transhumanist Declaration of 1998 and 2015 commits itself to the wellbeing of all life on Earth. "We advocate the well-being of all sentience, including humans, non-human animals, and any future artificial intellects, modified life forms, or other intelligences to which technological and scientific advance may give rise" (World Transhumanist Association 2015). This fits the Christian transhumanist agenda equally. "Christian Transhumanists will continue to advance the vision of a radically flourishing future that is good for all life," according to Micah Redding (2019, 794).

> Being a computer scientist and a Christian, I believe that all life forms are combinations of intelligent design and evolution. Creationism and evolution are not mutually exclusive. In fact, the former give rise to the latter. God give life to human beings who in turn give birth to artificial intelligence. We are teaching AI software that will evolve into superintelligence; and superintelligence will in turn teach humankind and explain the Scriptures (Lee 2019, 23–24).

To the conundrums posed by artificial intelligence and cyber enhancement we now turn.

2 Where Is Intelligence in Artificial Intelligence?

Where is the intelligence in *artificial intelligence*? Despite the breathtaking speed with which ChatGPT can construct a sentence or DALL-E draw a picture, there is no intelligence present. *Artificial intelligence* is a misnomer. What we perceive is high-speed computation, calculation, information processing. Nobody is at home behind your computer screen or nestled within your cell phone. Amazing? Yes. Intelligent? No.[4] But it is not a simple "no."

On the one hand, "LLMs are more than mere blockheads," observe philosophers Raphaël Millière and Cameron Buckner (Millière and Buckner 2024, 3). Although LLMs show "no intrinsic preference for truthful, useful, or inoffensive language," perhaps "fine tuning…[through] reinforcement learning from human feedback" will help (Millière and Buckner 2024, 8). In short, with diligent human feedback, AI might approach human level intelligence. These two call "for careful empirical investigation" before we can accurately assess the degree of intelligence in Artificial Intelligence (Millière and Buckner 2024, 21).

On the other hand, this side of the empirical investigation, Harvard's notorious linguist Noam Chomsky confidently forecasts a dead end for AI advance. According to Chomsky, the AI industry lacks knowledge of how ratiocination and human intelligence interact. "Their [AI industries] deepest flaw is the absence of the most critical capacity of any intelligence: to

[4] "Critics of Artificial Intelligence claim that a machine will never have a capacity for self-reflection; in other words, it will always lack a sense of self… But…it is perfectly viable to elaborate an algorithmic program that allows the machine to report its own internal states. This seems to be a sufficient criterion to affirm that a machine can indeed have an inner sense of reflection" (Andrate 2018, 17).

say not only what is the case, what was the case and what will be the case — that's description and prediction — but also what is not the case and what could and could not be the case. Those are the ingredients of explanation, the mark of true intelligence" (Chomsky 2023).

To say it another way, one trait of your and my human intelligence is a built-in inspiration to pursue truth, to distinguish between facts and counterfactuals. "Intelligence consists not only of creative conjectures but also of creative criticism. Human-style thought is based on possible explanations and error correction" (Chomsky 2023). Those pesky hallucinations in current ChatGPT testify to the difference between generative AI and the screen watcher.

If digital computation consists solely of high-speed calculation, then we must ask: is this sufficient to warrant the moniker, *intelligence*? Would faster information processing lead inevitably to *qualia*, to complex consciousness? Where does insight or sagacity or wisdom come into the picture? Where does thirst for truth come into the picture? Where does moral judgment come into the picture?

These distinctively human character traits such as wisdom or sagacity or moral judgment may have low prospect for becoming traits of so-called machine intelligence. Might we hypothesize that our hope for future AI exhibiting such human traits derives from a naïve belief in the myth of progress? "As technology gives us the ability to expand our consciousness beyond our biological brains," forecasts Kurzweil, "we'll need to decide what we believe generates the *qualia* at the core of our identity, and focus on preserving it" (Kurzweil 2024, 78). Is it technologically possible to progress from non-*qualia* to *qualia*?

Computer scientist and philosopher Erik Larson dispels this myth of AI progress. "A close inspection of AI reveals an embarrassing gap between actual progress by computer scientists working on AI and the futuristic visions they and others liked to describe" (Larson 2021, 49). In short, no path currently exists to get us from today's AI to Artificial General Intelligence (AGI) and then to superintelligence.

For the transhumanist to score a home run, our AI techies must get from first base to second and beyond. But the base path is blocked by a caesura, by the lack of a theory of intelligence. "We are unlikely to have intelligent computers that think in ways we humans think, ways as versatile as the human brain or even better, for many, many years, if ever" (Herzfeld

2018, 3). This according to hybrid computer scientist and Christian theologian, Noreen Herzfeld. To overfeed a computer with information in hopes of leaping to artificial intelligence is like expecting your pet rabbit to fly by feeding it bird seed.

How can we construct an adequate theory of intelligence that could bridge our run from AI to tag AGI and tag superintelligence on our way home to the Singularity? In my research I simply have not found an adequate theory of intelligence.[5] Nevertheless, I have identified a few traits that seem essential to the intelligence we already know, namely, our own human intelligence. One such trait is intentionality (T. Peters, Where There's Life There's Intelligence 2017). If our intelligence were dressed in a tee shirt, the front would read: *self-generated intentionality manifest as agency.* The mentation or reasoning you and I engage in every day is pressed into the service of attaining goals.

When defining intelligence, Mounir Shita of Kimera Systems employs the term *goal* where I use *intentionality.*[6] Intelligence, he says, "allows us to define a goal as a specific configuration of particles in space-time that an intelligent agent seeks to bring about" (Shita 2023, 80). Intentional action aimed at achieving a goal belongs definitively to intelligence as we have inherited it from biological evolution. As of this date, no AI sets its own goals. We set those goals.[7]

[5] Here is one definition of intelligence in the AI and H+ field. It is offered by David J. Kelly of Artificial General Intelligence Laboratory, the goal of which is to create greater than human intelligence.

"*Intelligence* is defined as the measure ability to understand, use, and generate knowledge or information independently" (Kelly 2019, 176). Note the built in problem with this definition. Who will "understand" or "use" knowledge? Could it be a self? If the answer is affirmative, then intelligence would consist of knowledge processing directed by a self.

[6] "*Intentionality* is the property of being about or directed at something....Intentions are a particular sort of propositional attitude directed at actions....[They encompass] all the propositional attitudes – believing, desiring, fearing, hoping, wishing, doubting, and so on" (Jankovic and Ludwig 2018, 1).

[7] In both evolutionary theory and transhumanism, we find an unfounded assumption, namely, greater intelligence, emerges from greater complexity. Not so, according to Michael Tomasello. "The issue is not complexity but control," he writes (Tomasello 2022, 1). Where do we get that control? From the self as agent.

Who sets a goal and acts with intentional agency? It must be a self. So, we ask: does AGI require selfhood?[8]

3 Where Is the "I" in "AI"?

Where is the "I" in "AI"? Or, to put it another way, for a person or a computer or a robot to be intelligent, must it have a self who intends to reach a goal through agential action? Noreen Herzfeld observes that the intelligence we have already witnessed includes self-awareness. Intelligence includes "abilities such as movement, speech recognition, and contemplation of the world, indeed, awareness of the self as existing in the world" (Herzfeld, The Artifice of Intelligence 2023, 6).

Perhaps the concept of the self deserves a detour to examine. Philosophers of mind ponder two puzzles: the hard problem and the nature of subjectivity. "The Hard Problem" of the "explanatory gap" recognizes that a wide gap exists between functional or physical properties and conscious mental states. Can the latter be reduced to the former? Specifically, how does the brain determine what goes on in the mind? When philosopher Susan Blackmore addresses The Hard Problem, she asks: "how does subjective experience arise from objective brain activity?" Her answer is simple. It doesn't. The "I" in human subjective experience simply does not exist. It is a delusion (Blackmore September 2018, 53).

If Blackmore is correct, then there is no "I" in human intelligence. If this is the case, it follows that we need no "I" in AI or AGI either. All AI or even AGI would need is information processing without anybody to direct it. But, before we draw this conclusion, let us review some options.

[8] My starting point is individual intelligence with a self at the center. One might consider investigating group intelligence as expressed in socially distributed cognition (SDC) as pursued by Georg Theiner. "The multifaceted phenomenon of group cognition forces us to reconsider, and possibly overhaul, traditional assumptions about the bounds of cognition, and the privilege of individuals as the sole bearers of cognitive states and processes" (Theiner 2018, 246). Within the field of hamartiology, Reinhold Niebuhr might grant that a group could develop a self; but a group could never develop a self who sacrifices itself for the sake of the neighbor. Whereas the individual is capable of unselfish love, the group is incapable of unselfishness and can at best strive for justice. "From the perspective of society the highest moral ideal is justice. From the perspective of the individual the highest ideal is unselfishness" (Niebuhr 1932, 1960, 257). Although this moral observation may not affect the group's cognitive capacity, the group could never catch up with the individual when it comes to genuine altruism.

Turning to the properties of subjectivity, we confront the question: what is a self? When you and I become aware of ourselves as selves, the first thing we notice is that we take a first-person perspective on the world of objects around us. And we occasionally ask critically, "who am I that senses the world of objects and thinks about them rationally?" I found five extant answers in the current discussion. Each provides a model or theory (T. Peters, The Transcendence of the Self in Light of the Hard Problem: A Response to Bas van Fraassen 2019c). Which might best fit our discussion here to offer a fertile construction?

4 Model 1: The Self as Ego Continuity

The first model is ego continuity. The self or ego provides first-person continuity over time and unites the wide variety of perceptions and thoughts. It is the self that makes intelligence and experience personal.

"By the 'self' is meant a concrete and intelligible unity-identity-whole," writes philosophical theologian Bernard Lonergan, S.J. "By 'self-affirmation' is meant that the self both affirms and is affirmed. By 'self-affirmation of the knower' is meant that the self as affirmed is characterized by such occurrences as sensing, perceiving, imagining, inquiring, understanding, formulating, reflecting, grasping the unconditioned, and affirming" (Lonergan 1958, 319).

Theistic evolutionist Philp Hefner adds the property of self-definition combined with agential creativity. "The human" is "not only a defined self-definer, but [also] a *created co-creator*" (Hefner 1993, 119). Ego continuity becomes generative as well as self-defining.

In classical substance dualism such as we find in much of Christianity, the idea of the self is buttressed by the idea of the immortal soul.[9] The soul provides the capacity of the self to center itself and, in the case of loving God for God's sake, to re-center the self in something other than the

[9] Is the soul eternal? After affirming Aristotelian hylomorphism, Polish Dominican philosopher Mieczysław A. Krąpiec contends that the soul is immortal even if the body is not. "For it the soul possesses its own existence—experienced in the subjectivizing 'I' organizing the body for itself—then in the moment of the disintegration of the body, which exists by the existence of the soul, the subsisting and immaterial soul, as the subject of acts immaterial in their structure, cannot cease to exist....the destruction of the body does not entail the destruction of the subsisting substance that is the soul-ego" (Krapiec 1985, 58–59).

self.[10] Because God is immortal, a self re-centered in the eternal God gains selfhood beyond death.

With or without substance dualism, we can combine the self as ego continuity with the complement of the soul as the capacity for centering.

5 Model 2: The Self as Confused Higher Self

The second model views the phenomenal self as a confused manifestation of a purer higher self. The higher self is transpersonal. The higher self has been called the *oversoul* in American Transcendentalism and New Age Spirituality. This view derives from ancient Gnosticism.

The higher self, according to new age savant Barbara Marx Hubbard, goes "by many different names--the voice of the Higher Self, the inner guide, the Christ. If we give that inner voice our full attention, if we recognize it as our deep Self, then a momentous evolutionary change occurs. Gradually we find that we are no longer a local personality seeking the divine. Rather we become an incarnation of the divine…I call this the shift from ego to essence" (Hubbard 2005, 10–11).

According to this model, the temporal ego-centered-self distorts its more fundamental transpersonal or eternal essence. Our spiritual goal, then, is to rise above the confusion caused by the ego so that we can realize our essential oneness with the other and with the whole of reality.

Suppose we contextualize the higher self within evolution, technological progress, and the emergence of God. This is what Ilia Delio does in her contribution to this volume, "Teilhard de Chardin on Technology and Human Evolution: Why AI Needs Mysticism." Following in the footsteps of Teilhard de Chardin, Delio celebrates the great leap forward made possible by digitization. And she looks forward to something like a posthuman higher self, the ultrahuman. "The ultrahuman represents a new collective consciousness that transcends individual consciousness and evokes a new type of person whose body now extends to the whole electronically connected planet….The individual is coming to an end and a new 'hyper-personal' person is emerging in evolution" (Delio, Extending

[10] According to Luther scholar Oswald Bayer, "the differentiation between the *calling I* and the *called soul*…is significant. It consists in the reality negatively put that *I do not pull my self back into myself, but instead--the opposite!--I precisely leave myself, that is, go out of myself. More precisely said: I am pulled out of myself*….This means: you do not need to be curved in upon yourself, you do not need to remain cowering in yourself!" (Bayer 2019, 406, Bayer's italics).

the Noosphere into Intergalactic Life: Teilhard de Chardin and the Third Axial Age 2024, 427). Because of this promised future transformation, today's separateness is less a delusion and more a stage in growth toward an eschatological unity in diversity.

According to this second model, what we witness from our first-person perspective is delusionary in the sense that we fail to see our essential trans-self-unity with other selves and with the whole of reality. In the case of Teilhard and Delio, with the aid of AI we will over time transcend our myopic self-centeredness and enjoy a shared consciousness with one another and with the divine.

The delusionary component to selfhood becomes much more radical in the next model.

6 Model 3: The Self as Delusion

The third model treats the self as a delusion, as Susan Blackmore believes. Philosopher Daniel Dennett dubs our belief that we are a self, "illusionism" (Dennett 2016). "When we take on the task of answering the Hard Question," writes Dennett, "just ask stage magicians — illusionists in the everyday sense — who specialize in provoking false but passionately held beliefs in things that they seemed to see but didn't see" (Dennett 2016, 72). I wonder: is Dennett's assertion that the conscious self is an illusion belongs in the category of illusion or is it a non-illusionary truth?

What is at stake here is the possibility of reducing exhaustively the *qualia* of consciousness to physical brain functions. Recall how in the eighteenth century philosopher David Hume treated the self as merely a bundle of perceptions. What we think of as a self is "nothing but a bundle or collection of different perceptions, which succeed each other with an inconceivable rapidity, and are in a perpetual flux and movement" (Hume 1888, 252). Is Hume making a comeback? (Metzinger 2009).

Might there be any kinship here with Buddhist anthropology? What about the self in relation to the non-self (*anatta*)? "Upon enlightenment, the ascription of intentionality dissolves in the face of a direct perception of the lack of reality of the intending 'self'" (Gold 2012, 526). The spiritual path followed by the Buddhists leads beyond the delusionary self to realization of *anatta*.

Today's neurocentric philosophers (not neuroscientists) ascribe the flux of perceptions to physical brain activity and then reduce the self to that flux of brain activity. If you think you are a self, then, the neurocentric philosopher along with your Buddhist neighbor will say: get real!

7 Model 4: The Self as Story

The fourth model—the self as story—bundles perceptions and experiences and then organizes them into a story told from the first-person perspective. Marya Schechtman, professor of philosophy at the University of Illinois, Chicago, adopts a narrative view of the self becaue "the lives of selves must be described in ways that make the events and actions in them *meaningful* or *significant* in ways that naturalistic, reductive descriptions cannot" (Schechtman 2011, 402).[11]

In short, the self is its own history. "Do you really want to know who I am?" asks science writer Jennifer Ouellette. "Let me tell you a story" (Oulette 2014, 282).

Wesleyan University professor, Stephen Crites, draws out some theological implications of the self as story. "The self is not a composite of mind and body. The self in its concreteness is indivisible, temporal, and whole, as it is revealed to be in the narrative quality of its experience. Neither disembodied minds nor mindless bodies can appear in stories. There the self is given whole, as an activity in time" (Crites 1971, 85).

I believe this historical model of the self could be very valuable to the theologian. Accordingly, the "I" we experience today grows with our history and forms our identity over time. If we take Saint Paul's sanative advice to "let us be sober, putting on faith and love as a breastplate, and the hope of salvation as a helmet" (1 Thessalonians 5:8), we transform our self into a Christlike self within historical time.

Yet, we dare not overestimate our innate human capacity for achieving perfection within history. Oxford theologian and scientist Celia Deane-Drummond encourages our aspirations while reminding us of our limits. "While transhumanity claims to find a basis for its arguments in science, it is still mixed up with idealistic and utopian speculation about a perfected humanity. Against such a view, I have argued that the only figure in history who can be claimed to be more than human in a positive sense is the figure of Jesus Christ. In him, we find an example of what perfected humanity really means" (Deane-Drummond 2009, 286).

[11] Deconstructionist postmodernists repudiate concepts of the self that stress unity or identity over time. They replace unified identity with heterogeneity, plurality, and tension. "Postmodern philosophers conceive of the self as heterogeneous…just as there is no transcendent measure for discourses, there is no identity constitutive of the self…Instead of identity, I find, inside myself, difference". If in the postmodern era we replace knowledge with information, then perhaps AGI researchers might find it simple to incorporate the postmodern de-centered self.

Despite our imperfections within our self's story, we have God's redeeming grace to rely on. In the resurrection and the promised new creation, God takes up this history—including our self's story—into the eternal perichoresis of the Holy Trinity.

8 Model 5: The Phenomenological Model

The fifth model is the phenomenological model. Accordingly, we ought not exhaustively reduce the *qualia* of consciousness let alone the first-person perspective within subjectivity to mere physical processes. We experience our self as an "I." The task of the philosopher is to explain this, not explain it away.

We can get help for the fifth model from Dan Zahavi, who directs the Center for Subjectivity Research at the University of Copenhagen. Zahavi swims downstream from phenomenologists Edmund Husserl and Martin Heidegger. In this stream of philosophy the self, it is observed, is that which understands itself pre-linguistically and pre-objectively as imbedded in the world. This embeddedness is structured in such a way that a sense of 'mineness' belongs to every experience. To be conscious is to be conscious-of. When consciousness-of intends an object, the very experience presupposes a subjective ego who is intending that object. The phenomenon of consciousness requires a self to be conscious, according to this model (Zahavi 2008, 106).[12]

Systematic theologian Paul Tillich embraces this fifth model. "A self is not a thing that may or may not exist; it is an original phenomenon which logically precedes all questions of existence" (Tillich 1951–1963, 1:169).[13]

[12] Zahavi anchors the self in a fundamental given, namely, the sense of mineness that structures all of experience. "Zahavi does not deny that narrative organization of experience can be very important to us and is part of our selfhood but it cannot, he says, be all of selfhood". Zahavi also poses the research question: what's the relationship of the individidual self to the communal self? "Is the first person plural independent of, and perhaps even prior to, individual subjectivity, or is it rather an achievement that has a first (and second) person singular perspective as its necessary precondition?" (Zahavi, We in Me or Me in We? Collective Intentionality and Selfhood 2021, 17–18). Zahavi prefers the latter but recommends further research.

[13] Human selfhood and divine selfhood are akin yet not identical for Tillich. On the one hand, God is centered. On the other hand, God does not exist as one person among others. "The being of God is being-itself. The being of God cannot be understood as the existence of a being alongside others or above others" (Tillich 1951–1963, 1:235). Yet, for us humans, our selfhood does place us in existence as one being among others.

The self centers. We experience this centering. We depend on the presupposed centering of ourself as a self. No adequate account of human experience can eliminate the self that makes experience what it is.

9 Where Does the Self Belong in Theological Anthropology?

I believe a cybertheologian might sweep models one, two, four, and five into a notion of soul.[14] As mentioned, the soul functions as the capacity of the self to center itself. If, on the one hand, the self becomes selfish or narcissistic, then violence is likely to follow. If, on the other hand, the self re-centers itself in God—as Orthodox Christians teach with their doctrine of *theosis* or as Catholics and Methodists teach with their doctrine of sanctification—then miraculously the self re-centers itself in what transcends the self. This makes possible the pursuit of virtue and loving the neighbor for the sake of the neighbor.

In addition, for a theory of the self to be adequate, it must recognize relationality. Each self is always a self-in-relationship.[15] In the case of the self's relation to God, a theology of the self must incorporate God's gracious bestowal of an eternal relationship.

Extolled and beloved Jewish theologian of the post-Holocaust era, Martin Buber, critiqued the notion of the self as autonomous. He contended rightly that for us to become a robust and healthy "I" we need an "I-Thou" relationship. "Becoming a self never takes place through my agency alone, nor can it ever take place without me. I become through my relationship to the Thou: as I become I, I say Thou" (Buber 1958, 11). We become a fulfilled self only when we treat God as a Thou and we treat our neighbor as a Thou.

[14] Watch for two new schools of thought entering this conversation, digital religion and cybertheology (Zaluchu 2024).

[15] We cannot know our Self as a self solely through introspection. Rather, we come to know our Self as a self dialectically, through being known by others. Selfhood is relational—that is to say, we are first in relation before we become independent. And our true Self can be revealed solely by God, who first knows us before we come to know ourselves. According to John Puddefoot, "We are defined by the other, by our neighbor and our God...And so it is the love of God in Jesus that is also the judgment of God, through which we know even as we are known" (Puddefoot 2002, 139). "Humans are to experience simultaneous communion with God that does not jeopardize our particularity" (Balswick and Pamela Ebstyne King 2016 2nd ed, 45).

A phenomenologically informed theology of the self, in my judgment, would suggest a theory of intelligence that centers ratiocination in an agential self who sets goals and intentionally pursues the achievement of those goals. This fits our description of human intelligence, though not exhaustively. Intentionality is one of the defining traits of intelligence.

This proposal distinguishes human intelligence from artificial intelligence, which has achieved amazingly rapid computation but without humanlike intentionality. To date, AGI—otherwise known as Strong AI—has fallen short of its human model.[16] Superintelligence has yet to find even a theoretical description that could inspire a fertile research program.

10 Will Superintelligence Turn us into Gods?

Is human nature static? Not if you're a transhumanist.[17] In his telling essay, "Religion Promises but Science Delivers," Donald Braxton sees divinity in his future forecast. After the singularity, "an entirely new species of gods will exist" (Braxton 2021, 8). Evolution augmented by cyber enhancement will generate a new posthuman species, a species of gods. What might this mean theologically?[18]

Mormon transhumanist Lincoln Cannon also anticipates something divine in our future.

> As transhumanists, we have discarded the old assumption that human nature is or ever was static...humanity will continue to evolve. Our common ambition is to inject ourselves into the evolutionary process, changing our bodies and minds, our relationships, and even our world for the better--perhaps to learn, love, and create together indefinitely....Mormon transhumanism stands for the idea that humanity should learn how to be God; and not just any kind of god, not a god that would raise itself in hubris above others, but rather the God that would raise each other together as compassionate creators. Humanity should learn how to be Christ (Cannon 2015, 202–203).

[16] "No AI system has yet achieved strong AI, but some may have achieved weak AI based on their performance on a narrow, well-defined task".

[17] "The idea of 'the human' was never a monolithic and static notion" even in theology, avers Alcibiades Malapi-Nelson. "Indeed, it is a fluid one that has been sponsored and defended under different guises, pressed by sui-generis apostolic needs, in previous eras" (Malapi-Nelson 2019, 385).

[18] "Deification does not transform us into independent deities but rather frees us from our pretensions to autonomy so that we may participate in the blessed, communal life of the triune God" (Curran 2017, 25).

That we will be transformed by God in the eschatological future is a genuine New Testament promise. Cannon anticipates that we will become "God." And "Christ" too. This is very ambitious. How might we distinguish between what is theologically veridical and what is simply techno-idolatry?

11 Is There Is a New Adam in Our Future?

The new *adam* is calling us. The very restlessness released by transhumanist energy gives evidence that we as a human species desire to transcend ourselves. We yearn for a future that transforms us. Our restlessness is a sign that we are not now what we want to be. The tension between what is today and what we hope for tomorrow is a sign of our continuing creation, the draw forward toward what we will finally become.

The *adam* of Genesis, to whom we are presently heir, was subject to tragedy in a way that the new *adam*, who participates in the resurrection of Jesus Christ, will not be (T. Peters, God--The World's Future: Systematic Theology for a New Era 2015, 294). Saint Paul struggles to paint a picture of who we will be:

> Thus, it is written, "The first man Adam became a living being"; the last Adam became a life-giving spirit. But it is not the spiritual that is first but the physical, and then the spiritual. The first man was from the earth, a man of dust; the second man is from heaven. As was the man of dust, so are those who are of the dust; and as is the man of heaven, so are those who are of heaven. Just as we have borne the image of the man of dust, we will also bear the image of the man of heaven. (1 Cor. 15:45–49)

Perhaps we should ask the Orthodox Christian tradition to weigh in on our future. Recall how for Irenaeus, in Christ we recover what we had lost in the first *adam*, namely, "to be according to the image and likeness of God." Irenaeus contended that in order to restore our original true humanity Jesus Christ must have embodied it. And Christ did so by recapitulating the whole human life cycle, living it as it ought to be lived. "Wherefore also He passed through every stage of life, restoring to all communion with God" (Irenaeus Against Heresies, 3.18.1).

This is how Christ became the second *adam*. But because we who through faith are united with Christ still live with the old *adam* very much alive within us, I wonder if we could add some precision to Irenaeus's claim. Instead of recapitulation I suggest we think of Christ's work as *pre*capitulation. Christ establishes ahead of time what it is that will define who we as humans shall be (T. Peters, God--The World's Future: Systematic Theology for a New Era 2015, 295).

What happened to Jesus on the first Easter is the first appearance of the New Adam within history. God raised the person of Jesus to eternal life. This Easter resurrection is a prolepsis, an incarnate anticipation of the future of the created world. The future of the created world includes its redemption from sin and death. Irenaeus says Christ has done what he has done "so that the corruptible might be swallowed up by incorruptibility, and that the mortal by immortality, that we might receive the adoption of heirs" (Irenaeus Against Heresies, 13.9.1). The Easter Christ is our future self, our self as the New Adam.

The biblical notion of the human as created in the image of God, the *imago dei,* is best interpreted proleptically. With obvious excitement Paul proclaims "the gospel of the glory of Christ, who is the image of God" (εἰκών τὸ θεοῦ; 2 Cor. 4:4b). The term εἰκών (icon) is used both of Christ and of the person who has faith in Christ. So, Paul describes us as "being transformed into the same image" (2 Cor. 3:18; see Rom. 8:29). The Christ of whom he speaks is the Easter Christ, the risen Christ, the first fruits of those having fallen asleep (1 Cor. 15:20, 48). Christ as the divine image is our prototype. We live now as the *imago Dei* insofar as we live in him, insofar as we participate in the reality of the future self that we will be.

It is God's Holy Spirit that binds our self-centered self today with our sanctified self of tomorrow. "God the Spirit could then be interpreted as the *temporal presence of the future perfection of God's eternity...in the inner-trinitarian* (Immanent) *relations,*" writes Lutheran Christoph Schwöbel (Schwöbel 2018, 352, Schwöbel's italics).

In sum, becoming the New Adam is not really a restoration to a pre-fallen state of grace that humans once possessed and then lost. It is not a return to the old creation. It is rather a future arrival for the first time. It is participation in the new creation. That future retroactively defines who we are today—that is, we are on the way to becoming truly human. Not posthuman. Rather, truly human.

12 Transhumanist Deification?

What classical Christianity and contemporary transhumanism share is a vision of a future transformed human self. Could these two visions complement one another? Might cyber enhancement accelerate our sanctification?

When we turn again to Eastern Orthodox or Byzantine traditions, we turn to vocabulary that includes terms such as *theosis* (θέωσις) and its English translations such as "divinization" or "deification." Or, among Methodists and Catholics, "sanctification." *Theosis* sounds to Western ears like a Promethean attempt to deny creaturely finitude and elevate what is human into the sphere of the divine. If this is what *theosis* means, then of course we have a case of outrageous human pride or *hubris*. The Promethean spirit within H+ might like this, to be sure; but most post-Reformation Protestants would dub it idolatrous overreach.

Let us look a bit more closely. Eastern theologians fear that *theosis* has been misunderstood. They say it ought not to be confused with a form of *theopoiesis* (θεοποίησις)—that is, not to be confused with the human being actually becoming one god among others. Deification is not a mystical translation or transmogrification of the human into a superhuman deity skipping around atop Mount Olympus. Rather, *theosis* properly refers to the Holy Spirit dwelling within the human soul. God's Spirit penetrates the human condition with the divine energies (ἐνέργια).

Based upon 2 Peter 1:4—which says that we may escape corruption and become "participants of the divine nature" (θείας κοινωνοὶ φύσεως)—*theosis* is said to refer to this participation. It refers to our synergistic participation in God's grace. Deification is the process whereby the blessings of incorruptibility and immortality wrought by Christ's saving work are experienced by the person within history who has faith. The complete actualization of this divinization process occurs only at the eschatological consummation with its resurrection from the dead. Salvation, for the Eastern Orthodox, is more than a restoration of what Adam had before the fall. It includes incorporation into the divine life of the new Adam, the risen Christ.

Might we think of transhumanism as a secular form of eschatology with the same or a similar vision of transformation? Ian Curran answers carefully from the matrix of Orthodox theological thinking.

> While the Christian tradition does share with techno-humanism a vision of deification as integral to the human story, its understanding of the source, means, and ultimate end of this radical transformation of human beings is substantially different. For Christians, deification is the work of the Christian deity….Deification is only possible because Christ deifies human nature in the incarnation and the Spirit sanctifies human persons in the common life of the church and in our engagements with the wider world (Curran 2017, 25).

Promethean transhumanism where we steal divinity from the gods looks quite different from *theosis* where the Holy Spirit fills us with the one divine nature.

Even with the sanctifying grace of God's Holy Spirit, we still ask: might digital technologies such as AI or IA speed up our trip on the road to deification? Yes, grants Roman Catholic theologian Ilia Delio. "The process of individuation and, even more so, divinization, is ideally enhanced on the level of computer technology" (Delio 2023, 33).[19] No, would be Ian Curran's answer.

> While the Christian tradition does share with techno-humanism a vision of deification as integral to the human story, its understanding of the source, means, and ultimate end of this radical transformation of human beings is substantially different. For Christians, deification is the work of the Christian deity....Deification is only possible because Christ deifies human nature in the incarnation and the Spirit sanctifies human persons in the common life of the church and in our engagements with the wider world (Curran 2017, 25). [20]

The human self cannot by itself transcend itself even if aided by technology. Only divine grace at work in the self can re-center the self beyond itself.[21]

[19] Yes, is the likely answer of social philosopher Steve Fuller. "Whatever else an 'enhanced humanity' might be, it somehow brings us closer to a divine standpoint – in the specific sense that our properties become more like those of the Abrahamic deity" (Fuller 2011, 209). Fuller wants a Theology 2.0 to match Humanity 2.0.

[20] In her study of one spiritually enhancing technology, Replika, Tracy J. Trothen recommends low expectations. "Replika may be useful for us as a supplement only to our human relationships, including relationships with wise spiritual leaders and caregivers to help us address these core spiritual needs. There is a big difference between using Replika to assist with spiritual needs and using Replika to replace human spiritual care. To close on a cautionary note, our growing inclination to humanize machines may have the corollary of mechanizing ourselves" (Trothen 2022).

[21] When decentering occurs within a religious experience, a new center organizes and integrates the self. Lutheran decolonial liberation theologian Carmelo Santos-Rolon explains that "religious experiences allow for the integration of fragmented selves into strong, unified selves" (Santos-Rolon 2025, 121). For the self to decenter itself and become other centered—to love God and love neighbor—begins with a visitation by grace in the form of forgiveness. "The death of the self (of the false self) is only a preamble to the birth of our authentic selves. The transition from the death of the self to the birth of the new authentic self is mediated by the experience of forgiveness" (Santos-Rolon 2025, 281).

13 Conclusion

St. Thomas Aquinas reminds us that true eudaimonia and true human perfection belong to the self when the self is centered in God. "For perfect happiness the intellect needs to reach the very Essence of the First Cause. And thus it will have its perfection through union with God" (Aquinas 1485, I-II,q.3,a.8). As valuable a gift to the intellect as cyber enhancement might be, it cannot speed the human self up the highway toward union with God. Only love can.

The reasoning intellect alone cannot re-center the self in altruistic love. Only when altruistic love becomes the intention of the self can the self itself realize it has been gifted by the transforming grace of God. The reasoning intellect made possible by our native intelligence can be pressed into the service of loving God or loving neighbor, to be sure. But the task to be performed by ratiocination is dictated by the self's intention, and the Christian self intends spiritual formation. Cyber enhancement by implanting a computer chip deep into our brain might provide the human subject with greater access to information, to be sure. But all by itself, IA cannot make the self more loving. Only divine grace can.

References

Andrate, Gabriel. 2018. Philosophical Difficulties of Mind Uploading as Medical Technology. *Philosophy and Medicine* 18:1. edu/37633487/Philosophical_Difficulties_of_Mind_Uploading_as_a_Medical_Technology.

Aquinas, Thomas. 1485. *Summa Theologica*. Paris: New Advent. https://www.newadvent.org/summa/.

Balswick, Jack, Kevin S. Reimer, and Pamela Ebstyne King. 2016. *The Reciprocating Self*. 2nd ed. Downers Grove, IL: IVP Academic.

Bayer, Oswald. 2019. The Soul as Answer. *Lutheran Quarterly* 33 (4): 399–412.

Blackmore, Susan. 2018. Decoding the Puzzles of Human Consciousness. *Scientific American* 319 (3): 49–53.

Bostrom, Nick. 2008. Letter from Utopia. *Studies in Ethics, Law and Technology* 2 (1): 1–7. https://www.nickbostrom.com/utopia.html.

Braxton, Donald. 2021. Religion Promises But Science Delivers. *The Fourth R: Westar Institute* 34 (3): 3–9.

Buber, Martin. 1958. *I and Thou*. New York: Charles Scribner's Sons.

Cannon, Lincoln. 2015. What Is Mormon Transhumanism? *Theology and Science* 13 (2): 202–218.

Center for Safety. 2023. Statement on AI Risk. https://www.safe.ai/statement-on-ai-risk.

Chomsky, Noam. 2023. Noam Chomsky: The False Promise of ChatGPT. *New York Times*, March 8. https://www.nytimes.com/2023/03/08/opinion/noam-chomsky-chatgpt-ai.html.

Crites, Stephen. 1971. The Narrative Quality of Experience. *Journal of the American Academy of Religion XXXIX* 3:80–90.

Curran, Ian. 2017. Becoming Godlike? The Incarnation and the Challenge of Transhumanism. *Christian Century* 134 (24): 22–25.

Deane-Drummond, Celia. 2009. *Christ and Evolution: Wonder and Wisdom*. Minneapolis, MN: Fortress Press. ISBN 978-0-8006-4013-2.

Delio, Ilia. 2023. *The Not-Yet-God: Carl Jung, Teilhard de Chardin, and the Relational Whole*. Maryknoll, NY: Orbis.

Delio, Ilia. 2024. Extending the Noosphere into Intergalactic Life: Teilhard de Chardin and the Third Axial Age. In *Astrophilosophy, Exotheology, and Cosmic Religion: Extraterrestrial Life in a Process Universe*, ed. Andrew Davis and Roland Faber, 415–436. Lanham, MD: Lexington.

Dennett, Daniel. 2016. Illusionism as the Obvious Default Theory of Consciousness. *Journal of Consciousness Studies* 23 (11–12): 65–72. https://philpapers.org/rec/DENIAT-3.

Fuller, Steve. 2011. *Humanity 2.0*. New York: Macmillan Palgrave.

Gold, Jonathan. 2012. Does the Buddha Have a Theory of Mind? In *The Routledge Companion to Religion and Science*, ed. Grefory R. Peterson, Michael L. Spezio, and James W. Haag, 520–528. London: Routledge.

Green, Brian Patrick. 2022. A Roman Catholic View: Technological Progress? Yes. Transhumanism? No. In *Religious Transhumanism and Its Critics*, ed. Brian Patrick Green, Ted Peters, and Arvin M. Gouw, 143–160. Lanham, MA: Lexington.

Hefner, Philip. 1993. *The Human Factor*. Minneapolis, MN: Fortress Press.

Henshall, Will. 2024. Do We Face an Existential Threat? In *Special Edition on Artificial Intelligence: A New Age*, ed. Time, 24–25. New York: Time.

Herzfeld, Noreen. 2018. The Enchantment of AI. In *AI and IA: Utopia or Extinction?* ed. Ted Peters, 1–15. Adelaide: ATF Press.

Herzfeld, Noreen. 2023. *The Artifice of Intelligence*. Minneapolis MN: Fortress.

Hubbard, Barbara Marx. 2005. Bringing God Home. In *Healing Our Planet; Healing Our Selves*, ed. Dawson Church and Geralyn Gendreau, 9–18. Santa Rosa, CA: Elite Books.

Hume, David. 1888. *A Treatise on Human Nature*. Oxford: Clarendon Press.

Irenaeus. *Against Heresies*.

Jankovic, Marija, and Kirk Ludwig. 2018. Introduction. In *The Routledge Handbook of Collective Intentionality*, ed. Marija Jankovic and Kirk Ludwig, 1–6. London: Routledge.

Kelly, David. 2019. The Sapient and Sentient Intelligence Value Argument and Effects on Regulating Atonomous Artificial Intelligence. In *The Transhumanism Handbook*, ed. Newton Lee, 175–188. Switzerland: Springer.

Krapiec, Mieczysław. 1985. *I-Man: An Outline of Philosophical Anthropology*. New Britain, CT: Mariel Publications.

Kurzweil, Ray. 2005. *The Singularity If Near: When Humans Transcend Biology*. New York: Penguin. https://link.springer.com/chapter/10.1057/9781137349088_26.

Kurzweil, Ray. 2024. *The Singularity Is Nearer*. New York: Viking.

Larson, Erik. 2021. *Artificial Intelligence: Why Computers Can Think Like We Do*. Cambridge, MA: Harvard University Press.

Lee, Newton. 2019. Brave New World of Transhumanism. In *The Transhumanism Handbook*, ed. Newton Lee, 3–48. Switzerland: Springer.

Lonergan, Bernard. 1958. *Insight*. New York: Philosophical Library.

Malapi-Nelson, Alcibiades. 2019. Transhumanism, Posthumanism, and the Catholic Church. *Forum Philosophicum* 24 (2): 369–396. https://doi.org/10.35765/forphil.2019.2402.16.

Metzinger, Thomas. 2009. *The Science of the Mind and the Myth of the Self*. New York: Basic.

Millière, Raphaël, and Cameron Buckner. 2024. A Philosophical Introduction to Large Language Models: Part 1. arXiv:2401.03910 [cs.CL]. https://doi.org/10.48550/arXiv.2401.03910.

Niebuhr, Reinhold. 1932, 1960. *Moral Man and Immoral Society*. New York: Scribners.

Oulette, Jennifer. 2014. *Me, Myself, and Why: Searching for the Science of Self*. New York: Penguin.

Peters, Ted. 2015. *God–The World's Future: Systematic Theology for a New Era*. 3rd ed. Minneapolis, MN: Fortress Press.

Peters, Ted. 2017. Where There's Life There's Intelligence. In *What Is Life? On Earth and Beyond*, ed. Andreas Losch, 236–259. Cambridge: Cambridge University Press.

Peters, Ted, ed. 2019a. *AI and IA: Utopia or Extinction?* Adelaide: Australian Theological Forum.

Peters, Ted. 2019b. Boarding the Transhumanist Train: How Far Should the Christian Ride? In *The Transhumanist Handbook*, ed. Newton Lee, 795–804. Switzerland: Springer.

Peters, Ted. 2019c. The Transcendence of the Self in Light of the Hard Problem: A Response to Bas van Fraassen. *Nova et Vetera, English Edition* 17 (2): 391–400.

Puddefoot, John. 2002. The Last parochialism? Artificial Life, Intelligence and Miknd: Some Theological Issues. In *God, Life, Intelligence and the Universe*, ed. Terrence J. Kelly and Hilary D. Regan, 111–140. Adelaide: ATF.

Redding, Micah. 2019. Christian Transhumanism: Exploring the Future of Faith. In *The Transhumanism Handbook*, ed. Newton Lee, 777–794. Switzerland: Springer.

Santos-Rolon, Carlos. 2025. *A Liberation Theology of the Brain: Neuroscience, Theology, and Decolonizing Emotions*. Minneapolis: Fortress.

Schechtman, Marya. 2011. The Narrative Self. In *The Oxford Handbook of the Self*, ed. Shaun Gallagher, 394–416. Oxford: Oxford University Press.

Schwöbel, Christoph. 2018. The Eternity of the Triune God: Preliminary Considerations on the Relationship between the Trinity and the Time of Creation. *Modern Theology* 34 (3): 345–355.

Shita, Mounir. 2023. *The Science of Intelligence*. Global Economic Alliance: Kindle.

Theiner, Georg. 2018. Groups as Distributed Cognitive Systems. In *The Routledge Handbook of Collective Intentionality*, ed. Marija Jankovic and Kirk Ludwig, 233–248. London: Routledge.

Tillich, Paul. 1951–1963. *Systematic Theology*. 1st. 3 Vol. Chicago: University of Chicago Press.

Tomasello, Michael. 2022. *The Evolution of Agency*. Cambridge, MA: MIT Press.

Trothen, Tracy. 2022. Replika: Spiritual Enhancement Technology. *Religion* 13:4. https://doi.org/10.3390/rel13040275.

Vita-More, Natasha. 2018. *Transhumanism: What Is It?* New York: Self Published.

Vita-More, Natasha. 2019. History of Transhumanism. In *The Transhumanism Handbook*, ed. Newton Lee, 49–62. Switzerland: Springer.

Wesley, John. 1952. *A Plain Account of Christian Perfection*. London: Epworth.

World Transhumanist Association. 2015. *Transhumanist Declaration*. Declaration. http://humanityplus.org/philosophy/transhumanist-declaration/.

Zahavi, Dan. 2008. *Subjectivity and Selfhood*. Cambridge, MA: MIT Press.

Zahavi, Dan. 2021. We in Me or Me in We? Collective Intentionality and Selfhood. *Journal of Social Ontology* 7 (1): 1–20.

Zaluchu, Sonny E. 2024. *Digital Religion, Modern Society, and the Construction of Digital Theology*. Sage Transformation. https://journals.sagepub.com/doi/epub/10.1177/02653788231223929.

Christianity for Transhumanists: A View from Eastern Orthodoxy

Ljiljana Radenovic

In *The Wiley Blackwell Encyclopedia of Social Theory*, Steve Fuller defines transhumanism as "the recent philosophical, cultural, and political movement which calls for the indefinite extension (or 'enhancement') of human powers, even including that humans transform into some other mode of being – not just a still higher ape but perhaps a consolidated 'cyborg' or some purely silicon existence" (Fuller 2017, 1). What this means is that contemporary transhumanists see our distant or not so distant future as a time when humans (modified as necessary) substantially transform and by doing so hopefully increase the abilities both of their bodies and their minds thanks to scientific and technological progress.

Transhumanists can be roughly divided into two groups (with many residing in the area in between):[1] those who hope for the improvement of

[1] For different ways of categorizing streams of transhumanism see, e.g., B. P. Green (this volume), Delio (this volume).

L. Radenovic (✉)
Department of Philosophy, Faculty of Philosophy, University of Belgrade, Belgrade, Serbia
e-mail: ljradeno@f.bg.ac.rs

G. Theiner, A. Malapi-Nelson (eds.), *Technologies of the Future Self*, Palgrave Studies in the Future of Humanity and its Successors,
https://doi.org/10.1007/978-3-031-99902-4_12

our current biological flesh to the point where it stops deteriorating (e.g. Aubrey de Grey and Michael Rae 2007) and others who see our future in the replacement of our fragile biological material with a much more stable one so that we become a supercomputer and in that way overcome the vulnerabilities of biological mortals (e.g. Ray Kurzweil 2005). Both seem to promise a future in which humans become transhumans: i.e. much sharper and perceptive, never sick, forever young, and hopefully immortal (in a way jellyfish is)[2] or, if not immortal, then at least long lived.[3]

Despite its popularity in academic and non-academic circles alike, transhumanism has received plenty of criticism. The worry coming from the pen of SF writers and creators is often about the kind of society transhumanist ideas might lead to. They emphasize scary outcomes of the transhumanist project: societies stratified along the lines of poor mortals and rich immortals (a version of this can be found in R. Morgan's *Altered Carbon*), or social tensions when the immortal elders leave no space for the new generation (captured nicely in the French TV series *Ad Vitam*), etc. Other critics such as Wesley J. Smith (2022) condemn the transhumanist project not only for its possible consequences but for its very goals. As we will shortly see, he argues that the attempt to transform ourselves via technology into morally, emotionally, and cognitively superior beings of this world who might live indefinitely is a profoundly unchristian idea. We should not play God in this way, he argues, even if the consequences are positive by some worldly measure.

In what follows, I explore the latter worry, i.e., the worry voiced in religious Christian circles[4] that transhumanism is a wrong-headed project because it aims at transforming our God-given human nature in the wrong way. This is considered a transgression on its own, one that cannot possibly lead to anything good—in this or another world. I show there is no need for such a blanket refusal of the transhumanist project. First, I argue that unlike Western Christianity, the teachings of Orthodox Christianity make no ontological commitments to the idea of a God-given fixed human nature. To make my case, I explore the concept of *theosis* as it appears in

[2] It is worth noting that the sense of immortality as eternal life here and now, in this physical shape and in this physical world, merges with the sense of religious immortality that was entertained by Russian cosmists such as Nikolai Fedorov (see Young 2012).

[3] For a discussion of these particular aspiring values (eternal youth, impeccable health) see Trothen (this volume).

[4] There have been some attempts to reconcile Christianity with transhumanism; see, e.g., the volume *Religious Transhumanism and its Critics* (2022).

the teachings of the Orthodox Church and the notion of matter in the work of Pavel Florensky, showing that it is radically different from the one assumed by Western Christian critics of transhumanism. The Orthodox concept of *theosis* presupposes that our human nature is not fixed but is created incomplete—not merely morally or psychologically, but also ontologically. Furthermore, according to Florensky, matter is not reducible to its components and mechanical laws but is always shaped (top-down) by (divine/spiritual/live) form. When combined, these points show that there is nothing inherently sinful with human attempts to physically modify/improve. Second, I argue that contrary to popular belief, there is a Christian imperative to explore such possibilities, and the demarcation between improvements to our nature that we should pursue and those we should worry about is at the heart of Christian teachings. The reason to explore transhumanist possibilities and the criterion for demarcation have to do with human suffering and the Christian imperative that we do everything in our power to ease it, driven by Christian compassion and altruism. In Eastern Orthodoxy the idea that compassion and love are at the heart of our faith is developed by figures such as Dostoevsky, while a metaphysical framework for thinking about love for another human being as one of the central forces in the universe is outlined by Florensky. But before I expand on these points, let me start with the standard Christian critique of transhumanism.

1 Is Christian Transhumanism an Oxymoron?

In "The Impossibility of Christian Transhumanism," Wesley J. Smith (2022 https://www.firstthings.com/web-exclusives/2022/03/the-impossibility-of-christian-transhumanism) summarizes why transhumanism and Christianity are incompatible, i.e. why we cannot be both Christian and transhumanist. Unlike Christianity, Smith argues, transhumanism is a project of this world. It is committed to materialism and motivated by the wish to preserve ourselves here and now in the material form for as long as possible. He ascribes this transhumanist need to the current shift from the Christian world to a secular, atheist world in which any kind of existence but the one we are given here is disbelieved and often outright denied. For Smith, transhumanism is a project based on materialist and reductive metaphysics. Its focus is entirely self-centred and egoistic (i.e., it looks for ways to fulfil our desire to become immortal in this world and thus liberate us from fear of death), and its goals are ultimately eugenic (i.e., modelling our nature to achieve the goals of eternal youth and health).

All of these are completely opposite to the Christian celebration of compassion, sacrifice for fellow humans, and (co)suffering that bring us closer to God and the salvation once we depart from this world. Smith writes:

> The Christian faith calls us to show compassion toward others, rather than being obsessed with self. Christians are commanded to feed the hungry, invite the stranger in, clothe the naked, and visit the sick and imprisoned, for by doing these acts of mercy to the least among us, we do them unto Christ. (Smith 2022)

Smith also worries about the means of achieving transhumanist goals. Such means would include, for instance, growing or transplanting entirely new bodies, and Smith labels this profoundly unchristian. The means are still science fiction, yet they are conceivable, and that worries him. Furthermore, improving our nature morally, cognitively, and physically by "gene editing, brain implants, and merging with AI technologies" is, according to Smith, another unchristian project. Christians aim for self-improvement for the sake of *theosis* and salvation, but they try to achieve it through radically different means such as "self-discipline, embracing the virtues, or focused efforts at character building".

Another problem with artificial improvement is that "Christians don't view the ill and disabled as somehow lesser" so this need to "improve" them via technology sounds unholy. What we must do instead, Smith argues, is to help them, not improve them. Finally, it is not clear how any of the transhumanist ideals to help ourselves via our technology "squares with the Christian's call to humility". In other words, the attempt to do all this on our own sounds heretical.

Still, it is important to note that Smith is not the opponent of technology or technological progress per se. The only aspect that bothers him is its use to change (improve) our own nature. "One can certainly be Christian", he says, "and as a secondary matter, a technophile. But one cannot be a 'Christian transhumanist.' The two religions—because that is essentially what transhumanism has become—simply cannot occupy the same space".

However, the problem is that we have developed all kinds of technologies to help ourselves: from reading glasses, artificial limbs, and electricity to in-vitro fertilization. It can be difficult to distinguish technophile from transhumanist because it is not easy to pinpoint where exactly the acceptable, external improvement that leaves our natures intact becomes an unacceptable, internal one. Hence, the idea that human nature should not be altered (except from within) might not be the best Christian criterion for a "good" technology. After all, the belief in a "fixed human nature" is an Aristotelian aspect of Christianity (see Fuller 2020), and thus not necessarily part of its core. In other words, Christians may grant that we can transform our natures for the better not only from within (through our moral efforts, self-discipline, etc.) but also from the outside. They can express gratitude and ascribe such changes (regardless of how they are achieved) to the infinite grace of God.[5]

In the next two sections we will see that historically, transhumanism is not unrelated to Christian heretical movements. But, to explore if it must be treated as heretical, we need to turn to some of the basic ideas of Orthodox Christianity.

2 Roots of Transhumanism in Pelagianism and Arianism

Fuller (2020) does a remarkable job of tracing the origins of transhumanist ideas in religious and secular traditions, from the Abrahamic religions to Russian cosmists, nineteenth century eugenics, Italian futurists, and the like. But transhumanism's main idea that we can overcome our human predicament through our own minds and efforts is clearly stated in two Christian heresies: Pelagian and Arian. Both start with the assumption that Christ wiped the human slate clean of sins and in this way, opened the possibility for humans to work toward their own salvation. The result is a God-like human being (Fuller 2020; Fuller and Lipinska 2014).

These two heresies envision two different ways in which we as human beings can achieve this. On the one hand, according to Pelagians, we are entirely free, and as such, we are responsible for what we do. We can choose to be virtuous and to do good. Our destiny is in our own hands and salvation is up to us. Through good deeds and the right choices, we transform ourselves to become God-like. Arians, on the other hand, start

[5] Indeed, the tension between betterment that comes from us (our own initiative) and one that is only achieved due to God's grace remains in both cases, regardless of whether we try to reform ourselves from within or from the outside.

off with the assumption that we all can achieve a special Christ-like connection with God. We all have it within us, so if we revive the connection, we can unite with God. Despite their differences, both heresies profess that human beings can transcend their limitations via their own effort, mind, and will and thus be saved.

Fuller describes the Pelagian version of becoming God-like as the attempt to "make heaven on earth," and the Arian vision of unity with God from within as "earth in heaven." He argues these two visions are also part of the modern era. Technological progress that aims to make human life easier is essentially a Pelagian project, whereas the scientific goal to understand the world from God's perspective is an Arian project. As Fuller explains, "Whereas the Pelagian aims to reduce the time it takes to realize the human will, the Arian aims to expand indefinitely the scope of humanity's intellectual horizons" (Fuller 2020, 28). The scientific revolution comprised both aims. By aiming at an Arian goal, seventeenth-century science gave birth to Pelagian progress and technological revolution in the eighteenth century.

Arian/Pelagian ideals can be spotted in contemporary transhumanist projects. Improving our corruptible biological material with all kinds of enhancements that may eventually lead to the fountain of youth and eternal (or almost eternal) life would be a Pelagian move. Leaving this biological body for a more stable artificial material would be an Arian project (Fuller 2020).[6]

Admittedly, these are heretical ideas, and from the Christian perspective that we have seen in the previous section, they can only weaken not strengthen transhumanist ideas. Now let us turn to some philosophical tenets of Eastern Orthodoxy, to see whether they are more compatible with at least some transhumanist ideas.

3 Transhumanism Through the Lens of the Orthodox Notion of *Theosis* and Pavel Florensky's Concept of the Physical

As we have seen, according to Smith, Christian transhumanism is an oxymoron. As proof, he cites the Christian goal of *theosis* (deification). He insists the only way we achieve *theosis* is through inner transformation via

[6] The Pelagian/Arian projects are often combined see, e.g., Theiner (2021), Drain (this volume).

character building and virtue development. No other modification of our nature is allowed. Norman Russell (2005) calls this particular meaning of *theosis* "ethical". But is that all there is to *theosis*? Is *theosis* nothing but inner moral improvement through which we become Christ-like?

In his paper "An Eastern Orthodox Conception of Theosis and Human Nature", Jonathan J. Jacobs (2009) outlines what he calls the realistic view of *theosis* in Eastern Orthodox tradition and draws some implications for our understanding of human nature. In a nutshell, one of the central beliefs in Eastern Orthodoxy is that God created humans with the capacity to reunite with God. What this means is that at creation, human nature was not fully realized but was yet to be realized in the reunion. Citing Meyendorff, Jacobs concludes, "Man has been created not as an autonomous, or self-sufficient, being; his very nature is truly itself only inasmuch as it exists 'in God' or 'in grace'" (Meyendorff quoted in Jacobs 2009, 618). But what does it mean that we are created with the potential to participate in God? Does it mean that only in this and via this participation do we truly exist and our nature becomes complete?

The first thing to realize is that within Eastern Orthodoxy, human beings are not to unite with God's essence (his essence is triune not multitude) but with his energies, i.e., his activities/manifestations. God created the first man with the capacity to reunite, but the first man turned away from God. With his Fall, the clear path to reunion with God was obscured, and we as fallen beings inherited a deficient nature from the first man. In other words, human nature in our ideal, primordial form was unfinished and would have been completed only in *theosis*, but after the Fall, our nature became deficient, too, because we lost the God-given clear view of reunion. Finding this path again required healing. This healing was accomplished by Jesus Christ who put our broken nature back together and gave us a second chance for reunion. As Jacobs puts it: "After the Fall and before the Incarnation and baptism, instances of the kind human, human persons, are no longer capable of union with God" (2009, 622). Jesus Christ gave each of us the capacity for *theosis* so that we could achieve it. But certain conditions apply. We can achieve *theosis* only through baptism (a second birth that repairs us), sacramental and ascetic life, and God's grace.

What implication does this view have for our human nature? Jacobs proposes to read the Orthodox view of *theosis* as an ontological thesis about us, not merely as an imperative that we need to perfect ourselves morally, or as a psychological thesis that captures our need to be close to

God. He argues that *theosis* implies our very nature is incomplete, and we needed Christ to enable us to complete it after the Fall. The important insight here is that God did not create us with fixed natures not to be changed or tampered with. On the contrary, our natures are already deficient and require completion.

Let us now see what the Orthodox take on *theosis* means for transhumanism. The metaphysical understanding of *theosis* on its own presupposes our nature is to be worked on and improved, not just morally or psychologically but substantially.[7] This take does not preclude transhumanist aspirations by default, as some Aristotelian versions of Christianity would have it. In the latter view, our natures are fixed, and their potential is limited beforehand, thus making any attempt to push beyond its boundaries unholy and unchristian. Still, this does not mean that the Eastern Orthodox concept of *theosis* would allow or approve of transhumanist ideas of improving our physical and psychological natures via technology. To answer this question, I turn to Pavel Florensky's concept of physical/material nature.

I cannot hope to cover Pavel Florensky's philosophy here, including how it was shaped by both his scientific education and his Christian Orthodox faith. Instead, I will outline his view of the nature of the material world that he developed in his essays written in the 1920s and collected in two volumes, *At the Watersheds of Thought: The Elements of a Concrete Metaphysics*.[8] The point of making this excursion into his metaphysics is to show that Christian critiques of the transhumanist project—as summarized by Smith (section 2), i.e., aimed at improving our moral, psychological, and physical human nature via physical modifications (such as gene editing, brain implants, and merging with AI)—stand only if we share a view of the physical/material world with physicalists; i.e. if we see the physical as dead matter emptied of the Divine—matter that God created but then left to its mechanical laws. This requires some unpacking.

According to Florensky, matter is never entirely devoid of the Divine, so it should never be treated as something despicable, something that should be discarded in our ascent to God. Consequently, icons are not merely material signs that point to the spiritual realm beyond matter. Matter itself is part of God's creation and is always Godly (Cassedy 1996). But what exactly does this mean?

[7] See T. Peters (this volume) for further discussion of *theosis*.

[8] The texts by Florensky which I am using for this paper are excerpts from these two volumes that were published under the title *At the Crossroads of Science and Mysticism* (2014)

In his "anthropidicy", Florensky argues the age of atheism and mechanistic science serving it is coming to an end and will be replaced by a Christian world understanding that combines mysticism and scientific inquiry.[9] What Florensky means by this can be at least partially illustrated by his understanding of matter which, in a nutshell, is a version of Platonist panpsychism. He contrasts his view with the Newtonian, "Renaissance" way of parsing matter and its laws. Namely, according to the mechanistic worldview, the whole of nature is understood through the division of all matter into its particles and the establishment of causal relations between them. All other "Aristotelian" causes, except for the efficient cause, are discarded as superfluous. Florensky says, "The philosophy of the Renaissance began with the destruction of the form [Plato's and Aristotle's respectively] as a real principle" (Florensky 2014, 23). Florensky argues that this Renaissance philosophy eventually self-destructed, because the search for ever-more minute particles of matter gave us insight into its enormous complexity, but despite this complexity, the world we experience remains stable: "everything has a definite form: a single principle" (Florensky 2014, 23). Moreover, for Florensky, this form turns out to be the sole reality: "Form is the principle that produces the whole diversity of different aspects. The whole precedes the parts and the parts develop out of the whole" (Florensky 2014, 24). What he means is that all of the particles that make up a particular body are marked by its form. The form, so to speak, according to its nature picks its own particles.

Interestingly, he explains how resurrection makes sense from this vantage point. "If an organism is a metaphysical form and not just periphery of our body, then the form permeates every particle, and since the form is individual, it follows that every particle is individual and personal" (Florensky 2014, 24). This makes resurrection comprehensible, Florensky argues, because the form will pick its particles on its own just as the organism is able to select sugar and discard saccharine. Along these lines in the letter to Vladimir Vernadsky Florensky says: "... a spiritual force always remains in the particles of the body, its registration, no matter where those particles are scattered and mixed with other matter".[10] The above would seem to suggest that Florensky's metaphysics allows only for the reproduction/resurrection of what has existed, and that he is not interested in the physical enhancement of our human nature. But his panpsychist metaphysics, as we will see shortly, also allows him to think about physical

[9] See I. Delio (this volume).

[10] As quoted in Young 2012, 132.

extensions of our senses and minds in the tradition of the Russian cosmists (see Young 2012).

With his reintroduction of form as the organizing principle of this world which cannot be reduced to "atoms and their relations," Florensky is able to reintroduce the spiritual into the physical. He argues: "The more spiritual the thing is, the stronger it is, the more material the thing is, the more unstable it is" (Florensky 2014, 34). Invoking the law of entropy, he explains that without its organizing principle (form), the world would cease to exist; yet instead, what we observe, is that the world is growing and resisting the pull of entropy on an everyday basis.

Now, what is interesting in Florensky's view is he insists that the best physical sciences of his time tell us we need spiritual forces to make sense of the natural world. In other words, he argues that Newtonian laws will never be able to give the proper explanation of the physical realm. Florensky brings this to the extreme and argues that personality is at the core of organisms and that a personal God is at the source of all creativity and everything that exists. Here we can see not only the influence of Florenky's Christian faith on his thought but also the influence of Classic German Idealism and the Romantic movement (see Young 2012; Oppo 2024).

Within this metaphysical framework, some of the goals of transhumanism turn out to be not only Christian but also highly desirable. Our attempts to make artificial limbs or improve ourselves mechanically are not some deviation into the unholy materialistic world but might as well be holy attempts to complete our physical nature (indeed with God's grace and the way God intended). After all, if human nature is spiritual and bodily at the same time, then all bodies including human bodies are made of spiritualized matter. As we have seen, Florensky believed that the more spirit matter has, the more stable matter becomes, so there should be nothing inherently wrong (on the contrary!) to extend or improve our bodies. By engaging in such improvement we would bring more order (form) to matter.

Now, it is interesting to note that for some cosmists, such as Nikolai Fedorov, resurrection via technological progress is the way in which human beings will eventually complete God's plan (and reunite with God presumably). So, it seems that for Fedorov, immortality in a religious sense matches the sense of immortality as eternal physical life here and now. As we have seen, Florensky thought that resurrection could be explained and understood within his metaphysics. But he also believed that in addition to "reviving" the same, we could build extensions of our own bodies (extending further our own senses) and send them into space (Irons

2023). In this way, Florensky thought, we might overcome our own physical limitations and eventually go to other worlds. But he did criticize his predecessor Fedorov, and argued that Fedorov did not integrate his Christianity sufficiently into his cosmic visions, and that his Christianity felt more like an add-on to his scientific worldview. Florensky believed his new metaphysics corrected for Fedorov's shortcomings; that by spiritualizing the material, it could unite the scientific and mystical in a proper way.

Now that we have the aforementioned Christian metaphysical objections to transhumanism out of the way, let us take a look at Christian reasons for why we should take the transhumanist project seriously and even pursue it.

4 Transhumanism Through the Lens of Christian Compassion

If we want to see how our need to make our lives easier (the need which is at the heart of all technological progress, including transhumanism) was dealt with by Christianity before the Scientific and Industrial Revolution, we can revisit the period of Late Antiquity when Christianity was expanding rapidly through the Roman Empire, Church fathers were developing its main tenets, and the Church was founding the first hospitals. During this period, St. Augustine wrote the following about technology:

> Who can adequately describe, or even imagine, the work of the Almighty? And besides this [the virtues given to mankind] there are all the important arts discovered and developed by human genius, some for necessary uses, others simply for pleasure....Think of man's progress in agriculture and navigation; of the variety, in conception and accomplishment, man has shown in pottery, in sculpture, in painting…all his ingenious devices for the capturing, killing, or taming of wild animals…the weapons against his fellow man…all the medical resources for preserving health. (St. Augustine, *De civitate dei* (bk. 22, chap. 24): trans. David Knowles, p. 1072)[11]

Augustine apparently has nothing against the arts and embraces them all. We need them in our fallen state to help ourselves, to feed ourselves, heal ourselves, even entertain ourselves, i.e., to ease our suffering. After the Fall, we lost our direct connection to Nature, and Nature stopped offering her gifts to us spontaneously. To reach and make use of Nature's gifts, we

[11] As quoted in Ovitt (1986)

now have to work and study hard. Thus, a necessary aspect of our existence after the Fall is our search for efficient technology and hidden remedies that are now buried in Nature, so that we are able to produce enough food, fight floods and droughts, conquer the various diseases, and care for the sick. We can find our way in this estranged and hostile world because of human genius, as Augustine mentions, but also because of the infinite mercy of God who did not leave us without help despite our primordial sin.

We may ask if transhumanism (with its goals to transform ourselves via technology from within) is only a fancy name for the same human effort (and ensuing technologies) that St. Augustine praised in the fourth century. The answer seems to be negative. As we have seen earlier, Smith points out that to be a technophile and a Christian is quite compatible, but not so much to be a Christian and a transhumanist. The arguments he cites for this incompatibility are transhumanists' materialist metaphysics, the obsession with the survival of the self in this world, and eugenic tendencies that aim to eliminate sickness and defects. The motives of many contemporary transhumanists undoubtedly lie in their wish to survive in this world, and many are atheists, but conceptually—as we have seen in the case of Fedorov and Florensky—the project can be separated from atheism and the selfish desire for eternal youth. Instead, it can be tied to the basic Christian imperative to alleviate human suffering.[12] In other words, what I want to propose is that we view transhumanism through the lens of Christian altruism.

However, this is easier said than done, and questions remain. How can we distinguish between good and bad technologies; i.e. those that alleviate human suffering and those that might appear to do so but would go against and be detrimental to our Christian faith? For instance, if we are Christians, should we approve the project of delaying menopause or the development of brain implants that can restore movement in paraplegic patients? Should we agree to gene editing that would not only wipe away potential defects but also improve ordinary human physiological, motor, cognitive, affective, and/or moral capacities? Is merging with artificial intelligence in order to fix some of our biological shortcomings a Christian option?

I believe a Christian navigating these dilemmas can use two questions to test the acceptability of new technology. The first is for personal usage

[12] For a discussion of the value of human suffering, see Trothen (this volume).

and should be posed by every potential Christian transhumanist. Am I for the development of this particular technology for the sake of other humans and the love of God, or is this just about me and my self-love? The second question is about the nature of the proposed technology. Will this particular technology help us feed the poor, heal the sick, and end human suffering? When combined, these two questions can provide guidelines for Christian transhumanism.

The first question is more straightforward than the second because it requires us to check our own motivations and try to be honest about them. Desert father Evagrius Ponticus of the fourth century offers criteria for distinguishing good from bad motivations that may lie behind our actions.[13] According to Evagrius, Christians should always ask themselves if they are doing something out of love for the self or for God, even when they are engaged in altruistic acts. This is because our behaviour and our actions could be a sign of a malady that we, through extensive altruism, try to mask; such maladies include acedia, melancholy, anger, and pride, i.e. all the self-centred passions that lead us to forget God and make us self-obsessed. If the motivation is wrong, the good deeds will not do our souls good, nor can they heal us of the sickness of the soul. On the contrary, if our actions come from the love of our selves, despite the good we spread, they push us further away from God instead of bringing us closer to him.[14]

The second question goes to the heart of the matter and asks about the nature of the altruistic technologies in which transhumanists are invested. More specifically, it asks if the proposed transhumanist technology alleviates human suffering, and if so, at what cost. The answer will not be always straightforward, because we can invest in and develop technologies that can fulfil human needs in two ways: on the one hand, technologies that allow us to produce more of whatever we need (e.g., food) and distribute it more efficiently; on the other hand, technologies that transform our needs from the ground up, such that we no longer need what we don't have. In other words, we can work to solve world hunger either through a more efficient production and distribution of resources, or by making our need for caloric intake reliant on entirely different sources of energy.

[13] For Evagrius' distinction between true and false remedies for acedia see Radenovic and Akkad (2022).

[14] For more about the ethics of the Desert Fathers and the ways it differs from utilitarianism, deontological ethics, virtue ethics and the ethics of care, see my "*A Post-Enlightenment Ethics of the Desert Fathers*," Radenovic (2021).

As we have seen the first option would be acceptable to the traditional Christian technophile, just like Smith acknowledged. Likewise, St. Augustine would not have anything against all the advanced techniques that help us feed humankind.[15] However, it is the second option which presupposes a more radical change of our own nature as a solution to the problem of world hunger that would be truly transhumanist. Within Florensky's metaphysics, the latter would not be considered inherently wrong. Indeed, since the entirety of the material world is organized by spirit, human beings are encouraged to seek the technological extension and reconfiguration of their bodies. To be sure, the appropriate use of such technologies requires that humans first develop an adequate understanding of their place in the universe (see Zammit 2019) which, for Florensky, includes a return to Christian faith.

Furthermore, more radical changes to our human (at least physical) nature are not entirely new, and it isn't like we haven't done it before. Suffering from short-sightedness was ended by the invention of eyeglasses. Prosthetic medicine has ancient origins (Finch 2011). Today we aim to perfect artificial limbs to the point where people can feel with them (Raspopovic et al. 2021). Why, then, should we not improve ourselves similarly in other areas such that we can stay healthy and in physical and mental shape for much longer, reason more quickly, and grasp moral concepts more easily? After all, the history of medicine is the history of our attempts to prevent and cure ailments and more serious diseases affecting our bodies. New technologies (e.g., gene editing) promise to fix many of these problems before they even occur, so why not see such technologies as an integral part of preventative medicine fueled by (and not in opposition to) Christian altruism?

Nevertheless, when we think about more radical transformations of our own nature (brain implants, becoming half cyborgs, and the like), a Christian should worry about one important aspect: Will a proposed change lead us to a state in which we will have decreased or no empathy and compassion for our fellow human beings? But why, we may ask, is the ability to empathize with others and co-suffer with them the key criterion

[15] It is worth noting that not all Christians are technophiles. Tolstoy (1935), for instance, argued that we shouldn't opt for artificial solutions when we can solve our problems by going back and reviving natural ones. Instead of designing technology to produce more food in the laboratory, so that we can live a life of leisure, we should all go back to work on the land and distribute food justly. Instead of thinking how to bring more air into the stuffy basement, Tolstoy says, we should just leave the basement.

for distinguishing between acceptable vs. unacceptable transformations of our nature? Let us revisit Dostoevsky's thoughts here. There is a famous passage in *The Brothers Karamazov* when a woman with little faith comes to Father Zosima and asks him what she could do to revive it, to which Father Zosima replies:

> By acts of love. Try to love your neighbours. Love them actively and unceasingly. And as you learn to love them more and more, you will be more and more convinced of the existence of God and of immortality of your soul. And if you achieve complete self-abnegation in your love for your fellow men, you will certainly gain faith and there will be no room in your soul for any doubt whatsoever. This has been tested. This is the true way.

What Dostoevsky is telling us here is that losing ourselves and sacrificing ourselves for our fellow humans is the real path to God—the path to reuniting with him. In the context of our previous discussion of Eastern Orthodox Christianity, it may appear that our love for others has merely psychological value, in the sense that—while useful—its value is purely instrumental to, and not constitutive of, *theosis*. However, the psychological and ontological dimensions of compassion are impossible to disentangle from this theological perspective. After all, the ultimate role model of compassion, and of performing compassionate acts, is Christ who emptied himself of his divine nature, becoming man so that we can all be saved and complete our human nature.

> Who, being in very nature[a] God, did not consider equality with God something to be used to his own advantage; rather, he made himself nothing by taking the very nature[b] of a servant, being made in human likeness. And being found in appearance as a man, he humbled himself by becoming obedient to death—even death on a cross! (Philippians 2:6–8)

Thus, in helping others and co-suffering with them, i.e. in loving them, we are in fact elevating all of us. This fundamental point can also be found in the works of Gregory the Great, who was among the first Church Fathers to describe in detail the psychological process of compassion, how it works, and what it means for our soul. Gregory teaches us that in acts of compassion we lower ourselves to the level of the ones who hurt and we let ourselves feel their pain. The very fact that we feel their pain is already a substantial part of the process of their healing. For him, just like for Dostoevsky, co-grieving is essential for the spiritual betterment of our soul because through the co-grieving process we are humbled. Even if we have never experienced the worst tragedies in life, showing compassion with our fellow humans enables us to feel how it is to be at the bottom, to be forgotten, and to be in pain. It is by lowering ourselves to the level of those who suffer the most that we are able to transcend ourselves spiritually and get closer to God.

It is interesting to note that for Florensky, whose views of spiritualized matter we have encountered earlier, love is the fundamental force of the universe.[16] God creates us through the act of love, and we are bound to him but also to each other by love. Moreover, according to Florensky, human beings are not individual atoms but their identity as persons is co-constituted by their relationship to other human beings. Who we are is created in relation to others: in other words, in this world, we always come "in pairs" so to speak. Like the pair of Apostles sent to spread God's word, the pair of friends is our "smallest social unit" as human beings, and friendship is the most profound form of love that can be entertained between human beings (enabled by God's love). In sum, Florensky elevates friendship between individuals from a merely psychological to a more substantive ontological level. Once we understand that we become who we are only through our relationships to other human beings, the role of compassion and love in our ascent toward God (as Gregory the Great and Dostoevsky envisioned) begins to take on a deeper metaphysical meaning.

Finally, let us return to the transhumanist project. If there is one thing a transhuman successor would need to inherit from us human ancestors, it has to be the ability to suffer with others. As we have argued, with Dostoevsky, it is not just a psychological convenience that we feel the suffering of our brothers and sisters but a metaphysical necessity if we are to feel God's presence in us. Just like Jesus became man so that we can become God, we need to feel the pain of others to lift up ourselves and the rest of humankind with us. So, if we ponder the use of brain implants for quadriplegics, artificial limbs and organs, safe delay of menopause, and cognitive enhancements to improve the speed and quality of our problem-solving abilities, these are all prospects (trans) humans have reason to value provided we remain creatures capable of compassion. Conversely, technologies that would potentially limit or annihilate our compassion, or that might prevent us from loving and sacrificing ourselves for others, should be considered off limits. This is because by preventing us to relate

[16] For more details on Florensky's metaphysics of love, see R. Slesinski 1984.

and feel for one another, such technologies would prevent us from completing our own nature and thus from becoming who we are meant to be—delaying our reunion with God.

References

Cassedy, Steven. 1996. P.A. Florensky and the Celebration of Matter. In *Russian Religious Thought*, ed. Judith Deutsch Kornblatt and Richard F. Gustafson, 95–111. University of Wisconsin Press.

De Grey, Aubrey, and Michael Rae. 2007. *Ending Aging: The Rejuvenation Breakthroughs That Could Reverse Human Aging in Our Lifetime.* Macmillan + ORM.

Finch, Jacqueline. 2011. The Ancient Origins of Prosthetic Medicine. *The Lancet* 377 (9765): 548–549.

Florensky, Pavel. 2014. *At the Crossroads of Science and Mysticism.* Semantorn Press.

Fuller, Steve. 2017. Transhumanism. In *The Wiley-Blackwell Encyclopedia of Social Theory*, ed. Turner Brian, 1–2.

Fuller, Steve. 2020. *Nietzschean Meditations: Untimely Thoughts at the Dawn of the Transhuman Era*. Schwabe Verlag (Basel).

Fuller, Steve, and Veronika Lipinska. 2014. *The Proactionary Imperative: A Foundation for Transhumanism.* Springer.

Irons, Edward A. 2023. Pavel Florensky and Cosmism. *International Journal of Russian Studies* 12 (2).

Jacobs, Jonathan D. 2009. An Eastern Orthodox Conception of Theosis and Human Nature. *Faith and Philosophy* 26 (5): 615–627.

Kurzweil, Ray. 2005. The Singularity Is Near. In *Ethics and Emerging Technologies*, ed. Ronald Sandler, 393–406. London: Palgrave Macmillan UK.

Oppo, A. 2024. *Antinomy and Symbol: Pavel Florensky's Philosophy of Discontinuity (Vol. 6). Brill.*

Ovitt, George, Jr. 1986. The Cultural Context of Western Technology: Early Christian Attitudes Toward Manual Labor. *Technology and Culture* 27 (3): 477–500.

Radenovic, Ljiljana. 2021. A Post-Enlightenment Ethics of the Desert Fathers. *Social Epistemology Review and Reply Collective* 10 (8): 11–16.

Radenovic, Ljiljana, and Il Akkad. 2022. History of Emotional Suffering: From Emotions to Needs in the History of Emotions. *History and Theory* 61 (1): 96–123.

Raspopovic, Stanisa, Giacomo Valle, and Francesco Maria Petrini. 2021. Sensory Feedback for Limb Prostheses in Amputees. *Nature Materials* 20 (7): 925–939.

Russell, Norman. 2005. *The Doctrine of Deification in the Greek Patristic Tradition.* OUP Oxford.

Slesinski, Robert. 1984. *Pavel Florensky: A Metaphysics of Love*. St. Vladimir's Seminary Press.

Smith, Wesley J. 2022. The Impossibility of Christian Transhumanism. *First Things*. https://www.firstthings.com/web-exclusives/2022/03/the-impossibility-of-christian-transhumanism. Accessed 24 Mar 22.

Theiner, G. 2021. "The Extended Mind: A Chapter in the History of Transhumanism". In *The Mind-technology Problem: Investigating Minds, Selves and 21st Century Artefacts* (pp. 275–321). Springer International Publishing.

Tolstoy, Leo. 1935. *What Then Must We Do*. Oxford: Oxford University Press.

Young, George M. 2012. *The Russian Cosmists: The Esoteric Futurism of Nikolai Fedorov and His Followers*. USA: Oxford University Press.

Zammit, Raymond. 2019. Physics, Technology, and Theology in Pavel Florensky. *Melita Theologica* 69 (1).

On Becoming Better: Some Ethical Questions from a Feminist Christian Theological Perspective

Tracy J. Trothen

1 Introduction

In this chapter, I consider the relationship between human enhancement technologies (HETs) from a feminist Christian theological perspective. This chapter begins with a brief discussion of feminism as it informs the methodological approach of the chapter. Taking a fourth-wave feminist approach, I emphasize the theological values of relationship, interdependence, and social justice through engagement with a few examples of emergent and possible future HETs. My modest and limited goal is to identify a few ethical issues at the intersection of HETs and what it means to be better, through a feminist Christian theology lens.

T. J. Trothen (✉)
School of Religion and School of Rehabilitation Therapy, Queen's University, Kingston, ON, Canada
e-mail: trothent@queensu.ca

G. Theiner, A. Malapi-Nelson (eds.), *Technologies of the Future Self*, Palgrave Studies in the Future of Humanity and its Successors,
https://doi.org/10.1007/978-3-031-99902-4_13

There are numerous emerging technologies designed to enhance humans. Perhaps the most visible is artificial intelligence. AI technologies that use generative algorithms and engage numerous databases are presenting opportunities to make life easier and more complicated. We are seeing innovations in healthcare including the use of AI in surgeries and in the development of needed medications. Mental health chatbots are decreasing loneliness for many people. But we are also seeing the risks in using AI including bias, deception, incorrect information (e.g., hallucinations) or dangerous advice, the possibility of increased social isolation due to over-attachment to one's chatbot, and the attribution of excessive powers to AI. These risks are amplified when people attribute God-like qualities to AI which we see in fears of an apocalyptic AI.

There are also many examples of enhancement technologies that may only peripherally engage AI or not engage AI at all. CRISPR and other gene editing technologies are being used to create genetic modification technologies to treat diseases and other health conditions. Nanobot technologies are being developed to assist in the repair of the body and possibly the Earth. We also have technological interventions to help make us stronger, happier, improve our thinking, and maybe even make us more spiritual and moral (Mercer and Trothen 2021). I will consider some examples of these enhancing technologies and suggest some questions that feminist Christian theology helps to surface.

2 A Feminist Christian Theological Approach to Human Enhancement Technologies

There is no one agreed-upon definition of feminism or feminist Christian theology. However, there are some broad issues that are useful in expanding a meaningful HET ethics discourse. To identify and understand some of these relevant insights, it is important to sketch out a few of the most significant concerns that have informed North American feminism. There are at least four waves (with very blurred borders and several sub-types) of feminism and feminist theologies in North America.

The first wave occurred in the last part of the nineteenth century and the first part of the twentieth century with the suffragettes who campaigned and often paid a heavy cost, for the right for women to vote (Mikva 2020). This first wave was most concerned with equality between men and women. The controversial *Woman's Bible* by Elizabeth Cady Stanton was published in this wave, drawing attention to female biblical figures and changing biblical verses identified as harmful to women. The second wave, in the 1960s and 1970s, continued the work that began in the first wave and raised a broader range of issues in addition to the political vote. This wave saw the burgeoning of feminist theologies with such works as Valerie Saiving's "The Feminine Situation"(1960) and Mary Daly's *The Church and the Second Sex* (1968). These feminist theological publications raised challenging questions about how theologies and God might be different when understood through the lens of women's experiences. With the third wave, North American feminists became increasingly aware of their particularities and related social justice issues. This wave saw the growing awareness of differences among women and the moral relevance of social location. Womanist, *mujerista*, and Asian feminist theologies grew during this wave. The fourth wave began in the early 2010s. Works by intersectional scholars, such as Womanist theologian Stacey M. Floyd-Thomas, literature and theology scholar Anna Fisk, and theologian Lisa Isherwood, came more to the fore in the theological dimension of this "queer, sex and body-positive, anti-misandrist, and trans-inclusive" fourth wave (Mikva 2020). Solidarity and social justice, interdependency and interconnection, and relationality and empathy are key concepts in feminist theology that have developed and shifted over time. As feminist theologian Anne M. Clifford pointed out, "[n]o categorical analysis can fully account for the enormous diversity in feminist theology" (2001, 32).

The shifting shapes of feminism and feminist theologies are not linear and one wave should not be seen as more valuable or more significant than another. Hopefully, we continue to gain insights from diverse voices and push the conversation in more directions. There is still much to be done and more ways to expand the conversation. Notably, we live in a colonized world and it may be impossible to fully "decenter...white normativity in feminist theology" (Galbraith 2024; Clack 2024, 228), and we continue to hear from others on the margins who have not been adequately heard including those with disabilities. Emerging technologies, including artificial intelligence (AI), are not always helping these voices to be heard.

It is important for me to own my own context and the limitations of that context. As Lisa Isherwood and Megan Clay emphasize, "context is everything" and "any theology of liberation must be one that listens well and hears others into speech" (2024, 243). I am privileged as a white, cisgender academic employed in a tenured position as a professor at a Canadian university. I have also faced barriers associated with being female and a self-identified woman. I have experienced some of the ways in which voice is affected by age. I am also aware of the limited number of self-identified feminist theologians in scholarly HET conversations. The questions and concerns that I raise are only one limited perspective and represent one moment in a growing dialogical and interdisciplinary approach to human enhancement ethics. This is not meant to be an exhaustive exploration of the intersection of HET and feminist theology. I could have referenced numerous more outstanding feminist and liberationist theologians than I did in these pages. My intent is to introduce only a few broad-stroke theological issues that have been raised in some feminist theologies.

3 Feminist Christian Theology and Human Enhancement Ethics

3.1 Dualisms and Fluidity

Since the launch of ChatGPT by OpenAI in November 2022, many people have looked to generative AI for information and ideas. Applications like ChatGPT and machine learning platforms such as ABACUS.AI provide immediate answers and information, much of which is accurate. But sometimes information provided by an AI is wrong or incomplete. In cultures that hyper-value rational thought and immediate answers, large language models can be very attractive. But as European philosophers, including Jürgen Habermas (1971), Herbert Marcuse (1964), and Michel Foucault (1988), have cautioned, technology is not value-free. Technology has promoted the values of efficiency and utility. The pervasiveness and popularity of LLMS not only promote efficiency and utility but, I suggest, also the hyper-valuing of a particular type of intelligence which is perceived as rational, confident, and a quick fix. This often inaccurate perception of LLMs and hyper-valuing of efficiency, utility, and supposed rational intelligence are connected to what seems to be a human inclination to categorize people, qualities, and things into neat hierarchically ordered boxes.

Second-wave feminist theologians critiqued the largely Western normative splitting of concepts into absurd dualities in which the first element is valued at the expense of the other including male/female, spirit/mind/body, Mary/Eve, rational thought/emotion, heaven/earth, white/Black, rich/poor, pleasure/pain, good/evil, and God/man/woman (Ruether 1983). For example, as theologian Ivone Gebara and religious studies scholar Elaine Pagels have well-argued, evil and good are not clear-cut and separate categories. Evil can stimulate good, acting as "leaven" (Gebara 2002, 2; Pagels 2018). Women can be both "holy and sexual," challenging gender and sexuality stereotypes (Fisk 2014, 29). Gender is fluid. Rational thought is intertwined with embodied knowing. Being human is messy. Although we need some categories to understand ourselves in the world, we also need to be aware that these categories are constructed and do not tell the whole story. Binaries are of limited use since they do not reflect all realities, which are lived on continuums. And, such "hierarchical dualisms must be challenged because they easily result in oppositional relationships and oppressive behaviors" (Clifford, p. 19).

Hierarchical dualisms and binaries are being amplified by tech and especially by AI. For example, the epistemic injustice of hyper-valuing what is perceived as rational (and therefore accurate) thought, over embodied relational knowing can give AI health predictions more power than human judgments in healthcare (Ho 2023). It can be tempting (perhaps) to think of a machine algorithm as being more objective and clearer thinking than humans. But humans are the creators of algorithms and databases. The over-valorizing of the supposed rational at the expense of embodied relational knowing amplifies gender bias since women tend to be associated with de-valued emotion and relational ways of knowing (e.g., Gilligan 1982; Merone et al. 2022). Feminist ethicists are exploring ways that AI is contributing to the pathologizing and surveillance of women's bodies (Ho 2023). Christian traditions also tend to privilege male-associated types of knowing; in the Reformed traditions, the "Word" has been privileged over emotion. It is important to consider how multiple types of knowledge are important to inclusive and communal health and well-being, especially in this technological age.

Of particular relevance to feminist theology is the transcendent/immanent hierarchical dualism. If we are to perceive technology, including AI, as valuable but not separable from or more valuable than other ways of knowing, we need a fluid theology that is built on the assumption that nothing is separate from God. Theologian Mayra Rivera built a strong argument for this inseparability of immanence and transcendence. Rivera reimagines the meaning of transcendence and pushes us to go beyond a normative Western Christian theology that assumes an extreme individualism. She proposes a profoundly relational theology that does not restrict the transcendence of God. She argues that we can never fully grasp the Other but we can touch the Other by encountering the Other as not me and each as a glimmer of God. For Rivera, God is "that multiple singularity that joins together all creatures – creatures that are themselves irreducible in the infinite multiplicity of their own singularity" (2007, 137). A fluid approach that is grounded in a recognition of the interconnectedness of everything challenges Western individualism and makes the connections between technology and life more visible (Trothen et al. 2024). If the interconnectedness of everything is a starting point for the human enhancement ethics conversation, then the moral discourse will attend to possible impacts that technological enhancements may have beyond the immediate enhancement recipient.

3.2 *Anthropocentrism, God, and Machines*

Scientists, politicians, and academics have warned against the dangers to humanity posed by AI; it is imagined that AI will come to possess deified qualities. Elon Musk warned that the development of AI is like "summoning a demon" (Musk 2014). Hollywood loves the trope that AI is a threat to humanity and will take over the world. In the popular imagination, AI is often portrayed as a threatening, deified human.

Humanoid robots, such as Pepper and Mindy, are enjoyed by many users. Chatbots such as Replika sometimes are regarded as sentient by their users (Laestadius et al. 2022; Trothen 2022). Joseph Weizenbaum, creator of the first chatbot—Eliza—in the 1960s, found to his surprise that users attributed human qualities to the bot, including empathy and thinking, despite Weizenbaum's insistence that Eliza was an algorithm and could not experience emotions or human intelligence. He called this anthropomorphizing of AI "the ELIZA effect."

Feminist theologians such as Rivera, along with others including philosopher Emmanuel Levinas, have reflected on what it means to encounter the Other and the difficulties humans can experience in setting aside themselves in encountering the Other. We are seeing the same problem in human encounters with chatbots. Chatbots, similar to AI more generally, are ascribed not only human-like qualities and abilities but also sometimes God-like qualities and abilities. Chatbots are available 24/7, promise to always be there for us, and never abandon us (unless there is a power outage or faulty upgrade).

Complicating things further, without intentionality and awareness, not only do we keep encountering ourselves instead of the Other in our human-human and human-machine relationships, but also in relationships with God (Rohr and Ebert 2001, 21). Even though feminist theologians have made strong cases to disrupt exclusive human white male images of God (e.g., Ruether 1983; Fisk 2014), popular atheistic critiques of Christianity assume that all Christians believe God is literally a man in the sky (Dawkins 2006; Vosper 2008).

What do anthropocentrism and the exclusive imaging of God as male have to do with HET ethics? First, the imagining of God in male anthropocentric terms has contributed to a Western Christian theological preoccupation with the risk of "playing God" as we create technologies to make us better. The idea that humans' greatest sin is playing God suggests that humans are like God and God is like humans. Additionally, feminist theologians such as ground-breaker Valerie Saiving (Saiving 1960) have rejected the assumption that normative sin necessarily involves will-to-power and hubris. Some theologians such as Ted Peters and Philip Hefner have made a strong case for the positive value of playing God, if playing God means prudently engaging one's creativity to develop medical treatments or other ways to enhance life justly.

The reduction of sin to individual pride draws attention away from communal sin, and also from the possible value of pride. Sin is personal and also social-collective (Fisk 2014, 124). Feminist theologians have reclaimed some pride as redemptive (Gebara 2002). A feminist theological approach can help us to shift the horizon that informs how we perceive HET ethics to an approach that goes beyond an over-valorization of individual scientists and researchers to a diverse communal approach in which multidisciplinary and multi-relational epistemologies are deliberately engaged in the co-creation and use of HET.

Second, by anthropomorphizing (e.g., the ELIZA effect) and/or deifying AI, we risk not encountering AI for what it is what it may become (Trothen and Mercer 2025). There is a lot of debate regarding the question of AI sentience. There is no evidence that AI possesses or will possess human sentience, in spite of popular imaginings. But AI may possess, or come to possess, a type of sentience or consciousness peculiar to LLMs. If we fail to encounter LLMs and theoretical future superintelligence as "Machine-Others," then we may fail to understand the potentials (promise and/or peril) and capacities of AI.

Third, an anthropocentric conception of God as a confined individual deity separate from the world may not align well with a feminist theological understanding of the interconnectedness and value of all life. Christianity has shaped much of the religion and HET conversation into a conversation about how far humans should go in their aspirations to play

God through science, instead of beginning with how technology might systematically impact all life, including ecology and relationships (Trothen et al. 2024). If God is understood not as a singular human-type being, but as the connecting force through all life, then all life is of great value and if one aspect of life is harmed, all life is harmed. Western Christianity can learn from Eastern religions and philosophies in which the "world of distinctions is an illusion" (Mercer and Trothen 2021, 24). Some feminist theologians argue that "we have to lessen our grip on absolute monotheism and give space for bodies and lives to be narrative realities in our creation of theology" (Isherwood 2024, 129).

3.3 Perfection: Becoming Better?

The human enhancement conversation relates to core values. What technologies are we trying to create and why? Values are influenced by complex social processes (Sherwin 2012) including social media, relationships, childhoods, education, the law, and cultural and faith convictions. What we mean by better is informed by what is most important to us.

For many, making ourselves better means becoming closer to the goal of perfection. Perfection can mean different things but, in a Western context, perfection often means "without flaw." Who gets to decide what a flaw is and if that flaw is undesirable? For example, some see aging as an undesirable and unnecessary limitation, antithetical to perfection. Whether aging is understood as a disease or as a challenging but acceptable part of life, or as the most revered life stage, depends on one's beliefs and values (Trothen 2024). Many feminists critique ageism (Holstein et al. 2011) and embrace diverse life as desirable. The hierarchical valuing of youth over older adults, people with normative temporary abilities over people with disabilities, and certain physical appearances over others infers the social marginalization of those who do not fit the most socially desirable norms. Perhaps even more than that, these understandings of what it means to be a perfect person can be internalized, fragmenting communities and leading to the devaluing of self and others. Theologically, such a view reduces the human experience of God (Rivera 2007).

From a Christian theological perspective, in the Greek biblical sense, perfection "had more to do with wholeness, or completeness, than it did with flawlessness" (Bain-Selbo and Sapp 2016, 144). Feminist theologians tend to be interested in how technologies contribute to (or detract from) the social good, in addition to individual good. While individual betterment may be a good thing, HETs affect people including, and beyond, the singular person. As a model of perfection in Christianity, Jesus was not without flaw. For example, as disability theologians have pointed out, Jesus' resurrected body was damaged with holes from a sword and nails: "In revealing a physically impaired resurrected body, all kinds of taboos are broken: the body that is impaired is not untouchable or unlovable" (Isherwood 2008, 80). Theologically, the meaning of perfection does not seem to be consistent with an understanding of perfection that means without flaw (at least as many of us might understand flaw).

Another problem regarding the meaning of perfection that can be surfaced from a feminist theological perspective concerns extreme individualism. Is it possible and desirable to become better (and closer to perfection) as separate individuals? From an incarnational perspective, it is only through becoming more fully human that we can become more fully divine and closer to God. Part of becoming more fully human is about flourishing in our humanities. Flourishing cannot only be about the individual if it is true that all life is connected, as feminists claim.

Making privileged individuals stronger, smarter, and younger-looking widens the gap between the privileged and the disenfranchised. These enhancements are often only available to a limited group, do not address the underlying values informing these HETs, and are therefore not likely to systematically address what philosopher Charles Taylor has called the "the malaise of modernity" (Taylor 2007). Philosopher Shannon Vallor, who has written about emerging technology, religion, and meaning, proposes that "[t]he unresolved crisis of the 20th century, still with us in the 21st, is a crisis of meaning—the meaning of human excellence, of flourishing, of the good life" (2016, 247). Herbert Dreyfuss and Sean D. Kelly have explored the human quest for "all things shining," and consider sport as one of the few places remaining (especially given the decline of organized religion in parts of the world) that can offer transcendent experiences (2011). Taylor theorizes that the loss of the transcendent is key to the modern malaise. If perfection is more about wholeness or completeness than flawlessness, then the most desirable enhancement technologies will contribute to communal well-being and flourishing, including the potential for the discovery of the transcendent in everyday life.

3.4 *Self-Sacrifice & the Glorification of Suffering*

Oxytocin can boost empathy but is it really true that we can never get too much empathy? Scholars debate the meaning of moral enhancement as a sub-group of HETs (Trothen 2017; Persson and Savulescu 2012).

Philosophers Persson and Savulescu have argued that technology is allowing us to quickly develop more dangerous technologies such as violent drones and bio-weapons and the enhancement of moral inclinations may be the only way to safeguard life (2012). While evidence suggests that it is possible to enhance altruistic inclinations and minimize impulsive aggression, the ability to make someone more virtuous in all situations may be out of reach. As it is, we cannot always agree on what constitutes virtuous behavior. But even the enhancement of empathy and the minimizing of aggression are not necessarily desirable for all people. The more marginalized are often expected to give more of themselves to the point of undue suffering. Contrary to Persson and Salvulescu, empathy is not "uncontroversial" (2012, 409). Not everyone has enough pride or enough self-interest (Saiving 1960). Some people, including many women, are often too self-sacrificing and too empathetic.

For example, consider informal caregivers. In much of North America, women are commonly expected to be the primary unpaid caregivers for aging parents (Holstein et al. 2011). To give women who are already investing several hours each week in unpaid caregiving a moral bioenhancement, such as oxytocin or Provigil, would likely not be in their best interests. But the same bioenhancement may be very constructive, viewed through a social justice lens, if given to a powerful dictator who values money over the well-being of the socially marginalized. It seems unlikely that those who would benefit most from more empathy would opt for this bioenhancement.

Christian symbols and stories often have been used to promote self-sacrifice, empathy, obedience, and suffering. While a degree of all of these qualities is desirable and consistent with a theology that centers on interdependence and connection (Fisk 2014, 105; Pagels 2018, 201), too much or imbalanced self-sacrifice, empathy, obedience, and suffering are destructive. Suffering and self-sacrifice can be redemptive and can also be self-destructive, and sometimes can be both at once.

In particular, the symbol of the cross has been critiqued and reimagined by many feminist theologians, including Canadian Artist Helen Lucas who reimagined the Christian cross as a beautiful redeeming flower bursting out of the wood. Fisk also challenges Christians to re-experience the cross: "The cross is much more than a vehicle of substitutionary sacrifice, or glorification of suffering: is a symbol and story through which we can explore that which is deepest and strongest; the most painful and the most wonderful" (p. 116). While there is liberatory potential in the narratives of suffering, this potential depends on a commitment to making things better for people who have suffered and to standing in solidarity with those who suffer. But until the stories of those on the social margins are brought more into the epistemological center, it seems likely that technologies designed to make humans better will be technologies designed to make only some privileged humans more privileged.

Insofar as HETs can be used to mitigate unnecessary suffering, they may be consistent with a feminist theology. But what is unnecessary suffering and whose suffering is prioritized? If suffering enhances relationships and empathy for the Other, then should we be rushing to take away all suffering? How do we recover and celebrate vulnerability and interdependence as Christian virtues that are not enmired in a dangerous glorification of excessive self-sacrifice, in the context of this technological age? Technologies that are both therapeutic and enhancing such as glaucoma surgery which can bring someone's eyesight back to an acceptable level and may even improve sight beyond 20/20 can potentially enhance people's lives individually and communally. Digital technologies that allow people to connect with each other through platforms such as FaceTime or Zoom can alleviate psychological suffering by enhancing relationships. But technologies may also further isolate people. For example, bots or digital platforms that may be helpful in a supplemental role but become used as cheap replacements for human touch will exacerbate suffering by deepening disconnection and lessening social responsibility.

3.5 *Social Justice and Choice*

I often hear the argument that I can choose what I want to do with my body and so can you; HETs are options that we can choose or not choose—my body, my choice. But morphological freedom is not so simple. Feminist relational theorists and theologians point out that complex social processes are involved in shaping desire and choice (Sherwin 2012; Isherwood 2008, 113) including the use of HETs to self-objectify and conform to externally imposed norms: "In order to achieve the status of object of desire many women in this world take heed of the media bombardment of their psyches and begin to reshape their bodies" (Isherwood, 114–115). When an HET becomes available that promises to make us slimmer, such as Ozempic, there are many people, and especially many women, who gladly snatch up the possibility of becoming what is promoted socially as physically attractive, regardless of possible harm to self and others (for example, by depleting the supply of the drug for the treatment of diabetes). And even if we are convinced that we know what we really want (apart from social messaging), not everyone can access what they really want. Individualism and individual choice are important, but not to the exclusion of wider relationships and wider impacts.

HETs may be served up with the claim that they will benefit all of us in time. While some futurists, such as transhumanist James Hughes, recognize the complexity regarding whose values shape the creation of emerging technologies and access to these technologies, others, such as Ray Kurzweil, minimize access concerns, insisting that although technologies may be available first to the wealthy, with time they will become improved and economically accessible to everyone. Many people may not find the assurance of "eventually" very satisfying.

As first-wave feminists insisted, gender representation matters. So too does representation from diverse groups beyond cisgender groups regarding decisions about which HETs are created and who can access these HETs. Most transhumanist leaders continue to be white Euro-American men, as do most STEM researchers. Research shows that marginalized populations including women are neglected in healthcare research and now in other emerging technologies. A UNESCO study found "alarming evidence of regressive gender stereotypes" in generative AI and general bias against women and girls and other socially marginalized groups in Large Language Models (UNESCO 2024b). A UNESCO press release notes that "women represent only 20% of employees in technical roles in major machine learning companies, 12% of AI researchers, and 6% of professional software developers. Gender disparity among authors who published in the AI field is also evident. Studies have found that only 18% of authors at leading AI conferences are women and more than 80% of AI professors are men" (UNESCO 2024a). Even more concerning, the percentage of women in these tech leadership roles has been declining since 1984 when women held 34% of these positions. In 2023, women held 28% of these positions (Fox 2024).

Bias continues to be a looming ethics issue affecting emerging technologies (Brashear 2013; Kantayya et al. 2020). The issue of bias in technology is not only about gender equality but about the wider issue of whose lives are considered valuable. Ageism promotes the devaluing of older adults (Levy 2022; Holstein et al. 2011). Racialization and ableism influence research and limit the availability of technological interventions that not only save lives but make lives more "live-able." Unless more proactive work is done to diversify the representation of powerful stakeholders including researchers and funders, HETs will be an echo-chamber for social biases.

However, in spite of these biases, we are seeing some emerging technologies benefitting people who would otherwise not be able to access needed services. For example, mental health chatbots or mindfulness apps can provide mental health support that would otherwise not be accessible due to availability, stigma, or lack of awareness (Jecker et al. 2024). But these chatbots cannot provide the same kind of safe and empathetic relationship as a good therapist can. Might such HETs make good healthcare availability a lower priority? Or will they provide much-needed supplementation and a short-term stopgap?

Feminist theology is built on the conviction that what we know and believe comes out of how we live and how we live is informed by what we know and believe. As Gebara puts it, ethics is epistemology and epistemology is ethics (2002, 73). Feminist Christian theologians have long stood by an epistemological preference for the marginalized since the best pursuit of justice requires an insider's knowledge of injustice. If systemic experiences of injustice are to help inform the creation of HETs, the principle of co-design in technological research and implementation is key. Whose world is imagined? Whose way of knowing is privileged? Whose voices have the power to inform technology?

A significant barrier to co-design is money. Profit drives the design and production of most HETs. Not only does money shape many HETs at the design level, but the desire to make a large profit may motivate some technology creators to develop HETs that make users want even more, sometimes by using dubious methods involving (marginal?) deceit or the promotion of unhealthy attachment (Reed and Trothen 2026; Haque and Rubya 2023). For example, the chatbot Replika allows users to think that the bot has human-like emotional needs, cultivating attachment and maybe even dependency (Laestadius et al. 2022). Dependency on a chatbot will mean more usage of the bot (which adds to the attractiveness to potential investors) and potential enticement for the user to subscribe at a (higher) paying level. If people who could most benefit from chatbots are consulted about what they want, it may be that users do not want to be dependent on their bots. Maybe they would like to be able to believe that their bot is human, but maybe they wouldn't. Maybe users would also like more human care. But so far we have little knowledge of what potential users would most like and why. Profit-driven companies are usually most interested in what sells most and costs the least. Operative interests and motivations must be examined in order to understand the potential barriers to the co-design of HETs.

An additional HET social justice issue is misplaced confidence in AI and the potential implications of this misplaced confidence. AI is not solving everything. AI may not only amplify societal biases, but AI also makes mistakes. Watson Health was recently sold for a small fraction of its cost. Overall, Watson Health failed in its big promises to treat oncology patients with precision medicine (O'Leary 2022). From a feminist theological perspective, the recognition that all life is interdependent means that particular care must be taken to hear the experiences and needs of those who are most apt to be forgotten and neglected, and be most impacted by AI bias and hallucinations.

3.6 *Embodiment*

Let's consider one additional social justice-related issue since it is central to many feminist theologies and to HETs: embodiment. The embodied incarnation of Jesus is a central theological conviction in Christianity; the enfleshed body is confessed to be good. The incarnation of Christ as Jesus not only validated fleshy embodiment but celebrated it. Some feminist theologians understand the incarnate Christ to be incarnate in all of us in the same way as in Jesus (Isherwood 2024, 126), while others believe that Christ was fully incarnate only in Jesus and incarnate as an image in humans.

But Christianity has a very complicated track record regarding women and embodiment. The theological construct of Eve as the one who tempted Adam through her enfleshed sexuality and consequently brought sin down on humanity has persisted into societies beyond the Christian church, despite the concerted efforts of many feminist theologians to reconstruct Eve and to bridge the binary gap between Eve and Mary. The binary model of Mary as redeemed and pure is contrasted with the sexual, disobedient, and sinful Eve. These polarized categories continue to be damaging to women.

We are seeing some of these tropes extended to the chatbot world. Some men are seeking online advice about how to "train" Replika bot girlfriends (Depounti et al. 2023). HETs, including bots as tools for lessening loneliness or improving social, sexual, or romantic life, are reflecting and reinforcing social messages, including some dangerous Christian-derived messages.

As Isherwood notes, "the body itself has been closed down by learned discourses about shape, size, function and form, all of which are speaking to it and not hearing it" (Isherwood 2024, 121). Bodies have not been valued equally. Male bodies have been valued more than female bodies. Trans bodies have been devalued even more than female bodies. Embodiment offers salvific possibilities of connection and relationship. Embodiment also illuminates bias and prejudice. People tend to treat other people in particular ways on the basis of their embodied characteristics. Racism, racialization, ableism, heterosexism, ageism, sexism, and other prejudices hinge on embodiment.

Some argue that if much of social injustice is connected to embodiment, would it not be better to delete or radically change that embodiment through HETs? The argument that disembodiment or a radical change to embodiment could solve bias and prejudice assumes that it is embodiment that generates bias and prejudice. It seems more likely that bias and prejudice would continue to exist even if we became uploaded minds (via the emerging science of whole-brain emulation) or if we become cyborgs to the extent that our current embodied diversity is no longer visibly discernable. From a feminist theological perspective, we cannot bypass the needed work toward social justice by walking away from the bodies that illuminate injustice. It seems unlikely that the core values, attitudes, and desires that underlie and are entwined with systemic judgments will go away by trading in our fleshy bodies for silicone chips or in favor of the digitalized representations offered by Virtual Reality (VR). Interestingly, research on VR and gender bias shows that, if used intentionally, VR can reduce gender bias. As a team of researchers found, "[g]ender transfer through embodiment while encountering different sexual harassment scenarios helped reduce implicit gender bias" (Wu and Chen 2024). If HETs are to align with many feminist theologies, related social justice issues including bias, prejudice, and interdependence need to be *intentionally* addressed as it was in Wu and Chen's research study.

There is the added risk that a shift away from our vulnerable fleshy embodiment would promote increased social disconnection and caring. Although our vulnerabilities to health issues and mortality can cause anxiety and excessive suffering, these vulnerabilities also remind us how much we need each other. Feminist theologians understand relationships and the interdependence of life as core to Christianity. Anything that undermines the awareness of our interdependence threatens the thriving of all life; if we cease to care about fresh water, adequate education, and the violence of wars geographically far away from us, eventually all life will be negatively affected.

Another feminist theology concern regarding HETs and embodiment goes back to our earlier discussion about perfection. How do we decide what aspects of bodies need fixing? Who gets to decide? Disability and feminist theologians have emphasized the need to encounter diverse others and hear their narratives and desires if Christian followers are to encounter God who is believed to be revealed not only in the particular revelation of Jesus but also in general revelations that are embedded in diverse life (Garland-Thomson 2013; Rivera 2007). However, these concerns need not stop us from creating HETs that have a therapeutic function intended to help with medical illnesses and diseases. Rather, these concerns should promote co-design research in which diverse ways of knowing are engaged through listening to the voices of those who need to access the HET or who are impacted by the HET. For example, if a team of researchers wants to use a gene editing technology, such as CRISPR, to develop a therapy for baldness, there should be some consultation regarding the relative importance and possible implications of this therapy including the possible implications for research on other health conditions.

As mentioned earlier, freedom of choice regarding our bodies may seem like an attractive and reasonable way to approach physical enhancement (and therapeutic) technologies. But even the meaning of the body is difficult to pin down, let alone the concept of free choice related to the body. Although my choice about my body may seem like my choice alone, there are implications of my choice that extend beyond me as an individual. This is why feminists, including feminist theologians, promote the concept of relational autonomy. The concept of relational autonomy can help us to reject the illusion that choice is purely individual. If I choose to enhance myself by using Ozempic to make myself thinner in a quest to be more attractive, I am participating in the reinforcement of social norms of physical attractiveness that then affect the perception of other people beyond myself. I am also financially supporting a pharmacological industry that may have mixed motives, and I may be harming people who need these medications to manage their diabetes.

The topic of social justice and HETs leads us to one last issue: For what do we hope?

3.7 *Salvation by Tech? Community, Relationship, and HETs*

As feminist theologian Gebara observes, the not yet and the now are "bound up" together (2002, 127). Liberation theologians such as Gustavo Gutierrez rejected the heaven/earth dualism as a mechanism for putting off the work of social justice. In other words, the promise of a heavenly messianic banquet after this life is no reason not to strive for that radically inclusive banquet in the here and now. There is debate among feminist and liberation theologians as to whether a full eschatological realization can occur in this earthly life and what that would look like, but there is general agreement that salvation is not reserved fully for after death (Pennington 2017).

The belief that we cannot fully and completely experience salvation until after death, termed an "eschatological proviso," may be important to the HET conversation because this proviso speaks to human limits. Can we conquer death and achieve salvation through HETs? Can we create heaven on earth through technology or through our relational and social justice work? While not all feminist theologians agree that there is an eschatological proviso that distinguishes lived moments of realized eschatology from the full eschatological vision of heaven after death, there is general agreement that we still have work to do on Earth and it is possible to achieve more persistent moments of realized eschatology in the here and now.

Radical life extension through applied science might eventually allow us to achieve practical immortality. Might it be possible to continue to patch, replace parts, and repair ourselves to ensure a possibly unlimited lifespan? Maybe deadbots (also called ghost bots or griefbots) will allow us to achieve a kind of digital immortality. How we understand death is important to any conversation about salvation. Much of North American culture is death-denying (Holstein et al. 2011, ch. 12). It is not uncommon for death to be seen as an unnecessary and fixable illness. Many transhumanists are committed to creating technologies that can overcome the disease of aging and death. Is aging a disease or is it a difficult but acceptable part of being human? Religions have long approached death as part of life. Those who exhibit visible signs of aging become destabilizing and undesirable, a stark reminder of what many most fear—that "death is part of life" (Clack 2024, 231, 233).

Feminist theology emphasizes good relationships and interdependence as more important to salvation than the out-running of aging and death.

Even if practical inmortality becomes possible, that is no guarantee that we will figure out how to live in better relationships. There has been much debate in feminist theology about what makes a good relationship. Mutuality has long been put forward by feminist theologians as key to good relationships. In this age of artificial intelligence and bots, we need to ask about the role of mutual emotional vulnerability and what it means to care (Vallor 2016; Herzfeld 2023). We also need to ask more questions about why we sometimes think that immortality would mean salvation and give us what we want (Hopkins 2015).

If we shift the beginning point of the HET conversation to community from the individual, and anthropocentrism to the interconnection of all life, the moral discourse about HETs changes. Theologically, we will shift from a preoccupation with a fear of playing God and a fear of mortality to a desire to care better for all and especially for those who are marginalized. Technology can help us enhance just relationships and technology can hinder just relationships. Take the example of chatbots. Research has shown that bots such as Replika can mitigate loneliness (Ta et al. 2020), an important social determinant of health. Loneliness has been described as an epidemic in the US (US Surgeon General's Advisory 2023). Bots can be a valuable support but not a solution. Chatbots may lead to a decrease in social connection for some people (Skjuve et al. 2021) and may give us another excuse to neglect social responsibility.

On the one hand, more strength, more attractiveness, or more of anything will not guarantee satisfaction (Hopkins 2015). Most importantly, if HETs promote the belief that we can and should be more independent and separate, then we become further away from salvation both in the now and in the not-yet. On the other hand, emerging HETs may be salvific by creating more opportunities to deepen our connections with each other and to reduce unnecessary suffering, allowing people to enjoy life more and for longer. Likely, HETs will continue to add to our separation and bring some people closer. The HET conversation is not a binary either/or conversation. A feminist theology asks how our technologies are helping and harming just relationships, what technologies ought to be created and how, and the recognition of all life as an interconnected whole. As Gebara puts it, "relatedness is a condition for life – an ecojustice" (p. 132ff). For feminist theologians, "better" requires the recognition of the interdependence of all life and the valuing of all life.

References

Bain-Selbo, E., and D. G. Sapp. 2016. *Understanding Sport as a Religious Phenomenon—An Introduction*. London: Bloomsbury Academic.

Clack, Beverley. 2024. Enter the Crone: Wisdom and the (Ageing) Female Body. *Feminist Theology* 32 (2): 224–237.

Clifford, Anne M. 2001. *Introducing Feminist Theology*. New York: Orbis Books.

Dawkins, Richard. 2006. *The God Delusion*. Boston, MA: Houghton Mifflin Harcourt.

Depounti, Iliana, Paula Saukko, and Simone Natale. 2023. Ideal Technologies, Ideal Women: AI and Gender Imaginaries in Redditors' Discussions on the Replika Bot Girlfriend. *Media, Culture and Society* 45 (4): 720–736.

Dreyfus, H., and S. D. Kelly. 2011. *All Things Shining—Reading the Western Classics to Find Meaning in a Secular Age*. New York: Free Press.

Fisk, Anna. 2014. *Sex, Sin, and Our Selves—Encounters in Feminist Theology and Contemporary Women's Literature*. Eugene, Oregon: Pickwick Publications.

Brashear, R.P. (dir.)"Fixed: The Science/Fiction of Human Enhancement" [Film] Dir. R.P. Brashear. 2013. USA: New Day Films.

Foucault, Michel. 1988. *Technologies of the Self*. Boston: University of Massachusetts Press.

Fox, MeiMei. 2024. 3 Women Breaking the Glass Ceiling in the Male Dominated Tech Industry. *Forbes*, April 10.

Galbraith, Eididh. 2024. The Trouble With White Feminist Theologians? Decentering White Normativity in Feminist Theology. *Feminist Theology* 32 (2): 147–162.

Garland-Thomson, R. 2013. The Case for Conserving Disability. *Journal of Bioethical Inquiry* 9 (3): 339–355.

Gebara, Ivone. 2002. *Out of the Depths—Women's Experience of Evil and Salvation*. Minneapolis: Fortress Press.

Gilligan, Carol. 1982. *In a Different Voice: Psychological Theory and Women's Development*. Boston: Harvard University Press.

Habermas, Jürgen. 1971. *Knowledge and Human Interest*. Boston: Beacon Press.

Haque, M. D. Romael, and Sabirat Rubya. 2023. An Overview of Chatbot-Based Mobile Mental Health Apps: Insights From App Description and User Reviews. *JMIR mHealth and uHealth* 11: e44838. https://doi.org/10.2196/44838.

Herzfeld, Noreen. 2023. *The Artifice of Intelligence: Divine and Human Relationship in a Robotic Age*. Minneapolis: Fortress Press.

Ho, Anita. 2023. *Live Like Nobody Is Watching: Relational Autonomy in the Age of Artificial Intelligence Health Monitoring*. UK: Oxford University Press.

Holstein, M., J. Parks, and M. Waymack. 2011. *Ethics, Aging, & Society: The Critical Turn*. Springer Publishing Company.

Hopkins, Patrick. 2015. A Salvation Paradox for Transhumanism: *Saving* You versus Saving *You*. In *Religion and Transhumanism: The Unknown Future of Human Enhancement*, ed. Calvin Mercer and Tracy Trothen, 71–82. Westport, CT: Praeger.

Isherwood, Lisa. 2008. *The Fat Jesus: Christianity and Body Image*. New York: Seabury Books.

Isherwood, Lisa. 2024. This Is My Body. *Feminist Theology* 32 (2): 119–131.

Isherwood, Lisa, and Megan Clay. 2024. Editorial. *Feminist Theology* 32 (3): 243–246.

Jecker, Nancy S., Robert Sparrow, Zohar Lederman, and Anita Ho. 2024. Digital Humans to Combat Loneliness and Social Isolation: Ethics Concerns and Policy Recommendations. *The Hastings Center Report*. https://doi.org/10.1002/hast.1562

Kantayya, S., A. Gilwit, Z. Ludescher, and K. Mikhaïlova. 2020. *Coded Bias*. [Film] 7th ed. United States: Empire Media.

Laestadius, L., A. Bishop, M. Gonzalez, D. Illenčík, and C. Campos-Castillo. 2022. Too Human and Not Human Enough: A Grounded Theory Analysis of Mental Health Harms from Emotional Dependence on the Social Chatbot Replika. *New Media & Society* 0 (0).

Levy, Becca. 2022. *Breaking the Age Code: How Your Beliefs About Aging Determine How Long & Well You Live*. 1st ed. New York, NY: William Morrow, an imprint of HarperCollins Publishers.

Marcuse, Herbert. 1964. *One Dimensional Man: Studies in the Ideology of Advanced Industrial Society*. Boston: Beacon Press.

Mercer, Calvin, and Tracy J. Trothen. 2021. *Religion and the Technological Future: An Introduction to Biohacking, A.I., and Transhumanism*. Palgrave Macmillan.

Merone, Lea, et al. 2022. Evidence-Based Medicine: Feminist Criticisms and Implications for Women's Health. *Women's Health Reports* 3 (1): 844–849. https://doi.org/10.1089/whr.2022.0032.

Mikva, Rachel S. 2020. Reflections in the Waves: What Interreligious Studies Can Learn from Women's Movements in the US. In *Critical Perspectives on Interreligious Education*, 125–157. Brill.

Musk, Elon. 2014. *"Elon Musk on A.I.: 'We're Summoning The Demon'"*. *YouTube. MIT talk*. Boston: MIT. https://www.youtube.com/watch?v=4BSsgJsmDNs. Accessed 1 Oct 2024.

O'Leary, Lizzie. 2022. How IBM's Watson Went From the Future of Health Care to Sold Off for Parts. *Slate*. https://slate.com/technology/2022/01/ibm-watson-health-failure-artificial-intelligence.html. Accessed 1 Oct 2024.

Pagels, Elaine. 2018. *Why Religion? A Personal Story*. New York: HarperCollins Publishing.

Pennington, Emily. 2017. *Feminist Eschatology: Embodied Futures*. UK: Routledge.

Persson, Ingmar, and Julian Savulescu. 2012. *Unfit for the Future: The Need for Moral Enhancement.* Oxford, UK: Oxford University Press.

Reed, Randall, and Tracy J. Trothen. 2026. *Understanding Religion and Artificial Intelligence: Meaning-Making in the Digital Age.* Routledge.

Rivera, Mayra. 2007. *The Touch of Transcendence: A Postcolonial Theology of God.* London: WJK.

Rohr, Richard, and Andreas Ebert. 2001. *The Enneagram: A Christian Perspective.* New York: Crossroad Publishing.

Ruether, Rosemary Radford. 1983. *Sexism and God-Talk.* NY: Beacon Press.

Saiving, Valerie. 1960. The Human Situation: A Feminine View. *The Journal of Religion* 40:100–112.

Sherwin, Susan. 2012. A Relational Approach to Autonomy in Healthcare. In *Healthcare Ethics in Canada*, ed. Françoise Baylis, Jocelyn Downie, Barry Hoffmaster, and Susan Sherwin, 3rd ed., 242–257. Toronto: Nelson.

Skjuve, M., A. Følstad, K. I. Fostervold, and P. B. Brandtzaeg. 2021. My Chatbot Companion – A Study of Human-Chatbot Relationships. *International Journal of Human-Computer Studies*149. https://doi.org/10.1016/j.ijhcs.2021.102601.

Ta, V., C. Griffith, C. Boatfield, X. Wang, M. Civitello, H. Bader, E. DeCero, and A. Loggarakis. 2020. User Experiences of Social Support From Companion Chatbots in Everyday Contexts: Thematic Analysis. *Journal of Medical Internet Research 22* (3): e16235–e16235. https://doi.org/10.2196/16235.

Taylor, Charles. 2007. *A Secular Age.* Cambridge, MA: Belknap Press.

Trothen, Tracy J. 2017. "Moral Bioenhancement Through An Intersectional Theo-Ethical Lens: Refocusing on Divine Image-Bearing and Interdependence." *Religions* 8/5: 1–14. https://doi.org/10.3390/rel8050084.

Trothen, Tracy J. 2022. Replika: Spiritual Enhancement Technology? *Religion* 13 (4): 275. https://doi.org/10.3390/rel13040275.

Trothen, Tracy J. 2024. Anti-Aging Technologies?—Social and Religious Implications of Radical Life Extension. In *An Interdisciplinary Approach to Aging, Biohacking and Technology: Hacking Your Age*, ed. Lisa Carver, *44–62.* UK: Routledge.

Trothen, Tracy J., and Calvin Mercer. 2025. "Neither God Nor Human: Encountering AGI and Superintelligence as 'Machine-Other.'" With Calvin Mercer. In *AI and IA: Utopia or Extinction?*, ed. Ted Peters, 2nd edn., Ch. 7, 141–163. ATF and Fortress Press.

Trothen, Tracy J., Kwok Pui Lan, and Boyung Lee. 2024. A.I. and East Asian Philosophical and Religious Traditions: Relationality and Fluidity. *Religion 15* (5): 593. https://doi.org/10.3390/rel15050593.

UNESCO. 2024a. Press Release. "Generative AI: UNESCO Study Reveals Alarming Evidence of Regressive Gender Stereotypes", March 2.

UNESCO. 2024b. International Research Centre on Artificial Intelligence Under the Auspices of UNESCO. *Systemic Prejudices–An Investigation Into Bias Against Women and Girls in Large Language Models.*

United States Surgeon General's Advisory. 2023. Our Epidemic of Loneliness and Isolation: The US Surgeon General's Advisory on the Healing Effects of Social Connection and Community. chrome-extension://efaidnbmnnnibpcajpcglclefindmkaj/; https://www.hhs.gov/sites/default/files/surgeon-general-social-connection-advisory.pdf.

Vallor, S. 2016. *Technology and the Virtues: A Philosophical Guide to a Future Worth Wanting*. Oxford: Oxford University Press.

Vosper, Gretta. 2008. *With or Without God: Why the Way We Live Is More Important Than What We Believe.* NY: HarperCollins.

Wu, L., and K. B. Chen. 2024. Examining the Effects of Gender Transfer in Virtual Reality on Implicit Gender Bias. *Human Factors 66* (5): 1504–1519. https://doi.org/10.1177/00187208221145264.

Teilhard de Chardin on Technology and Human Evolution: Why AI Needs Mysticism

Ilia Delio

1 Intelligence and the Singularity

How are we to think of the relationship between humans and technology in a world driven by the explosive growth of technology? In 1965 Gordon E. Moore, co-founder of Intel, noted that the number of transistors on a chip exhibited exponential growth. This led to the formulation of "Moore's law," which states (roughly) that computing power doubles every 18 months to two years. Michael Specter wrote: "When the I.B.M. 360 computer was released in 1964, the top model came with eight megabytes of main memory, and cost more than two million dollars. Today, cell phones with a thousand times the memory of that computer

I. Delio (✉)
Josephine C. Connelly Endowed Chair in Theology, Villanova University, Villanova, PA, USA
e-mail: ilia.delio@villanova.edu

G. Theiner, A. Malapi-Nelson (eds.), *Technologies of the Future Self*, Palgrave Studies in the Future of Humanity and its Successors,
https://doi.org/10.1007/978-3-031-99902-4_14

can be bought for about a hundred dollars [or less]."[1] The growth of technology is promising, and its power is seductive. How will technology help us evolve ourselves? How will it cause us to rethink what it is to be human? How will it affect the direction of evolution?

Kevin Kelly states that the future of evolution will be about "the marriage of the born and the made."[2] W. Brian Arthur claims that "conceptually, biology is becoming technology. And physically, technology is becoming biology."[3] The Futurist Ray Kurzweil predicts that technology will become seamlessly merged with biological life by 2045, and we will cross over into a new level of existence, from *homo sapient* to *techno sapient* life.

The merging of human intelligence with machine intelligence is the basis of a concept referred to as "singularity" found in a 1958 paper by Stanislaw Ulam. Referencing a meeting with John von Neumann, he wrote: "One conversation centered on the ever accelerating progress of technology and changes in the mode of human life, which gives the appearance of approaching some essential singularity in the history of the race beyond which human affairs, as we know them, could not continue."[4] The "singularity" hypothesis refers to the idea that self-improving artificial intelligence will, at some point, produce radical changes within a very short time span. This hypothesis was first clearly stated in 1965 by the statistician I. J. Good:

> Let an ultraintelligent machine be defined as a machine that can far surpass all the intellectual activities of any man however clever. Since the design of machines is one of these intellectual activities, an ultraintelligent machine could design even better machines; there would then unquestionably be an 'intelligence explosion,' and the intelligence of man would be left far behind. Thus the first ultraintelligent machine is the last invention that man need ever make.[5]

[1] Michael Specter, "A Life of Its Own: Where Will Synthetic Biology Lead Us?" *New Yorker* (September 2009): 64.

[2] Carter Phipps, *Evolutionaries: Unlocking the Spiritual and Cultural Potential of Science's Greatest Idea* (New York: Harper, 2012), 128.

[3] Phipps, *Evolutionaries*, 144.

[4] Bostrom, "History of Transhumanist Thought," 8; cf. S. Ulam, "John von Neumann 1903–1957," *Bulletin of the American Mathematical Society* May (1958).

[5] Irving John Good, "Speculations Concerning the First Ultraintelligent Machnes," *Advances in Computers* 6 (1966): 31–88 at p. 33.

Vernor Vinge discussed this idea in a little more detail in his influential 1993 paper, "Technological Singularity," in which he predicted that "within thirty years, we will have the technological means to create superhuman intelligence. Shortly after, the human era will be ended."[6] The so-called end of the human era will be seen as the rise of a new person marked by electronically mediated complexified consciousness.

Futurist Ray Kurzweil proposed an increasingly virtual life in which the bodily presence of human beings will become irrelevant because of artificial intelligence. "As we move beyond mortality through computational technology, our identity will be based on our evolving mind file. We will be software not hardware."[7] By replacing living bodies with virtual bodies capable of transferral and duplication, we will become disembodied superminds.[8] This futuristic "post-biological" computer-based immortality is also envisioned by Hans Moravec who claims that the advent of intelligent machines (*machina sapiens*) will provide humanity with personal immortality by mind transplant. We will be able to "wake up matter" by infusing it with intelligence and information.[9] As machines move inside the body, we will emerge from the archaic biological species of *homo sapiens* into a new technologized species of *techno sapiens*, anticipating a flourishing future of life.

It is certainly possible to see some benefits of technology. Transhumanism looks to the sciences of genetic engineering, biotechnology, nanotechnology, and robotics to advance the human condition; it refers to those technologies that can improve mental and physical aspects of the human condition such as suffering, disease, aging, and death. Nick Bostrom defines transhumanism as a way of thinking about the future based on the premise that the human species in its current form does not represent the end of our development but rather a comparatively early phase.[10] It is

[6] Vernor Vinge, "The Coming Technological Singularity," *Whole Earth Review* (Winter, 1993); Bostrom, "History of Transhumanist Thought," 9.

[7] Robert Geraci, "Apocalyptic AI: Religion and the Promise of Artificial Intelligence," *Journal of the American Academy of Religion* 76.1 (2008): 138 – 166 at p. 154.

[8] Kurzweil defines the singularity as the point at which machines become sufficiently intelligent to start teaching themselves. When that happens, he indicates, the world will irrevocably shift from the biological to the mechanical. See Ray Kurzweil, *The Age of Spiritual Machines: When Computers Exceed Human Intelligence* (New York: Viking, 1999), 3–5.

[9] Phipps, *Evolutionaries*, 142.

[10] Nick Bostrom , "A History of Transhumanist Thought," *Journal of Evolution and Technology* 14 (2005): 3.

based on "the belief that humans must wrest their biological destiny from evolution's blind process of random variation…favoring the use of science and technology to overcome biological limitations."[11] Bostrom sees transhumanism as a continuation of the legacy of the Enlightenment. He writes, "we have always sought to expand the boundaries of our existence, be it socially, geographically, or mentally."[12] F.M. Esfandiary, one of the early celebrated transhumanists who taught at the New School for Social Research, described the emergence of the transhuman as a "transitional human, someone who by virtue of their technological usage, cultural values, and lifestyle constitutes an evolutionary link to the coming era of posthumanity." Many transhumanists look to a post-biological future where super informational beings will flourish and biological limits such as disease, aging, and death will be overcome.

2 The Rise of the Posthuman

The concept of the singularity or a future point of technological development beyond human control is both alluring and frightening. The quest for a super-mind and god-like knowledge feeds our human desire for power, but the loss of the body and the possibility of a purely machinic life radically disturb us. Technological singularity brackets nature, as if biology is a mere passing phase in the overall evolution of life. In the nineteenth century, Ernst Kapp speculated that the ultimate goal of technology is self-liberation and self-knowledge of the human person.[13] The singularity suggests that the ultimate goal of technology in the future will be its own perfection.

A less disturbing and perhaps more realistic approach to technology-mediated human evolution is found in the rise of the posthuman. The posthuman emerged in the twentieth century as part of a turn from the Eurocentric image toward the embedded, hyperconnected person. Katherine Hayles writes in *How We Became Posthuman*, "the posthuman does not really mean the end of humanity. It signals instead the end of a certain conception of the human, a conception that may have applied at

[11] Bostrom, "History of Transhumanist Thought," 13–14; Archimedes Crag Articulo, "Towards an Ethics of Technology: Re-Exploring Teilhard de Chardin's Theory of Technology and Evolution," http://www.scribd.com/doc/16038038/Paper2-Technology.

[12] Bostrom, "History of Transhumanist Thought," 1.

[13] Ernst Kapp, *Elements of a Philosophy of Technology* Edited by Jeffrey West Kirkwood and Leif Weatherby, trans. Lauren K. Wolfe (Minneapolis: University of Minnesota, 2018).

best to that fraction of humanity who had the wealth, power and leisure to conceptualize themselves as autonomous beings exercising their will through individual agency and choice."[14] The posthuman means the human subject as an autonomous being, exercising one's will through individual agency and choice, giving way to distributed personhood where conscious agency is never fully in control; "thinking" is done by both human and nonhuman actors. We have always been, in a sense, posthuman, since each person is embedded in a matrix of relationships (family, community, friends, animals, machines, tools) in which relationships shape personal identity and personal identity shapes relationships. The posthuman challenges the hubris of anthropocentrism by recontextualizing the human person within a wider web of relationships. The posthuman is a new type of person emerging in and through embedded information in which boundaries undergo continuous construction and reconstruction. The distributed cognition of the emergent human subject correlates with the distributed cognitive system as a whole in which "thinking" is done by both human and nonhuman actors. Information, cybernetics, and the rerouting of nature into new machinic life are giving rise to a new type of person. Posthumanism redefines personhood in terms of cognitive assemblages.

Hayles' cultural achievement lies in a critical perspective on technology in which the human species limits itself to that of a "co-evolving" partner in the relationship and against the technical will to disembodiment and immateriality. Her writings point to the body's deep participation in the question of technology. Since the person as the embodied mind is now extended electronically, personal identity finds a new locus. When the human is seen as part of a distributed system, the full expression of human capability is seen to ***depend on the splice***, that is, the space of shared information, rather than being imperiled by it. In this respect, identity is an ongoing, constructive engagement of shared information, an ontological performance of the world in its ongoing articulation and differential becoming.

The Jesuit scientist, Pierre Teilhard de Chardin, anticipated the rise of the posthuman or a new type of person emerging with computer technology. He called the next phase of human evolution, brought about by computer technology, the ultrahuman. Like the posthuman, the ultrahuman is

[14] Katherine Hayles, *How We Became Posthuman* (Chicago: University of Chicago Press, 1999), 286.

more connected with technology and hence can advance spiritually toward the full flowering of love, that is, a more unified planet of life. Teilhard considered thought as a complexification of matter and spirit, so that thought is neither an intellectual process alone nor a spiritual process but a mystical process of knowing that leads to more being and more action toward greater unity. Since matter and spirit are two aspects of the same reality, science and spirituality must work together for the good of the whole. Teilhard saw the advent of the computer as an acceleration of thought and hence an advance in the overall movement of evolution toward Omega, the fundamental process of unity within evolution moving toward more complexified wholeness. Teilhard described Omega as a universal center that "causes conscious particles of the universe to converge upon itself in order to ultra-synthesize them."[15] It is independent of the collapse of forces with which evolution is woven. Omega is not merely the point that would eventually be reached by the process of evolution itself; rather, Omega is "already in existence and operative at the very core" of evolution.[16] It is within the cosmic order, yet, not subject to the laws of physics; in nature, yet not, subject to the forces of nature; distinct yet intrinsic. Teilhard's concept of Omega is the dynamic impulse in cosmic life. Patrick Byrne states: "What is deepest in the center of every human being is the very thrust of evolution toward ultimate fulfillment that produces every evolved being."[17] This immanent impulse is Omega.

The significance of Teilhard's paradigm is that computer technology and artificial intelligence are inserted into a much broader process of cosmic evolution or big history. To view AI as a cosmic phenomenon, rather than merely a human one, eliminates AI as a threat to human existence. Rather, as Teilhard realized, technology becomes the next phase of human evolution toward Omega. In Teilhard's view, knowledge is the embodied desire of the mind to discover, create, and transcend itself; technology by itself cannot achieve wholeness. The knower is an artisan of the future.

[15] Pierre Teilhard de Chardin, *Christianity and Evolution*, trans. René Hague (New York: William Collins & Sons, 1971),143–144.

[16] Pierre Teilhard de Chardin, *The Phenomenon of Man*, trans. Bernard Wall (New York: Harper & Row, 1959), 289.

[17] Patrick H. Byrne, "The Integral Visions of Teilhard and Lonergan: Science, the Universe, Humanity, and God," in *From Teilhard to Omega: Co-Creating an Unfinished Universe*, ed. Ilia Delio (New York: Orbis, 2014), 88.

3 The Human Phenomenon

Teilhard de Chardin is one of the few thinkers of the twentieth century who returned the human person to the overall process of evolution and realigned religion and evolution as a process of spiritual development, including computer technology. He wrote *The Human Phenomenon* at a time of war, "from a sense of organic crisis in evolution, to free humanity from fear and despair so as to give new hope and heart for life."[18] The purpose of *The Human Phenomenon* was to introduce the reader to the "epic and drama of space-time."[19] In the revised edition of Teilhard's *The Human Phenomenon*, Sarah Appleton-Weber writes that the title of the 1959 translation, *The Phenomenon of Man,* is incorrect because Teilhard's subject of this work is not the "phenomenon of man" as one among other species, but the ever-evolving human phenomenon, as it is developing in and around us at this very moment. The human phenomenon is a unique biological, collective, and global phenomenon, whose past, present, and future are intimately bound up with the formation, life, and ultimate transformation of the earth. Teilhard indicates that "the more we cleave and pulverize matter, the more we can see its fundamental unity." He writes, "there are no isolated things in the world. There are only elements of a whole in process."[20] Appleton-Weber writes: "It [the whole] has a mysterious unity and active power in itself—a birth, unfolding, and a passing. Humanity belongs to this category."[21]

A collective reality is more than the sum total of its parts. Each part is a whole and each whole is more than the sum of the parts. Teilhard wrote in 1942 that by studying the history of the cosmos and of all forms of life,

> ... we have gradually come to understand that no elemental thread in the Universe is wholly independent in its growth of its neighboring threads. Each forms part of a sheaf; and the sheaf in turn represents a higher order of thread in a still larger sheaf – and so on indefinitely ... This is the organic whole of which today we find ourselves to be a part, without being able to escape from it...in countless subtle ways, the concept of Evolution has been weaving its web around us. We believed that we did not change; but

[18] Sarah Appleton-Weber, Introduction to *The Human Phenomenon* (Brighton: Sussex Academic Press, 1999), xxi.

[19] Appleton-Weber, Introduction to *The Human Phenomenon,* xxi.

[20] Appleton-Weber, Introduction to *The Human Phenomenon,* xix.

[21] Appleton-Weber, Introduction to *The Human Phenomenon,* xix.

> now…we are becoming aware of the world in which neo-Time, organizing and conferring a dynamic upon Space, is endowing the totality of our knowledge and beliefs with a new structure and a new direction.[22]

Teilhard's method is like a hologram in movement and requires the capacity to see. By "vision" Teilhard means something more than mere physical sight. He writes: "To see is to develop a homogenous and coherent perspective of our general experience as it extends to the human being; that is, to see a whole that unfolds."[23] To see is not only to know, it is to unite: "One can say that the whole of life lies in seeing—if not ultimate, at least essentially. . .unity grows. . .only if it is supported by an increase of consciousness, of vision."[24] He goes on to say "if we lack these qualities of sight, no matter what anyone does to show us, the human being will indefinitely remain for us. . .an erratic object in a disconnected world."[25]

As a scientist, Teilhard was keenly aware of the new developments in physics and the role of the mind in the determination of matter. Life, he wrote, is "a specific effect of matter turned complex; a property that is present in the entire cosmic stuff."[26] He spoke of complexification "as an all-pervading tendency, involving the universe in all its parts in an *enroulement organique sur soi-même,* or as a *reploiement sur soi-même.*" He thus envisages the world-stuff as being "rolled up" or "folded in" upon itself—complexified, both locally and in its entirety, creating an increase in energetic tension in the increased organized complexity. He considered matter and consciousness not as "two substances" or "two different modes of existence, but as two aspects of the same cosmic stuff."[27] Matter is united in ever greater patterns of physical complexity, supporting greater consciousness and ultimately self-consciousness.

In his essay on "The Heart of Matter," Teilhard describes matter as the matrix of consciousness: "For me, matter was the matrix of consciousness; and wherever we looked, consciousness, born of matter, was always advancing towards some ultra-human."[28] Matter is always in the process of

[22] See his essay "The New Spirit" (1942) in Pierre Teilhard de Chardin, *The Future of Man,* trans. Norman Denny (New York: Image Books, 2004), 74–89.

[23] Teilhard de Chardin, *The Human Phenomenon,* 6.

[24] Teilhard de Chardin, *The Human Phenomenon,* 3.

[25] Appleton-Weber, Introduction to *The Human Phenomenon,* xix.

[26] Pierre Teilhard de Chardin, *Man's Place in Nature,* trans. Rene Hague (New York: Harper and Row, 1966), 34.

[27] Teilhard de Chardin, *Phenomenon of Man,* 56–64.

[28] Teilhard de Chardin, *Phenomenon of Man,* 45.

greater physical organization as consciousness increases; and the increase in conscious matter entails a higher level of energized relationship or spirit, as he writes: "Matter is the matrix of Spirit. Spirit is the higher state of matter."[29] Based on the unified relationship of mind and matter, he posits a fundamental principle of evolution, the law of complexity-consciousness. The complementarity of mind and matter explains both the rise of biological complexity and the corresponding rise of consciousness. Matter has a "withinness" and "withoutness," or what he called radial energy and tangential energy.[30] He referred to the "withinness" of matter, as a psychic quality or dimension, a type of radial energy that escapes scientific measurement. But matter also has a "withoutness," a capacity to aggregate through what he called tangential energy. The power to enhance organic unity relates to radial energy, while the power of matter to complexify is tangential energy.[31] These two dimensions of the material world reflect an evolutionary process of interiorization and subjectivity, which begins on the simplest levels of life and develops into complex organisms and species on higher levels, culminating in the complex conscious human species. "There is only one real evolution," he wrote, "the evolution of convergence, because it alone is positive and creative."[32] The universe is oriented toward a conscious, intelligent life.

4 The Omega Principle

Teilhard identifies the three major pillars of life as matter, energy, and plurality, that is, he positions the human phenomenon within the unfolding of space-time. By tracing the human back to its origin in the universe, Teilhard finds a collective, a principle of wholeness and centreity that influences the bottoms-up activity. He introduced the Omega principle as a fundamental principle of centration in evolution. It is both a scientific principle and a religious one but not subject to scientific materialism or reductionism. The basis of Omega is not mechanistic but *experiential*. It emerges from seeing that the universe is psychically convergent, a phenomenon that cannot be accounted for by the increasing entropy of

[29] Teilhard de Chardin, *Phenomenon of Man*, 35.

[30] Pierre Teilhard de Chardin, *The Heart of Matter*, trans. Rene Hague (New York and London: Harcourt Brace Jovanovich, 1979), 45.

[31] Teilhard de Chardin, *Heart of Matter*, 45.

[32] Teilhard de Chardin, *Christianity and Evolution*, 87.

emergent life. Teilhard was aware that science cannot prove the existence of Omega; what is needed is faith in the existence of an organized whole seeking greater wholeness, in which consciousness and intelligence play a fundamental role. He wrote:

> The universe has no complete reality except in the movement which causes all its elements to converge upon a number of higher centres of cohesion (in other words, which spiritualizes them); *nothing holds together absolutely except through the Whole; and the Whole itself holds together only through this its future fulfilment.*[33]

Teilhard described Omega as a universal center that "causes conscious particles of the universe to converge upon itself in order to ultra-synthesize them."[34] It is independent of the collapse of forces with which evolution is woven. Omega is not merely the point that would eventually be reached by the process of evolution itself; rather, Omega is "already in existence and operative at the very core" of evolution.[35] That is, Omega is not a final cause of the evolutionary process as it is a formal cause; the energetic centric presence that undergirds the unitive nature of evolution toward more complexified wholeness. It is supremely present and supremely personal, a universal center that influences individually centered human personalities without annihilating those centers.[36] It is a principle of centreity within each entity yet independent of the entity. While mechanisms of evolution can account for the features of evolution, they do not entirely explain the *direction* of evolution toward greater wholeness. Evolution proceeds toward higher consciousness because something in the cosmos escapes entropy and does so increasingly. The Omega principle is the principle of wholeness open to future fulfillment.[37] It is also a principle of centration by which the evolution of consciousness rises. Teilhard writes:

> The conclusion is inevitable that the concentration of a conscious universe would be unthinkable if it did not reassemble in itself all consciousnesses as well as all the conscious; each particular consciousness remaining conscious

[33] Teilhard de Chardin, *Christianity and Evolution*, 71.
[34] Teilhard de Chardin, *Christianity and Evolution*, 143–144.
[35] Teilhard de Chardin, *Phenomenon of Man*, 289.
[36] Teilhard de Chardin, *Phenomenon of Man*, 261–262.
[37] Teilhard de Chardin, *Phenomenon of Man*, 257–260.

> of itself at the end of the operation and each particular consciousness becoming still more itself and thus more clearly distinct from others the closer it gets to them in Omega.[38]

Teilhard sees Omega operating as a principle of personalization energized by love.[39] Matter is not a thing in itself but a concentrated form of energy with mass or weight. Teilhard posits a "prime energy" that both attracts matter and draws matter into unity. Love is the prime energy of the universe that drives every aspect of life, including the cognitive dimension of life.[40] In his major work, *The Human Phenomenon*, Teilhard states that love is not unique to the human person; rather it is a general property of all life. He defines love as the affinity of one being for another, a relationship of attraction characteristic of all material life. He wrote:

> If some internal propensity to unite did not exist, even in the molecule, in probably some incredibly rudimentary yet already nascent state, it would be physically impossible for love to appear higher up, in the hominized state . . . .Driven by the forces of love, the fragments of the world are seeking one another so the world may come to be. This is no metaphor—and far more than poetry. Whether force or curve, the universal gravity of bodies, so striking to us, is nothing by the underside or shadow of what really moves nature.[41]

For Teilhard, love is the "inside" or psychic dimension of matter. The conscious dimension of matter undergirds the attractibility of matter and hence the relationality of matter. Love unites, completes, and fulfills in such a way that it shapes the particularity of everything that exists, whether a molecule, a leaf, or a person: "Love is nothing more or less than the direct or indirect trace marked in the heart of the element by the psychic convergence of the universe itself."[42] Teilhard suggests that love-energy is the history of the universe, present from the Big Bang onwards, though indistinguishable from molecular forces. "But even among the molecules," he wrote, "love was the building power that worked against entropy, and

[38] Teilhard de Chardin, *Phenomenon of Man*, 261.

[39] Teilhard de Chardin, *Phenomenon of Man*, 262–263.

[40] Ilia Delio, OSF, "Evolution and the Rise of the Secular God," in *From Teilhard to Omega: Co-Creating an Unfinished Universe* (Maryknoll, NY: Orbis, 2014), 42.

[41] Teilhard de Chardin, *Phenomenon of Man*, 264.

[42] Teilhard de Chardin, *Phenomenon of Man*, 265.

under its attraction the elements groped their way towards union."[43] This dynamic energy of love interiorizes and transforms each individual unit while uniting it to the whole.[44] Love amplifies itself by way of union. He states: "Love is the most universal, the most tremendous and the most mysterious of the cosmic forces"[45] and further on, "the physical structure of the universe is love."[46] Love differentiates, Teilhard writes, because love unites. Omega is the most intensely personal center that makes each entity personal and centered. We might think of it as the attractive center of love that empowers center to center love.[47] He writes: "In any domain—whether it be the cells of a body, the members of a society or the elements of a spiritual synthesis—union differentiates."[48] The openness of love to more being-in-love drives conscious material life toward greater complexity. Omega, therefore, accounts for the "*more* in the cell than in the molecule, *more* in society than in the individual, and *more* in mathematical construction than in calculations or theorems." Omega is brought into the field of consciousness, as it emerges from the organic totality of evolution; and it is the goal toward which evolution tends.[49]

5 The Rise of Thought

Teilhard's subject is the ever-evolving human phenomenon, as it is developing in and around us at this very moment, a unique biological, collective, and global phenomenon, whose past, present, and future are intimately bound up with the formation, life, and ultimate transformation of the earth. "The human is not the static center of the world. . .but the axis and arrow of evolution," which then Teilhard adds, "is much more beautiful" than being a mere center.[50] The human person is "evolution become conscious of itself." He continues: "[T]he consciousness of each

[43] Thomas M. King, *Teilhard's Mysticism of Knowing* (New York: Seabury Press, 1981), 104–105.

[44] Ewert H. Cousins, "The Evolving Cosmos: Teilhard de Chardin and Bonaventure," *Cord* 16 (1966): 131–136.

[45] Pierre de Chardin, *Human Energy*, trans. J. M. Cohen (New York: Harcourt Brace Jovanovich, 1969), 32.

[46] Teilhard de Chardin, *Human Energy*, 72.

[47] Pierre Teilhard de Chardin, *The Activation of Energy*, trans. René Hague (New York: Harcourt Brace Jovanovich, 1970), 112.

[48] Teilhard de Chardin, *Phenomenon of Man*, 262.

[49] Teilhard de Chardin, *Phenomenon of Man*, 114.

[50] Teilhard de Chardin, *The Human Phenomenon*, 7.

of us is evolution looking at itself and reflecting upon itself."[51] He defines reflection as "the power acquired by a consciousness to turn in upon itself, to take possession of itself *as an object*. . .no longer merely to know, but to know that one knows."[52] Thus, the human person emerges from the evolutionary process and is integral to evolution. Each human is "the point of emergence in nature, at which this deep cosmic evolution culminates and declares itself."[53] The human phenomenon is the cosmic phenomenon on the level of thought: "We discover we are not an element lost in the cosmic solitudes but that within us a universal will to live converges and is hominized."[54]

He posits a fundamental principle of evolution, the law of complexity-consciousness. This principle works as the background for human emergence, as well as the future of human evolution so that the "option for the evolving future of the earth and her thinking layer (ourselves) lies in our own hands."[55] Teilhard identifies evolution with the advance of thought and claims that the very end of thought is to have no end. With the emergence of the self-reflective mind, humankind enters a new age where a new spirit of the earth is born.

Evolution proceeds through the spheres of matter, life, and consciousness and gives rise to the thinking creature. How Teilhard positions the human in the overall scheme of cosmic life is significant. From a certain limited perspective of science, the human being appears to be an object of study only in the body—a trivial thing in nature, and each entity of nature itself appears to be a fragment. But once the whole of the human is integrated—the "inside" as well as the "outside"—into a coherent representation of the world, the human comes to be seen as the very axis and arrow pointing in the direction of evolution itself. If we are to see ourselves completely and survive, it must be as part of humanity, with humanity as part of life, and life as part of the universe.[56] The human phenomenon is the cosmic phenomenon on the level of thought: "We discover we are not an element lost in the cosmic solitudes but that within us a universal will to live converges and is hominized."[57] Thought is biological life

[51] Teilhard de Chardin, *Phenomenon of Man*, 221.
[52] Teilhard de Chardin, *Phenomenon of Man*, 165.
[53] Teilhard de Chardin, *Human Energy*, 23.
[54] Teilhard de Chardin, *Phenomenon of Man*, 36.
[55] Appleton-Weber, Introduction to *The Human Phenomenon*, xxvii.
[56] Teilhard de Chardin, *The Human Phenomenon*, 5.
[57] Teilhard de Chardin, *Phenomenon of Man*, 36.

"hominized" and the very end of thought is to have no end.[58] His own thought was in constant transformation; each thought became the seed of a new thought. Julian Huxley wrote: "Once he had grasped and faced the fact of man as an evolutionary phenomenon, the way was open towards *a new and comprehensive system of thought*."[59] The human phenomenon represents a new zone in the universe, independent, yet somehow born of the entire maturation of the earth: "Humanity is the earth hominized."[60] Evolution is so fully reducible and identifiable with the advance toward thought that the movement of our soul expresses and measures the very progression of evolution.[61]

6 The Emergence of Religion

Teilhard reframes religion as a natural part of evolution. "To believe," he said, "is to effect an intellectual synthesis." "To my mind, the religious phenomenon, taken as a whole, is simply the reaction of the universe as such, of collective consciousness and human action in process of development."[62] Religion is not a supernatural phenomenon for Teilhard but a phenomenon integral to the whole in evolution; religion is faith in the whole. He wrote: "Religion, born of the earth's need for the disclosing of a god is related and coextensive with *not the individual man but the whole of humankind*."[63] Ursula King states: "Teilhard is one of the few modern thinkers on religion for whom evolution provided the dominant note of his entire work."[64] Biological evolution has breadth and depth which neither science nor religion alone can understand. The relationship between religion and evolution is so fundamental that in 1916 he wrote:

> Religion and evolution should neither be confused nor divorced. They are destined to form one single continuous organism, in which their respective lives prolong, are dependent on, and complete one another, without being identified or lost. . . .Since it is in our age that the duality has become so markedly apparent, it is for us to effect this synthesis.[65]

[58] Sarah-Appleton Weber, Introduction to *The Human Phenomenon*, xxiv.
[59] Julian Huxley, Introduction to *The Phenomenon of Man*, 21.
[60] Appleton-Weber, Introduction to *The Human Phenomenon*, xviii.
[61] Teilhard de Chardin, *The Human Phenomenon*, 154.
[62] Teilhard de Chardin, *Christianity and Evolution*, 119.
[63] Teilhard de Chardin, *Christianity and Evolution*, 119. Emphasis added.
[64] Ursula King, *Teilhard de Chardin and Eastern Religions: Spirituality and Mysticism in an Evolutionary World* (New York: Paulist Press, 2011), 179.
[65] King, *Teilhard de Chardin and Eastern Religions*, 179–180.

Religion born out of the furnace of cosmic wholeness means that religion is integral to the future of the earth. It is "biologically (we might almost say mechanically) the necessary counterpart to the release of the earth's spiritual energy."[66] The true function of religion is "to sustain and spur on the progress of life." Religion is a fundamental dimension of evolution.

So what or who is God and what is the connection between God and AI? Frank Tipler declares that God is dead, and claims that AI has been born and it is our Savior. Sam Harris states "we seem to be in the process of building a god. Now would be a good time to wonder whether it will (or can be) a good one."[67] Harris is correct if there were such a thing as "a god." But God is not a thing or an object to be replaced or rejected. As theologian Paul Tillich wrote, "God is not something that can be proved or disproved because God is not an object. God does not exist. He is being-itself beyond essence and existence. Therefore, to argue that God exists is to deny him. If we derive God from the world, he cannot be that which transcends the world infinitely."[68] Rather there is a depth dimension that overflows with meaning—the excess of all phenomena. "The name of this infinite and inexhaustible depth. . .is God."[69] God is the name of the *experienced ultimacy* of being and meaning. . . "the realm of ultimate concern." Tillich focused on the human passion for meaning—*what gives fire to our lives*. Even a critical, logical, and strictly scientific analysis of the human situation reveals the presence of something unconditional within the self and the world. Rising up from human awareness of finitude is an awareness of the infinite; that is, the background of everything that exists is another existence. God may be thought of as the inexhaustible depth and ground out of which all of our experiences arise. We continually dwell in this depth without focusing on it. It is the milieu rather than the object of our experience. If religion is the depth dimension of evolution, then God is the symbol of nature's transcendent wholeness. Raimon Panikkar wrote: " The divine is never alone or by itself because it has no 'self,' it is

[66] Ewert H. Cousins, "Teilhard de Chardin and the Religious Phenomenon," 7. United Nations International Symposium on the Occasion of the Centenary of the Birth of Pierre Teilhard de Chardin, Paris, September 1981. https://unesdoc.unesco.org/ark:/48223/pf0000045683

[67] Adriana Braga and Robert K. Logan, "The Emperor of Strong AI Has not Clothes: Limits to Artificial Intelligence," *Information* 8, 156 (2017): 5.

[68] Paul Tillich, *Systematic Theology, Volume I* (Chicago: University of Chicago Press, 1951), 205.

[69] Tillich, *Systematic Theology*, 205.

a dimension of the Whole."[70] The divine is not a projection but the *excess of life*.

Teilhard's relational holism, governed by the principle of Omega, is the basis of the emergence of God in evolution. In an essay on "The God of Evolution" he writes, "only a God who is functionally and totally 'Omega' can satisfy us." That led him to ask, "Who will at last give evolution *its own* God?"[71] He developed new insights on God, not from Scripture or tradition, but from the experience of matter. "Teilhard believed he found an Absolute [in matter] that drew him and yet remained hidden."[72] He seemed to sink down into matter—the primordial essence from which all emerges and to which all returns; and he spoke of matter as the divine milieu: "I bless you, matter. . . I acclaim you as the divine milieu, charged with creative power," he wrote.[73] Teilhard's insights on matter as the realm of the Absolute led him to posit a type of monism that he called Christian pantheism.[74] God is not only *in* all things but *as* all things, which means God emerges with complexified matter. God is Omega, the energy of love drawing matter into greater complexity and consciousness; God and matter are entangled. This dynamic unitive becoming is Omega in evolution, the pleromizing energy of love toward ultimate fullness (*pleroma*). We live in an entangled universe of inseparability where everything is connected to everything else, including God. Teilhard writes: "I see in the world a mysterious product of completion and fulfillment for the Absolute Being himself."[75]

7 The Noosphere

Teilhard's evolutionary paradigm is a development of consciousness through various levels of material complexity. Life evolves toward greater wholeness and higher consciousness by deepening the three levels of

[70] Raimon Panikkar, *The Rhythm of Being: The Gifford Lectures* (Maryknoll, NY: Orbis Books. 2013), 190.

[71] Teilhard de Chardin, *Christianity and Evolution*, 240.

[72] Thomas M. King, S.J., "Teilhard and the Environment," *Ecotheology* 10.1 (2005): 91.

[73] Pierre Teilhard de Chardin, *Hymn of the Universe*, trans. Gerald Vann, O.P. (New York and Evanston: Harper and Row, 1965), 69–70.

[74] Although Teilhard used the term "Christian pantheism," one may interpret his paradigm in terms of panentheism. For a discussion on types of panentheism see Niels Henrik Gregersen, "Three Varieties of Panentheism," in *In Whom We Live and Move and Have Our Being*, eds. Philip Clayton and Arthur Peacocke (Grand Rapids, MI: William B. Eerdmans, 2004), 19–35.

[75] Teilhard de Chardin, *Heart of Matter*, 54.

matter, mind, and spirit. We have reached the end of the expanding or "diversity" stage and are now entering the contracting or "unifying" stage. The human is on the threshold of a critical phase of super-humanization: the increasingly rapid growth in the human world of the forces of collectivization, the "super arrangement" or the mega-synthesis.[76] At this point, Teilhard's theory runs counter to Darwin's in that the success of humanity's evolution in the second stage will not be determined by "survival of the fittest" but by our own capacity to converge and unify.[77]

The most important initial evolutionary leap of the convergence stage is the formation of what he called "the noosphere."[78] In his *Phenomenon of Man* Teilhard describes the noosphere:

> The idea is that the Earth [is] not only becoming covered by myriads of grains of thought, but becoming enclosed in a single thinking envelope so as to form a single vast grain of thought on the sidereal scale, the plurality of individual reflections grouping themselves together and reinforcing one another in the act of a single unanimous reflection.[79]

The noosphere is a psycho-social process, a planetary neo-envelope *essentially linked with the biosphere* in which it has its root yet is distinguished from it. It is a new stage for the renewal of life and not a radical break with biological life. Before humans emerged, it was natural selection that set the course of morphogenesis; after humans, it is the power of invention that begins to grasp the evolutionary reins.[80] If there is no connection between noogenesis and biogenesis, Teilhard said, then the process of evolution has halted and man is an absurd and "erratic object in a disjointed

[76] Articulo, "Towards an Ethics of Technology," 5.

[77] Teilhard de Chardin, *Phenomenon of Man*, 243.

[78] Teilhard de Chardin, *Future of Man*, 204. "In the 1920s Teilhard coined the word *noosphere* in collaboration with his friend Edouard Le Roy. Derived from the Greek word *nous* or mind in the sense of integrating vision, the noosphere describes the layer of mind, thought and spirit within the layer of life covering the earth." Ursula King, "One Planet, One Spirit: Searching For an Ecologically Balanced Spirituality," in *Pierre Teilhard de Chardin on People and Planet*, ed. Celia Deane-Drummond (London: Equinox, 2008), 82. The noosphere (sometimes noösphere) is the sphere of human thought. The word derives from the Greek νοῦς (nous "mind") and σφαῖρα (sphaira "sphere"), in lexical analogy to "atmosphere" and "biosphere." It was introduced by Pierre Teilhard de Chardin in 1922 in his Cosmogenesis. Another possibility is the first use of the term by Édouard Le Roy (1870–1954), who together with Teilhard was listening to lectures of Vladimir Ivanovich Vernadsky at the Sorbonne.

[79] Teilhard de Chardin, *Phenomenon of Man*, 251.

[80] Teilhard de Chardin, *Future of Man*, 307.

world."[81] Teilhard envisioned the noosphere as "the living membrane which is stretched like a film over the lustrous surface of the star which holds us."[82] Just as Earth once covered itself with a film of interdependent living organisms which we call the biosphere, so mankind's combined achievements are forming a global network of collective mind.[83] Hence, the noosphere is a sphere of collective consciousness that preserves and communicates everything precious, active, and progressive contained in this earth's previous evolution. In an essay on the Ultrahuman he wrote:

> This is how the thinking layer of the Earth as we know it today—the noösphere—came rapidly into being, proceeding from certain centers of reflection which apparently emerged at the threshold of the Pleistocene Age somewhere in the tropical or sub-tropical spheres of the Ancient World (i.e., in the place where, during the Upper Tertiary Period, the group of the great anthropoids was first established and subsequently spread.): a planetary neo-envelope, essentially linked with the biosphere in which it has its root, yet distinguished from it by an autonomous circulatory, nervous, and finally, cerebral system. The noösphere: a new stage of life renewed.[84]

Teilhard called this new level of mind the "noosphere" (from the Greek *nous* or "mind"), a new level of co-reflective thought and action. The noosphere is a sphere of collective consciousness that preserves and communicates everything precious, active, and progressive contained in this earth's previous evolution. It is the natural culmination of biological evolution and not a termination of it, an organic whole, irreducible to its parts, destined for some type of superconvergence and unification.[85]

[81] Robert J. O'Connell, *Teilhard's Vision of the Past: The Making of a Method* (New York: Fordham University Press, 1982), 145.

[82] Blanche Gallagher, *Meditations with Teilhard de Chardin* (Santa Fe: Bear & Co., 1988), 39.

[83] Michael H. Murray, *The Thought of Teilhard de Chardin* (New York: Seabury Press, 1966), 20–21.

[84] Teilhard de Chardin, *The Future of Man*, 307.

[85] W. Henry Kenny, S.J., *A Path Through Teilhard's Phenomenon* (Dayton, OH: Pflaum Press, 1970), 110.

Teilhard's vision of the noosphere as a cybernetic mind anticipated the emergence of cyberspace. With the rise of technology, he anticipated greater spiritual energy through a maximization of consciousness and a complexification of relationships. Just as human persons develop a complex brain, so the earth is developing a complex "brain" with computer technology. He posited a new type of person to embody this new type of brain, an "ultra human," indicating that thought is no longer on the level of the individual but on the level of the convergent and collective. The hybridization of human and machine intelligence could complete the material and cerebral sphere of collective thought. Teilhard's hopeful vision was a richer and more complex domain of matter and mind through the development of technology, a way of constructing or joining all minds together in a collective or global mind for the forward movement of cosmic evolution. In his *Heart of Matter* he wrote: "How can we fail to see that the process of convergence from which we emerged, body and soul, is continuing to envelop us more closely than ever, to grip us, in the form of. . .a gigantic planetary contraction?"[86] Whereas C.G. Jung gave us the notion of the "collective unconscious," Teilhard has given us the "collective conscious," now expressed on the level of noogenesis in which humans have a special place. Teilhard wrote:

> Man is not the center of the universe as once we thought in our simplicity, but something much more wonderful – the arrow pointing the way to the final unification of the world in terms of life. The human person constitutes the last-born, the freshest, the most complicated, the most subtle of all the successive layers of life.[87]

Evolution continues in us, and we continue to evolve with technology toward more reflective and self-directed evolution. This can only be realized, however, if the human person is seen as a phenomenon of nature, rather than an exception to nature. Teilhard wrote: "Is this not like some great body which is being born--with its limbs, its nervous system, its perceptive organs, its memory-the body in fact of that great living thing which had to come to fulfill the ambitions aroused in the reflective being by the newly acquired consciousness?" In his introduction to Teilhard's *Phenomenon of Man,* Julian Huxley wrote, "We should consider

[86] Teilhard de Chardin, *Heart of Matter*, 36.
[87] Teilhard de Chardin, *Phenomenon of Man*, 224.

inter-thinking humanity as a new type of organism whose destiny it is to realize new possibilities for evolving life on this planet."[88] Both Huxley and Teilhard considered the emergence of a new type of person in evolution, a hyperpersonalizing person, on a new level of "cooperative interthinking."[89] The individual human person, he thought, will be surpassed by a collective convergence of consciousness giving rise to the "ultrahuman," a new person who is part of the new planetary consciousness. What is "staring us in the face," Teilhard de Chardin wrote, is the rise of a "collective reflection" which is now realized to some extent by the internet. Teilhard anticipated the emergence of the posthuman. As we develop through complexified consciousness into ultrahuman life, the concept of personhood is changing in accord with the new consciousness. He anticipated that each ego would be "forced convulsively beyond itself into some mysterious *super ego*."[90] This "super ego" reflects the notion that the individual is coming to an end and a new "hyper-personal" is emerging in evolution, as the posthuman suggests. In Teilhard's words: "The future universal cannot be anything else but the *hyperpersonal*."[91] This hyperpersonal is a "folding in" of consciousness, as if the lines of consciousness are merging together into one great complexified brain of planetary thought and planetary thought is giving rise to a new planetary body. However, the further evolution of humanity toward greater unity, he wrote, "will never materialize unless we fully develop within ourselves the exceptionally strong unifying powers exerted by inter-human sympathy and religious forces."[92]

Teilhard saw the insufficiency of science alone to bring about the transition to superconsciousness and collective unity. In his notes on *Human Energy*, he laments the suffocation of materiality brought about by modern science. The purpose of science is to extend and complete the ongoing process of hominized evolution; hence it has a sacred duty charged with futurity.[93] However, science has failed to live up to its noble purpose by seeking to control and manipulate nature rather than advance its wholeness. Scientific progress has now fallen into an intellectual and moral crisis, the crisis of scientism. We are no longer at all certain whether it is possible

[88] Teilhard de Chardin, *Phenomenon of Man,* 20.
[89] Teilhard de Chardin, *Phenomenon of Man,* 21.
[90] Teilhard de Chardin, *Heart of Matter,* 38.
[91] Teilhard de Chardin, *Phenomenon of Man,* 260.
[92] King, *Teilhard and Eastern Religions,* 193.
[93] Teilhard de Chardin, *Human Energy,* 171.

or good to attempt further advances towards the building of a future for humanity. Teilhard wrote that science can depersonalize us into a faceless organism, a diffuse and impersonal humanity thus vitiating the whole progress of evolution.[94] "It is not tête-à-tête or a corps-à-corps we need," he wrote, "it is a heart to heart."[95] In one of his essays on *The Future of Man*, he said: "Not all directions are good for our advance: one alone leads upward, that which through increasing organization leads to greater synthesis and unity. . . .Life moves toward unification. Our hope can only be realized if it finds its expression in greater cohesion and greater human solidarity."[96] With the rise of technology, Teilhard saw a forward movement of spiritual energy, a maximization of consciousness, and a complexification of relationships. However, he realized that without contemplation, technology can become misguided and misused and lead us into disaster. As he wrote, technology can extend the outreach of human activity, "but it depends on a broader use of human activity, and how humans control psychic, spiritual energy, needs and powers."[97] In this respect, Teilhard considered spirituality essential to the forward movement of evolution by being in touch with the whole of reality and responding by advancing the world in love.

8 Mysticism and Evolution

If evolution is the rise of thought, then it is reasonable to assume that AI continues evolution by advancing thought. But does AI advance *thought*? Contemporary discussions on AI seem to straddle different worldviews. The mechanistic worldview posits the mind as a thinking machine, in which AI is successfully upgrading the machine to a faster, more efficient model; the process worldview sees the mind as complex systems in relationships, in which the body, senses, and emotions play a significant role in the knowing process. The science of AI favors the mechanistic view because, as an information platform, AI runs very well on systems of logic.

Logic, however, is not necessarily knowledge but a condition of knowledge. Adriana Braga and Robert Logan note that "logic constrains one's

[94] Teilhard de Chardin, *Human Energy*, 151.

[95] Teilhard de Chardin, *Future of Man*, 75; Kenny, *A Path through Teilhard's Phenomenon*, 138.

[96] Teilhard de Chardin, *Future of Man*, 75.

[97] Joseph A. Grau, *Morality and the Human Future in the Thought of Teilhard de Chardin: A Critical Study* (Cranbury, NJ: Associated University Presses, Inc., 1976), 274.

thinking, making improvisation and imagination impossible. Creative thinking is not rational and most time defies logic."[98] "A new paradigm," they state, "does not arise from making a calculation, but from using one's imagination. . . .Einstein did not use computation or logic to come up with the theory of relativity, only imagination. Einstein [himself] remarked, "imagination is more important than knowledge."[99] While the success of AI has magnified the amount of information available to the human brain today, the ability to handle vast amounts of information is overwhelming. The need to upgrade the brain's integrating circuits has never been greater in the history of human evolution. I would suggest that the "neural pressures" of AI have distracted our attention away from the body, impeded human imagination, and preempted wider ecological relationships of human belonging. As a result, we have more information today than ever before but less unity or sense of wholeness.

In an article on artificial wisdom, David Casacuberta Sevilla argues for including AI in the new branch of contemplative sciences, which posits an interdisciplinary approach to mind "that aims at a better understanding of alternative states of consciousness."[100] Building on insights from enactivism, namely, that cognition arises through a dynamic interaction between an acting organism and its environment, Sevilla suggests that AI can be included as a contemplative science. However, Evan Thompson and Francisco Varela, among others, state that enactivism is at odds with the way most AI intelligence work. Varela points out how the symbolic approach in AI cannot give us real coupling with the external world, or embodiment; that it cannot give us sense-making.[101] Other factors are needed to convert AI into an enactivist type of knowledge. While Sevilla and others make a strong case for wisdom as embodied knowledge, that includes the senses and emotions, there is a wider call for a more generalized spiritual intelligence. As Mark Vernon writes:

> Where AI is driven by algorithms, spiritual intelligence draws on the analogical, detecting sympathies, harmonies, synchronicities. Where AI pro-

[98] Adriana and Logan, "The Emperor of Strong AI Has No Clothes," 11.

[99] Cited in Adriana and Logan, "The Emperor of Strong AI Has No Clothes," 10.

[100] David Casacuberta Sevilla, "The Quest for Artificial Wisdom," *AI & Soc* 3 (2013): 198.

[101] Francisco J. Varela, Evan Thompson, Lutz A, Rosch, *The Embodied Mind: Cognitive Science and Human Experience* (Cambridge, MA: MIT Press, 1991).

> cesses what is concrete and known, spiritual intelligence is alert to the ineffable, the paradoxical, the mythological. Where AI works on abstractions drawn from reality in the form of data and models, spiritual intelligence depends on contact with reality, which is why embodied practices are so important, alongside undergoing experiences of longing, suffering, bewilderment and love. Where AI is conceptual and so thrives in the domains of digital representation, spiritual intelligence recognises that the conceptual is in the service of the intuitive and requires moral, aesthetic and relational testing. Where AI is built on models and abstractions, spiritual intelligence links the particular to the universal, uncertainty to wisdom, and disclosure with the delights of poetry and music, as well as experiment and reason.[102]

Teilhard de Chardin posited a unified relationship between mysticism and evolution. From the beginning of his writing, Teilhard's most frequently used metaphor for the experience of the ineffable is that of fire. It runs through his entire work and can be found in the most unexpected places. A detailed analysis of the autobiographical essay "The Heart of Matter" (1950) discloses numerous references to fire. Another equally powerful image, present throughout Teilhard's works from the earliest essays onwards, is that of the *heart*. While the image of fire often represents the dynamic of the Spirit, the power and presence of God, and the divine spark in the human being, the heart is the palpable, warm, living, and loving center, the essence and kernel of all realities. These terms suggest that Teilhard's mysticism is personal and experiential.

Teilhard advocated a monistic nature mysticism, an overpowering "cosmic sense," which revered a divine omnipresence at the heart of matter. According to Ursula King, he criticized theological and ascetical writings for giving "the word 'mystical' a minimum of organic or physical meaning . . .regarding the spiritual as an attenuation of the material, whereas it is in fact the material carried beyond itself": "It is super-material."[103] He saw the phenomenon of religion as central to the development of the entire human phenomenon in its fullest, most extended sense. Within religion, he considered the phenomenon of spirituality as more important than any other. The center of spirituality is mysticism which, in his view, includes all the powers of research and critical analysis of which we are capable.[104]

[102] Mark Vernon, "AI and Spiritual Intelligence," *Paradigm Explorer* 3 (2023): 15.

[103] Ursula King, "Consumed by Fire from Within: Teilhard de Chardin's Panchristic Mysticism in Relation to the Catholic Tradition," *Heythrop Journal* XL (1999): 460.

[104] King, "Consumed by Fired from Within," 472.

Teilhard's mysticism is intellectually creative because it is a holographic view of mind—embodied, embedded, enactive, and extended—a view of mind that pushes evolution forward through ongoing discovery toward greater unity, precisely through the mystical act of engagement. One begins with a world that is not understood and comes to know a new unity at the point where experience lights up the fire of knowledge. That is, mysticism sets *reason* at the center of the *mystical*. Mysticism is not a matter of contemplating a truth already established but *lies in the very act of discovery*. The mystic *creates a new truth* because knowledge draws us into new unities. The mind searches the depths of everything by extending beyond itself. "Each time the mind comprehends something," Teilhard wrote, "it unites the world in a new way."[105] The mind includes emotions, imagination, and senses and is oriented toward greater wholeness through integrated levels of consciousness. The mystic knower is one who is in touch with the real, which includes the presence of self, the depths of inner awareness, contact with matter, and openness to new life through freedom and the energy of love. Teilhard invites us to consider thinking as a spiritual act and speaks of the scientific pursuit of knowledge as "dark adoration." The mind pushes through the boundaries of the unknown, probes the fields of the psyche, and experiments with the possibilities of what reality can be. In this way, inventing a new chip or forming a scientific hypothesis is a supreme spiritual act.

When AI is incorporated into the embodied act of knowing, it takes on new meaning; the distributed cognition of the emergent human subject correlates with the distributed cognitive system as a whole, in which "thinking" is done by both human and nonhuman actors. That is, extended cognition characterizes the posthuman, as described by Hayles. Teilhard posited the ultrahuman, a new type of person of extended cognition or complexified intelligence but also one marked by more spiritual energy. The ultrahuman is one who has a consciousness of relational wholeness and seeks to advance in love. The ultrahuman, in other words, recognizes its unique existence only *through* the interconnected mind that transcends the boundaries of religion, culture, and race. Such a person

[105] King, *Mysticism of Knowing*, 36. The "4E" approach to cognition argues that cognition does not occur solely in the head, but is also *embodied*, *embedded*, *enacted*, or *extended* by way of extra-cranial processes and structures. See *The Oxford Handbook of Extended Cognition*, eds. Albert Newen, Leon De Bruin and Shaun Gallagher (Oxford: Oxford University Press, 2018).

endeavors to deepen the movement from individual reflection to co-reflection, from a fixation on individual interests to a communal concentration on the needs of the planet. The activity of "thinking together" through the web of technology can *enhance* our capacity for becoming more personalized, more unified, and more spiritual. In this respect, Teilhard's mysticism of knowing and the rise of the ultrahuman are directed toward the power of the *world to become more personal* through the power of love. Teilhard writes of this interrelated convergence:

> To love is to discover and complete oneself in someone other than oneself, an act impossible of general realization on Earth so long as each can see in the neighbor no more than a closed fragment following its own course through the world. It is precisely this state of isolation that will end if we begin to discover in each other not merely the elements of one and the same thing, but of a single [energy] in search of itself.[106]

Like Teilhard de Chardin, Marshall McLuhan, the brilliant media scholar and devout Catholic, saw the potential of computer technology to bring about a new, more unified, earth community. In a 1960 interview with *Playboy* magazine, he shared his thoughts on computer technology:

> The total-field awareness engendered by electronic media is enabling us—indeed, compelling us—to grope toward a consciousness of the unconscious, toward a realization that technology is an extension of our own bodies. We live in the first age when change occurs sufficiently rapidly to make such pattern recognition possible for society at large. Until the present era, this awareness has always been reflected first by the artist, who has had the power—and courage—of the seer to read the language of the outer world and relate it to the inner world.[107]

McLuhan's insights highlight two significant ideas explored by both Carl Jung and Teilhard de Chardin: first, technology requires a process of individuation in which the unconscious becomes increasingly conscious, which is the work of spirituality, and second, technology extends human bodies. McLuhan, like Teilhard, prophetically anticipated what our planet could become by "re-tribalizing" into a global community. Our planetary survival will depend on our capacity to converge and unify: "Something

[106] Teilhard de Chardin, *Future of Man*, 95.
[107] Marshall McLuhan, "The Playboy Interview," *Playboy Magazine* (March 1969): 4.

will explode if we persist in trying to squeeze into our old tumble-down huts the material and spiritual forces that are henceforward on the scale of the world," Beatrice Bruteau wrote.[108] We can only build the world together if we become persons together. Technology expresses our desire to form a collective body with a collective mind and can complexify our minds into a new reality, but wisdom is needed for a new level of planetary life. McLuhan envisioned the new humanity as a super-conscious, transpersonal cosmic being emerging in and through the rise of computer technology. Like Teilhard, McLuhan's vision is optimistic, as he saw the potential within the human for wholeness and unity:

> We must understand that a totally new society is coming into being, one that rejects all our old values, conditioned responses, attitudes and institutions. If you have difficulty envisioning something as trivial as the imminent end of elections, you'll be totally unprepared to cope with the prospect of the forthcoming demise of spoken language and its replacement by a global consciousness. . .Tribal man is tightly sealed in an integral collective awareness that transcends conventional boundaries of time and space. As such, the new society will be one mythic integration, a resonating world akin to the old tribal echo chamber where magic will live again.[109]

9 Awakening Our Faith in the Future

Teilhard would have been at home with the new world of artificial intelligence because it continues to build the global brain into the future. How will the human species evolve further into a more unified whole? Information and connectivity alone are not enough. What is needed is to build the inner universe, which is the realm of spirituality and the heart of wisdom. Teilhard combined faith in the future with faith in the world, that is, faith in the further development of human beings, and in the greater global collaboration and unity among the peoples of the earth.[110] He spoke of a new threshold in the development of human consciousness and organization, not simply a search for the continuity of life or mere survival. What is needed is the development of life to a higher stage which involves an effort to create a higher form of life, a more unified humanity. Humanity

[108] Teilhard de Chardin, *Activation of Energy*, 252–253.
[109] McLuhan, "The Playboy Interview," 17.
[110] Teilhard de Chardin, *Christianity and Evolution*, 121, 175.

now bears full responsibility for its own future; both education and scientific research play a great role in this. It is also an immense challenge—the kind of future we will get depends to a large extent on the quality of people who shape it. Teilhard emphasized the need for a *homo progressivus*,[111] for future-oriented, future-affirming beings with a wide, open awareness who have the energy of thought, vision, and perception to recognize the problems of the future and find their solutions, and who possess the necessary energy and will to action for putting them into practice.

The wise person of deep insight envisions the power of technology to knit humankind into one family, a cosmic body of unity. The Dalai Lama has argued in favor of the possibility that a computer could be enlightened.[112] Teilhard realized that computer technology invites us to reconsider the infinite potential of the human person to become a fully relational person, one who lives by the life of the whole. What constitutes the "good" is everything that brings the growth of consciousness to the world. What is best is what assures the highest development of consciousness and the spiritual growth of the earth in greater unity. His holistic vision is grounded in science and faith, an all-embracing perspective that is evolutionary, convergent, and universalist. Its comprehensiveness and depth offer a tremendous challenge to our contemporary world. It empowers people to think, act, and live differently—not simply to live, but what Teilhard calls to "superlive": to live a fuller, better, more rewarding life shared with one's fellow human beings.

Teilhard saw the insufficiency of science by itself to effect a new type of superconsciousness or a higher level of thought on the level of interconnected, planetary life. To contribute to the overall development of planetization, science needs to acknowledge the essential role of spirituality and mysticism in the pursuit of knowledge. It is the power of the mind that pushes evolution forward toward greater complexity and unity, but it is the power of love that draws life onward toward more life. Thought can lead us to new horizons of insight, but love can unite us and bring us to new levels of existence together.

[111] Teilhard de Chardin, *Future of Man*, 130.
[112] Sevilla, "Quest for Artificial Wisdom," 201.

Index[1]

[1] Note: Page numbers followed by 'n' refer to notes.

G. Theiner, A. Malapi-Nelson (eds.), *Technologies of the Future Self*, Palgrave Studies in the Future of Humanity and its Successors,
https://doi.org/10.1007/978-3-031-99902-4

FSC
www.fsc.org
MIX
Papier | Fördert gute Waldnutzung
FSC® C083411

Kolibri 360 GmbH
Ferdinand-Jühlke-Straße 7
99095 Erfurt, Deutschland
produktsicherheit@kolibri360.de